ECOSYSTEM PROCESSES IN ANTARCTIC ICE-FREE LANDSCAPES

PROCEEDINGS OF AN INTERNATIONAL WORKSHOP ON POLAR DESERT
ECOSYSTEMS/CHRISTCHURCH/NEW ZEALAND/1-4 JULY 1996

Ecosystem Processes in Antarctic Ice-free Landscapes

Edited by
W. BERRY LYONS
Department of Geology, University of Alabama, Tuscaloosa, Alabama, USA
CLIVE HOWARD-WILLIAMS & IAN HAWES
*National Institute of Water and Atmospheric Research, Christchurch,
New Zealand*

A.A.BALKEMA/ROTTERDAM/BROOKFIELD/1997

Photo cover: Kathy A.Welch

The texts of the various papers in this volume were set individually by typists under the supervision of the editors.

Authorization to photocopy items for internal or personal use, or the internal or personal use of specific clients, is granted by A.A. Balkema, Rotterdam, provided that the base fee of US$1.50 per copy, plus US$0.10 per page is paid directly to Copyright Clearance Center, 222 Rosewood Drive, Danvers, MA 01923, USA. For those organizations that have been granted a photocopy license by CCC, a separate system of payment has been arranged. The fee code for users of the Transactional Reporting Service is: 90 5410 925 4/97 US$1.50 + US$0.10.

Published by
A.A. Balkema, P.O. Box 1675, 3000 BR Rotterdam, Netherlands (Fax: +31.10.4135947)
A.A. Balkema Publishers, Old Post Road, Brookfield, VT 05036-9704, USA
(Fax: 802.2763837)

ISBN 90 5410 925 4
© 1997 A.A. Balkema, Rotterdam
Printed in the Netherlands

Ecosystem Processes in Antarctic Ice-free Landscapes, Lyons, Howard-Williams & Hawes (eds)
© *1997 Balkema, Rotterdam, ISBN 90 5410 925 4*

Table of contents

3 *Aquatic environments*

4 *Human impacts*

Ecosystem Processes in Antarctic Ice-free Landscapes, Lyons, Howard-Williams & Hawes (eds)
© *1997 Balkema, Rotterdam, ISBN 90 5410 925 4*

Preface

It is just over 20 years since the publication of the landmark volume "Polar Deserts and Modern Man" (Smiley & Zumberger 1974). At that time the major issues facing polar deserts were resource development and transportation. The section on climatology did not even mention "Climate Change" and terrestrial ecosystem research was in the descriptive phase. How perspectives have changed in two decades. In Antarctica, the Minerals Convention has, for the moment, put aside discussions on mineral resources, while the Madrid Protocol to the Antarctic Treaty has placed a major emphasis on restricting the impact of man on ecosystems and landscapes. Similarly, environmental awareness in the Arctic is playing a major role in influencing human activities. Climate change, and the effect of ozone depletion on UV levels over the polar regions are now major topics of research, particularly with the polar amplification factor pertaining to both of these effects. Ecosystem research in polar landscapes has become increasingly process oriented and increased understanding has undermined naive assertions that these are simple ecosystems.

This volume brings together some of the papers presented at the 1996 International Workshop on Polar Desert Ecosystems held in Christchurch, New Zealand. The workshop was, sadly, the last international scientific forum attended by Emeritus Professor Sir Robin Irvine whose opening address to the workshop we are honoured to publish. The title of the address "There can be no contentment without proceeding" aptly sums up the drive and high level of energy displayed by scientists of polar regions in all disciplines. That Antarctic science is rapidly on the move was illustrated at this focused workshop, with specialists in the study of ice-free polar ecosystems from the USA, Great Britain, Australia, New Zealand, Germany, and Canada providing a stimulating series of papers and workshop debates. Some of these are published in this volume.

The volume is divided into four parts; the overviews, terrestrial environments, aquatic environments and human impacts.

Two of the overview papers have a broad bipolar focus, with an emphasis on spatial variation in ecosystem structure and function and addressed temporal variability and global ozone depletion. The other plenary paper reminds us of the extent and speed of natural variability in climate in the region. The sensitivity of polar ecosystems to climate change is a recurrent research theme, and is the underlying rationale behind a number of SCAR international programmes. Polar desert landscapes are dependent on ice melt to provide water and are among the most sensitive of systems to change, since the balance of this phase change is so sensitive to small fluctuations in radiation or temperature.

One of the central themes of Part Two is the overwhelming importance of water to organisms in the system. As in their warmer counterparts, plant and animal growth in polar deserts is driven by the availability of water. The paper by Schroeter *et al.*, clearly shows that sub-zero temperatures alone do not prevent growth. A second theme is that of colonisation, often linked to water availability by the need for a suitable habitat for growth for colonisation to proceed. Exciting new work on the genetic diversity of polar desert organisms is shedding light on the process of colonisation.

The papers in Part Three illustrate how the concept of antarctic ecosystems as "simple" is no longer valid. Complex interactions between a diverse set of microbial components characterise Antarctic inland waters and recent work on aquatic viruses, (Kepner *et al.* this volume) further extends this complexity. In the preface to the Book "High Latitude Limnology" Vincent & Ellis-Evans (1989) pointed out that one of the most significant areas still awaiting investigation was that of catchment-lake interactions. Several papers here contribute to this topic (Nyogi *et al.*; Howard-Williams *et al.*) and the recent major effort of the Long Term Ecological Research Programme in the Taylor Valley funded through the US NSF will make considerable advances in the area in the next few years. A major issue identified in this workshop was the potential importance of benthic communities, long overlooked in lacustrine systems.

In Part Four, the four papers provide contrasting sets of data and experiences on human influence in ice free landscapes. Human activity in inland Antarctica is restricted to a few isolated places but the great scientific and wilderness values of the region has led to a recognised need for special restrictions (Vincent 1995). It is particularly refreshing to find that representatives of National Antarctic logistic agencies are interacting so closely with the Scientific community as to contribute two of the papers in this section. The final chapter in Part Four represents the summary of the interactive workshop discussions which were held in two parallel sessions, Natural Change and Human Impacts.

The volume is aimed at scholars and postgraduate students involved in polar science and hopefully gives a flavour to the huge breadth of work available to researchers in the area. It is our wish that some of the papers presented here will stimulate younger scientists to work in polar ice free landscapes. The work is exciting, the fragility of these ecosystems and their sensitivity to change make the work highly topical, and we have yet to find a better working environment on this planet.

The editors wish to acknowledge the considerable help with the production of this volume and the organisation of the 1996 workshop. Carol Whaitiri, the workshop secretary, provided a friendly e-mail service world wide to ensure the smooth planning of the workshop and that participants were welcome in Christchurch. Carol, and Drs Julie Hall and Anne-Maree Schwarz helped with the day to day organisation. Drs Mike Timperly, David Walton and Anne-Maree Schwarz acted as workshop rapporteurs. The primary responsibility for word processing and formatting was also Carol's, and she is thanked again for the long hours put into this work. Monique Baars provided a valuable copy editing service in the final stages. As with other multi-authored volumes this book reflects a compromise between uniformity and diverse opinions of the authors. The editors accept responsibility for mistakes and omissions. The US National Science Foundation, The New Zealand Foundation for Research, Science and Technology, and the National Institute for Water and Atmospheric Research (NZ) are thanked for sponsorship.

W B Lyons
C Howard-Williams
I Hawes
(Editorial Panel)

References

Smiley, T.H. & J.H. Zumberger, 1974 *Polar Deserts and Modern Man* The University of Arizona Press, Tucon. 173 p.

Vincent, W.F. (Ed.), 1995. Environmental management of a cold desert ecosystem: The McMurdo Dry Valleys. Desert Research Institute, University of Nevada, U.S.A. Special Publication. 57 p.

Vincent, W.F. & J.C. Ellis-Evans, 1989. *High Latitude Limnology*. Kluwer Academic Publishers, Dordrecht, the Netherlands. 323 p.

Ecosystem Processes in Antarctic Ice-free Landscapes, Lyons, Howard-Williams & Hawes (eds)
© *1997 Balkema, Rotterdam, ISBN 90 5410 925 4*

There can be no contentment but in proceeding: Opening address to the International Workshop on Polar Desert Ecosystems

Sir Robin Irvine
Management Board of the Antarctica New Zealand, Christchurch, New Zealand

In opening this workshop I have decided to range widely, certainly to emphasise the importance of international collaboration in Polar studies, and also to make some remarks about science in general.

I have a theme for my address, "There is no contentment but in proceeding". To explain this sentence I have to go back to my own days of medical research and to an interest in kidney transplantation at a time when this procedure, now a common-day event, was still in the area of research. One of the doyens of the transplantation field in the 1950s and 60s, whom I was privileged to meet, was Peter Medawar, a professor at the age of 32, a Fellow of the Royal Society of London at 34, and a Nobel Prize winner at 45. At the age of 54 he suffered a crippling stroke. Days before this catastrophe he gave the Presidential Address at a meeting of the British Association. It is the concluding paragraph of this address that I want to impinge on your memories. He said:

'We cannot point to a single definitive solution of any one of the problems that confront us - political, economic, social or moral, that is having to do with the conduct of life. We are still beginners, and for that reason may hope to improve. To deride the hope of progress is the ultimate fatuity, the last word in poverty of spirit and meanness of mind. There is no need to be dismayed by the fact that we cannot yet envisage a definitive solution of our problems, a resting place beyond which we need not try to go. Because he likened life to a race, and defined felicity as the state of mind of those in front of it, Thomas Hobbes (a 17th century writer) has always been thought of as the arch materialist, the first man to uphold go-getting as a creed. But this is a travesty of Hobbes' opinion. He was a go-getter in a sense, but it was the going, not the getting he extolled. As Hobbes conceived it, the race had no finishing post. The great thing about the race was to be in it, to be a contestant in the attempt to make the world a better place, and it was a spiritual death he had in mind when he said that to forsake the course is to die. 'There is no such thing as perpetual tranquillity of mind while we live here,' he told us in Leviathan, 'because life itself is but motion, and can never be without desire, or without fear no more than without sense. There can be no contentment but in proceeding.' I agree" said Medawar.

Peter Medawar died in 1987 and carved on his gravestone in a Sussex Churchyard are those telling words "There can be no contentment but in proceeding."

There has to be a time when each one of us, thinks seriously about why he or she is interested in the Polar regions and why we want to go there. Of course there is the excitement and the beauty of the regions themselves and the unique nature of life in the Arctic and

---

[1] Sir Robin Irvine died in September 1996 - a friend and motivator of the highest quality international research efforts in the Antarctic

Antarctic, but there is also the science. And when we think about it, basically the science is the reason why we are all here. We are here because "there can be no contentment but in proceeding."

In relation to this theme I have time to ask only two questions. Why is international research important in polar science, and why do research anyway? The answer to the first question is simple; to the second it is clear-cut, but the answer has to be considered in a logical manner.

Surely polar research is one of the best examples of a scientific field where international collaboration is essential. Rising international concern for the global environment, and the need for investigations of global processes, along with the increasing costs associated with polar research, have provided unique opportunities for scientists to form collaborative ventures. This Workshop is all about international collaboration. Ice-free polar desert ecosystems, be they in the Arctic or the Antarctic, can only be studied in depth if scientists from all interested nations work together. The scope of this workshop, the interest you show in each others work and the passion you show in finding the answers to questions about your ecosystems indicate very strongly to me that international collaboration will be the basis of bipolar studies, and in particular the study of ice-free polar desert ecosystems.

But I don't want to spend much time on Polar Science itself, I would rather take this opportunity to answer the second question, why do research anyway? Why is there no contentment but in proceeding?

It is a true saying, but nevertheless worth stating, that the harmonious development of any society depends to a very large extent on research. It is indispensable for the solution of the problems created by the technological developments of the modern world. It has to be remembered, however, that the new appears as a minority point of view and hence is unpopular. The function of research is to give new ideas a home.

It is true, too, that fundamental research is necessary in every society. Without it, a society could lose its active link with the development of knowledge, and thus find itself cut off from the source of its vitality. The expansion of knowledge is the greatest human achievement of our time. Knowledge is the essential element on which any modern society depends for its survival, for its prosperity, and for its hope of improving its condition - not only its economic condition, but its social condition as well. Research if pursued for its own sake, i.e. with the object of advancing knowledge and understanding and to satisfy curiosity, requires not only training and high ability on the part of the researcher, and easy access to the means for his or her research (libraries, apparatus and materials), but also a supportive atmosphere which is not authoritarian or censorious and which is conducive to able minds to range over any part of knowledge which they deem worthy of attention. This is one of the reasons why the polar regions are ideal environments for many researcher workers and why research in the Arctic and Antarctica should be encouraged at all levels.

In general terms, this is an age in which scientific knowledge is altering our lives with accelerating speed - on a scale and to a depth which are unprecedented. It is an exciting age, but it is also frightening because it presents a bewildering array of problems and challenges. We do have to do some revolutionary rethinking about our attitudes, our opportunities, and even the most familiar problems like social responsibilities.

In keeping to my theme of "there can be no contentment but in proceeding" I have ranged from the need for international collaboration in Polar science, to the importance of research. I have only had time to touch on these topics and I hope, to stimulate your thinking. Best wishes for a highly successful, rewarding, and relaxing Workshop.

1 Overviews

Ecosystem Processes in Antarctic Ice-free Landscapes, Lyons, Howard-Williams & Hawes (eds)
© *1997 Balkema, Rotterdam, ISBN 90 5410 925 4*

Polar desert ecosystems in a changing climate: A north-south perspective

Warwick F. Vincent
Département de Biologie et Centre d'Etudes Nordiques, Université Laval, Québec, Canada

ABSTRACT: Aridity and extreme cold characterize much of the Arctic as well as the south polar region. In northern parts of the Canadian Arctic Archipelago, the Eurasian high Arctic and Greenland, the average annual precipitation is at or below 150 mm water equivalent. Vegetation in these regions is sparse and discontinuous, often covering <5% of the ice-free ground. These Arctic desert ecosystems share certain environmental traits with their counterparts in Antarctica, for example high sensitivity to year-to-year variations in climatic forcing, and large between-site variations controlled by water availability. However, there are also fundamental differences; for example, in food web complexity and the degree of isolation from source biota at lower latitudes. Global climate change is likely to have strong effects on polar desert ecosystems because of the greater magnitude of environmental change at high latitudes; the general sensitivity of desert environments to climate variation; and the precarious freeze-thaw balance which influences many aspects of Arctic and Antarctic ecosystems. In both polar regions, lakes provide a downstream, integrated measure of overall landscape processes. A north-south comparison of polar desert lakes reveals many similarities including low allochthonous inputs of carbon from the surrounding catchment, extreme transparency to ultraviolet radiation, and in their potential value as sensitive indicators of global change.

1   INTRODUCTION

Polar desert landscapes are characterized by barren catchments in which sparse, discontinuous patches of vegetation are limited by water, temperature, nutrient supply and a brief growing season. According to Aleksandrova (1988), the concept of a 'cold desert' was first put forward by S. Passarge in 1920 to describe the largely unvegetated land at the highest northern latitudes. The term 'Arctic desert' was subsequently applied in 1935 by Gorodkov to the vegetation of the Russian high Arctic 'nival zone' (Fig. 1) and later extended by Vera Aleksandrova and others to the more general term 'polar deserts' to include the ice-free land of Antarctica.

The exact climatic definition of polar desert regions varies between authors and reflects their centre of field studies. The Russian literature favours the criteria of mean temperatures less than 2°C during the warmest month and precipitation in the range 100 to 250 mm (Aleksandrova 1988); North American ecologists apply the term 'polar desert' to western and northern parts of the Canadian Arctic Archipelago (Fig. 2) in which mean temperatures for the warmest month are up to about 5°C, with annual precipitation at or below 150 mm. In the ice-free desert regions of Antarctica, the warmest monthly mean air temperatures are within the range -2 to +2°C, with total annual precipitation often well below 100 mm (Schwerdtfeger 1970). Bliss & Matveyeva (1992) estimate that the Arctic polar desert extends over *ca.* 850 000 km², accounting for about 43% of the ice-free land in the high Arctic (the circumpolar region north of continuous tundra). This compares with an overall total of 332 000 km² of exposed rock and soil in Antarctica (Drewry *et al.* 1982).

There is now a considerable and rapidly growing literature on polar deserts, and increasing interest in these environments for research and resource exploitation, including ecotourism. This latter activity has accelerated greatly over the last decade with increasing access to sites in the Ross Sea region, Antarctica (e.g. 200 to 300 tourists per year in the McMurdo Dry Valleys since 1993; Vincent 1996), and in the high Arctic where a variety of national parks have been created, for example Ellesmere Island National Park Reserve in the northern polar desert region of the Queen Elizabeth Islands (Fig. 2). The combination of the most pronounced global environmental changes (rising solar UV-B radiation and increased air temperatures) in the polar regions, and the sensitivity of desert systems to climatic forcing, suggests that polar deserts may be preferred sites within the biosphere for monitoring global change. The biological and geochemical processes in these ecosystems operate under extreme conditions, and research on these 'natural experiments' is generating new insights into a variety of environmental phenomena that operate at all latitudes. The recognition of the global importance of these sites, in combination with the increasing ease of access by researchers and tourists, highlights the need for an improved understanding of their structure and dynamics. Such information is required to not only harness the scientific value of polar desert ecosystems, but also to ensure their long term conservation and protection.

The present article summarizes some of the major features of the Arctic desert environment, with comparison to deserts of the south polar region. The climatic and biological characteristics of polar desert ecosystems are first examined, and then their potential sensitivity to global climate change is explored.

2   CLIMATE

The ice-free desert regions of Antarctica have long been held to represent a climatic extreme relative to all other parts of the biosphere. On the basis of the mean annual temperature regimes there are strong similarities between polar desert environments in the Arctic and Antarctica;

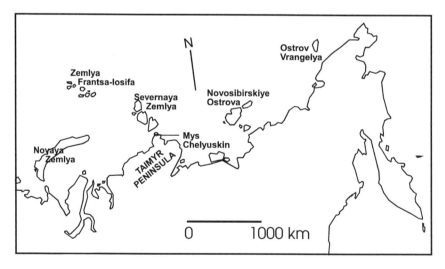

Fig. 1. Key sites in the Eurasian polar desert region. Mys Chelyuskin is located at latitude 77°45'N, longitude 104°20'E.

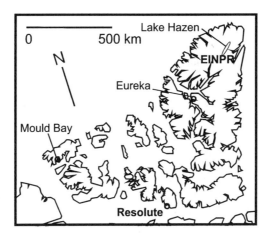

Fig.2. The Queen Elizabeth Islands in the Canadian Arctic Archipelago, showing the location of sites mentioned in the text. EINPR - Ellesmere Island National Park Reserve. Resolute is located at latitude 74°41'N, longitude 94°54'W.

e.g. -20.4°C at Lake Hazen, Ellesmere Island (Fig. 2) and -19.8°C at Lake Vanda in the McMurdo Dry Valleys, Antarctica. However, such a comparison may be misleading for ecological purposes because the mean annual temperature includes much of the year in which there is little plant, animal or even microbial activity. A more appropriate climatic separation of biological relevance is on the basis of total annual precipitation and the mean daily temperature during the warmest month. This latter temperature measure is more likely to reflect conditions during the peak growing season, and is known to be a correlate of certain biological characteristics. For example, Chernov (1995) has shown that in western and middle Siberia the number of nesting bird species correlates with July mean daily temperature, and similarly that the number of beetle species on the Taimyr Peninsula is a close log function (r = +0.99) of July mean temperatures. Although much of the precipitation falling on these arid environments is rapidly lost by evaporation (Bliss *et al.* 1984), some of the winter snowfall may eventually be made available for biological processes, and annual totals are probably more biologically meaningful than summer precipitation.

A plot of total annual precipitation versus mean daily temperature for the warmest month (January or February for Antarctica, July for the Arctic) confirms the extreme position of the Antarctic deserts relative to analogous types of environment in the high Arctic (Fig. 3). Precipitation measurements are extremely difficult to obtain under the low snowfall and blowing snow conditions which characterize these regions, however a set of careful measurements are available for Vanda Station in the McMurdo Dry Valleys. The Vanda Station point on the graph in Fig. 3 represents the 'wettest' year on record (1974), and the snowfall values were converted to mm water equivalent using the average water to snow density measured during that study of 1:12.6 (Bromley 1985). Even given this above-average value for total precipitation, the Dry Valleys are clearly drier than any Arctic site monitored to date, with the possible exception of Lake Hazen. The anomalously low value for the latter site is based on limited data collected during the International Geophysical Year, and is considered by some authors to be an underestimate (see below). The summer air temperatures at Vanda Station, are similar to many sites in the high Arctic, particularly islands of the Siberian Arctic Archipelago.

There is a striking contrast between climates in the McMurdo Dry Valleys LTER (Long Term Ecological Research Site, a global network of 19 ecosystem research and monitoring sites maintained by the US National Science Foundation) and the northern-most LTER site, Toolik Lake in Alaska (latitude 68°30'5'N) where precipitation totals 203 mm, and the mean July air

5

temperature is 12.2°C. Much lower precipitation values (104 mm) but still relatively warm temperatures (7.1°C) characterize Barrow, a US International Biological Program site on the Alaskan coast (latitude 71°N). Both of these Alaskan sites support tundra communities (shrub tundra at the Toolik LTER) with a much richer diversity and biomass of plants than in the high Arctic desert. Further to west and north, however, in the Siberian Arctic Archipelago, summer mean air temperatures plummet to values at or below those of the McMurdo Dry Valleys, with a trend of decreasing temperatures towards the west, from 5.6°C on Ostrov Vrangelya (Wrangel Island) to 0.6°C on Ostrov Rudol'fa (latitude 81°N, longitude 57°E), the northernmost island of Zemlya Frantsa-Iosifa (Frantz Joseph Land; Fig. 1). The ice-free parts of these regions as well as the northern tip of the Russian mainland, Mys Chalyuskin on the Taimyr Peninsula (July mean of 1.7°C) are characterized by polar desert terrain. Much further to the west in Novaya Zemlya temperatures as well as precipitation begin to rise (6.7°C, 243 mm). The climate of Svalbard similarly reflects the increasing influence of the Gulf Stream (4.5°C and 378 mm at Isfjord), and this region is typically classified as semi-desert rather than true polar desert. These relatively moist conditions characterize much of Greenland, with extremely high precipitation on the southeastern side (7.6°C and 1940 mm at Torgilsbu). Even at the very north of Greenland mean July temperatures are around 4°C and precipitation high (204 mm at Nord); the lowest precipitation recorded at low altitude is on the western coast, for example 186 mm at Upernavik and 125 mm at Thule.

In the Canadian Arctic Archipelago, particularly to the north and west, the climate is much drier and colder than coastal Greenland with July temperatures around 4 to 5 °C and precipitation below 150 mm, for example 67 mm at Eureka. The western sector of the Queen Elizabeth Islands, for example the Forsheim Peninsula lying in the rain shadow of Ellesmere and Axel Heiberg Islands, is especially dry and has been referred to as the barren wedge; Rea Point on Melville Island has been identified as the driest location in North America (Van der Leeden *et al.* 1990). All of these sites however, have total precipitation that is much higher than

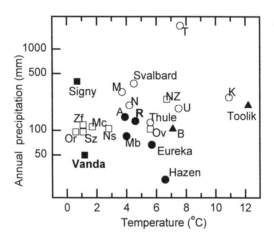

Fig 3. Total annual precipitation (water equivalent) versus mean daily temp-erature for the warmest month at Arctic and Antarctic sites. Antarctica: Vanda Station (Continental Antarctica) and Signy Island (Maritime Antarctic). Canadian high Arctic: Alert (A), Eureka, Lake Hazen, Mould Bay (Mb), and Resolute (R). Eurasia: Mys Chelyuskin (Mc), Novaya Zemlya (Nz), Novo-sibirskie Ostrova (Ns), Ostrov Rudol'fa (Or), Severnaya Zemlya (Sz), Zemlya Fransta- Iosifa (Zf), Svalbard (Sv). United States Arctic: Barrow (B), and Toolik Lake. Greenland: Kapisigdlit (K), Myggbukta (M), Nord (N), Thule, Torgilsbu (T) and Upernavik (U). Data sources are Bigl (1984), Bromley (1985), Van der Leeden *et al.* (1990), Putnins (1970), Schwerdtfeger (1970), Vincent (1988) and Vowinckel & Orwig (1970).

in the McMurdo Dry Valleys. In the Lake Hazen area, a thermal desert effect gives rise to prolonged periods of warm summer temperatures (summer daily maxima up to 15°C) but the annual precipitation is extremely low at this site (25 mm, although the reliability of these measurements has been questioned; Bigl 1984).

A major feature of desert climates in general is the magnitude of year-to-year variability in weather conditions. Because precipitation is so low, a single storm event can completely change the availability of water from one year to another. At Thule on July 1957, for example, 47.5 mm, almost the entire season's precipitation, fell in one day (Wilson 1967). For the three complete years for which 12 months of weather data are available for Vanda Station in the McMurdo Dry Valleys area, the precipitation (in terms of snowfall) was 82 mm (1969), 7 mm (1970) and 115 mm (1974); 80% of the, 1973 precipitation (89 mm, well in excess of the 1969 and 1970 totals) fell during a 7-day period in March (Bromley 1985).

3   MICROBIAL ECOSYSTEMS

Much of the ice-free land in the high Arctic is devoid of vegetation and is exclusively dominated by microscopic life-forms. Cyanobacteria and other micro-organisms form a black patina over rock surfaces (Aleksandrova 1988), while other species penetrate the cracks of rocks to form chasmolithic communities (e.g. on the exposed dolomite sections at Resolute). Phototrophic as well as heterotrophic species may penetrate deep within certain rock types to form structured cryptoendolithic communities beneath the surface (e.g. in the exposed sandstones of Ellesmere Island) or form sublithic communities under translucent rocks. Analogous lithophytic communities of each of these types have been well described from Antarctic desert ecosystems (Broady 1981; Nienow & Friedmann 1993). The microbial ecology of soils in desert ecosystems is still poorly understood in either the north or south polar zones, but in both regions cyanobacteria appear to play an important role (Alexandrova 1988; Vincent 1988; Wynn-Williams 1990), including in the nitrogen economy of certain environments (e.g. Henry & Svoboda 1986). In the Arctic as well as Antarctica, lakes, ponds and streams are a major focus of microbiological activity (see below).

4   PLANT COMMUNITY STRUCTURE AND DYNAMICS

Discontinuous patches of vegetation occur throughout the deserts of both polar regions, but are strictly confined to sites of favorable moisture content. In Antarctic deserts, the plant communities occur as small, isolated patches of cryptogams (algae, moss and lichens), with more extensive cover at sites well supplied by meltwater (e.g. the Canada Glacier moss flush in the McMurdo Dry Valleys; Schwartz et al. 1992). In barren regions of the Queen Elizabeth Islands (Fig. 2), the sparse vegetation is dominated by vascular plants; for 23 sites the vascular plant cover averaged 1.8%, visible cryptogams 0.7%, and bare rock and soil 98% (Bliss et al. 1984). The Russian polar desert is characterized by a greater ground cover, higher species diversity and by dominance of mosses and lichens rather than vascular plants. For example, Zemlya Alexandra, an island at the western end of the Frantsa-Iosifa Archipelago, contains an extensive area of ice-free land in which 25% is completely devoid of plants. Another 35% of the ground contains scattered plants with negligible total cover and 10% is occupied by lakes and permanent snowfield; however over the remaining 30% of land there are well developed communities in which the vegetation cover ranges from 3 to 15%, and up to 100% at a small number of sites (Aleksandrova 1988). These latter communities are mostly moss-lichen associations with less than 10% flowering plants. They are more comparable in biomass, diversity and vegetation type with the semi-desert rather than desert plant communities in the Canadian high Arctic (Bliss et al. 1984).

4.1   Biodiversity

The diversity of Arctic plant communities drops substantially with increasing latitude, however even the north polar deserts contain a diverse assemblage of vascular plants by comparison with Antarctica. Some 400 species have been recorded in the high Arctic, compared with about 900 species in the entire Arctic (Bliss 1979). Only two vascular plants are native to Antarctica (a grass *Deschampsia antarctica* and a cushion-forming angiosperm *Colobanthus quitensis*) and although their populations appear to be expanding (Smith 1994), both are at present restricted to the maritime zone.

The Arctic flora also contains about 600 moss species, >150 liverworts and 2000 lichens; this compares with 250 to 300 mosses, 150 to 175 liverworts and 300 to 400 lichen species from Antarctica (Longton 1988). The polar desert ecosystems of each region contain only a subset of this total diversity. For example, about 30 mosses, 1 liverwort and 125 lichens occur in continental Antarctica (Longton 1988); 102 mosses, 115 lichens and 55 liverworts have been recorded on Zemlya Frantsa-Iosifa (Aleksandrova 1988); about 166 moss species occur in northern Ellesmere Island (Longton 1988).

In part the difference in floral diversity between the north and south polar deserts reflects the greater severity of climate in Antarctica, but it is also a function of the biogeographical isolation of Antarctica. In the northern hemisphere there is a continuity of land into the Arctic allowing continuous opportunities for colonization from the more diverse communities of the south. The persistence of glacial refugia in northern high latitudes in the past (e.g. Beringia) and present (e.g. sites on Ellesmere Island) is also likely to have contributed to the relatively high diversity of Arctic communities.

4.2   Standing crop and productivity

The north-south gradient in biodiversity is accompanied by major changes in other biological properties of the plant communities. According to the estimates of Bliss & Matveyeva (1992), the plant stands in polar deserts have standing stocks and primary productivities some two to three orders of magnitude less than values for vegetation in the low Arctic (Table 1).

The availability of liquid water has a major influence on the production characteristics of desert plant communities of the Arctic as well as Antarctica. Kennedy (1993) argues that water supply rather than biogeographical isolation or low temperatures *per se* is the overall limiting factor for the distribution and abundance of organisms in the Antarctic terrestrial environment; for example in Ablation Valley (a 40 km^2 desert area on Alexander Island, Antarctica) the vegetation is largely restricted to seven moss patches occurring on north-facing slopes supplied by groundwater seeps. Similarly, mosses form extensive mats in snowflush sites in the Canadian Arctic desert, but these habitats cover less than 0.1% of the overall landscape (Bliss 1979). Much of Ellesmere Island National Park Reserve contains clay and rock barrens, with sparse vegetation; however in the vicinity of rivers, streams and ponds and at other moist sites such as wet slopes of mountains, there are luxurious stands of plant communities dominated by sedges and moss (Soper & Powell 1985). The two order of magnitude higher productivity values for wetlands relative to desert sites under the same high Arctic climate regime (Table 1) underscores the strong controlling influence of water. Such effects are also apparent at the microhabitat level in north and south polar deserts; for example the occurrence of moss cushions around and beneath stones (Kennedy 1993) and the moss and lichen colonization of desiccation cracks in patterned ground (Aleksandrova 1988).

5   ARCTIC DESERT FAUNA

The Arctic desert contains a complex animal community structure with functional components

Table 1. Characteristics of Arctic desert vegetation compared with other major plant communities in the north polar zone. Cryptogams (algae, mosses and lichens) are important vegetation components in all communities. Compiled from the information given in Bliss & Matveyeva (1992).

Vegetation type	Vascular dominants	Standing crop (kg m^{-2})	Net production (g m^{-2} year^{-1})
High Arctic			
Polar desert	Herbs (*Draba, Papaver,Saxifraga, Puccinellia*	0.02	1
Polar semi-desert	Cushion plants and herbs (*Alopecurus, Draba, Papaver, Saxifraga*, prostrate *Salix* spp.)	1.2	35
Wetlands (mires)	Sedges and grasses (*Carex stans, Dupontia fisheri, Eriophorum* spp.)	2.4	140
Low Arctic			
Wetlands (mires)	Sedges and grasses	4.6	220
Tussock	Sedges (especially *Eriophorum vaginatum*) dwarf shrubs (heaths)	7.4	225
Shrub tundra	Low shrubs (*Betula, Salix, Alnus*) and sedges (*Carex, Eriophorum*)	3.1	375
	Tall shrubs (riparian)	5.8	1000

that are completely absent from Antarctic desert ecosystems, for example top carnivores (e.g. the Arctic wolf, *Canis lupus*) and mammalian herbivores (e.g. muskox (*Ovibos moschatus)*; lemmings (*Lemmus*); and voles (*Microtus*)). Faunal biodiversity in the Arctic is greatly increased in summer by migratory species from the south, such as many birds; the extreme isolation of Antarctic deserts precludes such immigrants, apart from humans and a few bird species (particularly the Antarctic skua *Catharacta maccormicki* and the snow petrel *Pagodroma nivea*). Arctic animal populations have a broad range of effects on ecosystem properties including carbon fluxes, vegetation dynamics and nutrient recycling (Remmert 1980); these 'top-down' controls are without parallel in Antarctic desert ecosystems.

Certain invertebrate groups are common to both polar regions. As with other components of the biota, the invertebrate fauna of Arctic deserts is depauperate by comparison with lower latitudes in the north polar zone, but is much richer than in the Antarctic. Only one insect species is native to Antarctica, the dipteran *Belgica antarctica*, but it is restricted to the maritime zone in conditions that are much less harsh than in the polar desert environment. In the Arctic some 3000 insect species have been recorded with a few hundred occurring in the desert regions. Certain orders are completely lost across the Arctic climate gradient from south to north; for example, the number of ground beetle taxa (Carabidae) drops from 60 in the southern shrub tundra of the Taimyr Peninsula, to 15 to 20 in the tundra, two to six in the high Arctic tundra, and to complete absence in the polar desert (Chernov 1995).

Similar south-north trends are observed in the soil micro-invertebrates. Chernov (1995) reports that on the Taimyr Peninsula the diversity of nematodes drops from 160 species at tundra sites to 50 species at Mys Chalyuskin. This compares with six genera and 11 species of nematodes which have been reported from the Antarctic continent; in the McMurdo Dry Valleys, two nematode species (*Scottnema lindsayae* and *Eudorylaimus antarcticus*) dominate under a wide variety of conditions (Powers *et al*. 1995).

The trend of diminished biodiversity at high latitude is accompanied by a shift towards dominance by less advanced, generalist species that have broad environmental tolerances (Chernov 1995). For example, the number of families of collembolans (an invertebrate group also represented in Antarctic deserts) drops from more than eight in mixed forest habitats to three in the Arctic desert; there are about 10 species in the Taimyr polar desert and most of these are generalists, occurring in most habitat types. The advanced, diversified order of birds, the Passeriformes, drops off markedly in number of taxa with distance north, with a shift towards dominance by the phylogenetically much lower order Charadriiformes (plovers) containing generalist species which eat both plant and animal matter. Within the insects there is again a loss of specialists, with a shift towards a phenotypically variable group of dipterans, the infraorder Tipulomorpha (craneflies). This effect results in 'super-dominance' where a few species dominate in many different communities.

6   GLOBAL CHANGE

6.1   *Environmental change in the polar regions*

There is increasing evidence that biologically significant changes in the Earth's climate have occurred this century, and that these continuing trends are most pronounced in high latitude regions of the planet. These long-term environmental trends include changes in air temperature, precipitation, soil temperature, moisture and albedo. They also include the continuing rise in ultraviolet radiation, specifically in the UV-B waveband which is the photochemically most active region of the solar spectrum (Vincent & Roy 1993). Increased atmospheric pollution and human activities in the Arctic and Antarctic are additional factors contributing to long term environmental change in the polar regions.

Global climate models predict that the greatest effects of atmospheric CO_2 increases will be experienced at high latitudes. For example, in a CO_2-doubling scenario, summer-time temperatures around Hudson Bay rise by $7^{\circ}C$, with a 50% decrease in soil moisture (McBean 1994). However, the currently available data sets show that there are major regional differences in the pattern of temperature change across the Arctic, with warming trends in Alaska and Siberia, but pronounced cooling trends in the eastern Canadian Arctic and Greenland (Chapman & Walsh 1993). Several authors have drawn attention to the potential effects of global warming on increased precipitation in the high Arctic (resulting from less sea ice cover) and the positive feedback effects on regional albedo and cooling trends (e.g. Svoboda 1994).

The predicted changes in soil temperature and moisture have implications for many of the biological communities and processes in polar ecosystems, but the nature of any effects remains controversial. Bliss & Matveyeva (1992) suggest a scenario whereby global warming causes increased winter precipitation, later snow melt and thus a later start for plant growth in the Arctic, with a negative or only small effect on the plant cover in polar deserts relative to the low Arctic. Other authors argue that changes in soil moisture will have their greatest effects on species diversity in the Arctic polar desert and in other high Arctic environments in which recruitment from seeds is more important than in the low Arctic (Callaghan & Jonasson 1995).

Stratospheric ozone depletion and the associated rise in UV-B radiation have been especially pronounced over Antarctica. For example, at McMurdo Station (in the vicinity of the Dry Valleys polar desert) the total atmospheric ozone concentration in 1993 fell from 275 Dobson Units (DU) in late August to 130 DU in early October (Johnson & Deshler 1994). Ozone depletion has been much less marked over the north polar region, but long term rises in UV-B are now apparent; for example, spring-time levels of UV at the wavelength 310 nm appear to have increased 10-20% in the subarctic between the late 1970s and 1995 (International Arctic Science Committee 1995). Current models of global UV predict that this rising trend will continue until about the turn of the century, with a recovery towards pre-ozone depletion levels over the subsequent 50 years (Madronich *et al.* 1995).

Desert environments in general contain simplified communities that are strongly regulated by climatic forcing. In his seminal work on ecosystem theory, Margalef (1968) argued the importance of community structure and cybernetic feedback processes in controlling the dynamics of temperate latitude ecosystems. When it came to desert ecosystems, however, he conceded that "desert ecologists working in arid countries where weather fluctuations exert a controlling influence on poorly organized communities, would hardly accept as a suitable basis for ecological theory (these) points of view". Given their responsiveness to the physical environment, deserts might be expected to provide better indicators of climate change than other ecosystems where biological interactions predominate and act as a buffer against external forcing.

Water availability is a primary controlling variable for biological processes in all desert environments, including polar deserts (Section 4.2 above). In polar regions this effect is compounded by the precarious balance between freezing and melting (Vincent 1988). Community structure and processes in the high Arctic and continental Antarctica may therefore be especially sensitive to long term climate change. Desert environments, however, also experience high interannual variability in precipitation (compounded by the annual variations in meltwater production in the polar zones) and the resultant year-to-year variability in their biological properties could obscure long term trends.

Lakes, ponds and streams are to be found in the desert environments of both polar regions, and they provide key monitoring sites for the early detection of global change effects. These inland aquatic ecosystems share many characteristics including strong seasonality of radiation, persistent low temperatures, low nutrient inputs and the pervasive influence of ice formation and decay. There are also important differences between the north and south polar zones, for example the presence of fish and the more diverse zooplankton and zoobenthos in Arctic freshwater ecosystems. In the aquatic environments of Arctic as well as Antarctic deserts, microbial food webs play a dominant role in the biological flux of carbon and energy. As in polar desert soils, cyanobacteria appear to be major elements of high latitude lake and stream ecosystems (Vézina & Vincent, in press, and references therein).

Global climate change is likely to impact the limnology of polar desert lakes in a variety of ways such as changes in the duration and thickness of snow and ice-cover (thus underwater light availability), in their thermal and associated chemical regimes (e.g. oxygen dynamics), and in near-surface mixing and stratification properties (e.g. the transition from cold-monomixis to dimixis). Lakes in general reflect the integrated effects of their surrounding catchments, and therefore global change may also influence desert lakes indirectly through climate-induced shifts in catchment vegetation or hydrology. The Arctic desert waterbody Lake Hazen (Fig. 2), may be especially interesting in this regard because its substantial water volume and residence time may smooth out year-to-year differences, and thereby reduce the large interannual variability noted above.

Global climate change is likely to impact the limnology of polar desert lakes in a variety of ways such as changes in the duration and thickness of snow and ice-cover (thus underwater light availability), in their thermal and associated chemical regimes (e.g. oxygen dynamics), and in near-surface mixing and stratification properties (e.g. the transition from cold-monomixis to dimixis). Lakes in general reflect the integrated effects of their surrounding catchments, and therefore global change may also influence desert lakes indirectly through climate-induced shifts in catchment vegetation or hydrology. The Arctic desert waterbody Lake Hazen (Fig. 2), may be especially interesting in this regard because its substantial water volume and residence time may smooth out year-to-year differences, and thereby reduce the large interannual variability noted above.

Recent research has drawn attention to the importance of dissolved organic carbon (DOC) in natural waters in controlling the spectral penetration of underwater light (particularly UV radiation) and its responsiveness to change in catchment processes (Schindler *et al.* 1996).

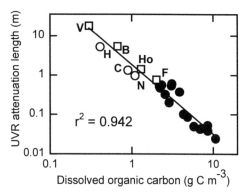

Fig. 4. UV transparency versus DOC in polar desert and subpolar lakes. Open symbols: desert lakes in the Dry Valleys, Antarctica (V - Vanda, B - East Bonney, Ho - Hoare, F - Fryxell) or in the Canadian high Arctic (H - Hazen, C - Char, N - North). Closed symbols: lakes in northern Québec at the subarctic treeline. Attenuation length is the reciprocal of the attenuation coefficient for 320 nm radiation measured in the upper water column; data are from Laurion *et al.* (1997) or unpublished records. Note the logarithmic scales

Polar desert lakes are unusual in that their catchments contain little vegetation and the input of UV-screening humic and fulvic acids is therefore small. A comparison of polar and subpolar lakes (Fig. 4) emphasizes two aspects of this effect: (1) the tight controlling influence of DOC on underwater UV penetration across this broad range of freshwater ecosystems, and (2) the similarity between Arctic and Antarctic desert lakes in occupying the extreme low DOC/high-UV transparency end of this relationship. Over the DOC range 0 to 4 g C m^{-3}, small changes in the concentration of coloured dissolved organic matter ('gelbstoff' or 'yellow substance') derived from the surrounding drainage basin can give rise to major non-linear variations in the spectral composition of underwater UV (Laurion *et al.*, in press). Bio-optical measurements such as UV spectral attenuation therefore offer potentially sensitive, integrative measures of the desert lake-plus-catchment response to global climate change. Such measurements could be conducted from aircraft or satellites, with remote sensing of the spectral fluorescence or absorbance signature of DOC. Furthermore, the ability to predict DOC from the record of fossil diatoms contained within lake sediments suggests that historical reconstructions may ultimately be possible, thereby allowing the interpretation of current changes within the context of climate-related variations in the past (Vincent & Pienitz 1997).

7   CONCLUSIONS

Terrestrial environments in the high Arctic are characterized by a cold, arid climate, often with sparse vegetation limited by water supply. These Arctic desert ecosystems share a number of climate and ecosystem properties with the polar deserts of Antarctica including low biological production rates and high sensitivity to environmental forcing. Biodiversity is greatly reduced by comparison with the terrestrial ecosystems at lower (including subpolar) latitudes, but to a much greater extent in continental Antarctica where many of the functional components of north polar desert communities (e.g. vascular plants, migrant birds, mammalian carnivores and herbivores) are completely absent.

Several features of polar desert ecosystems suggest that they may be especially sensitive to global change; these features include the measured and predicted magnitude of climate change at high latitude, the precarious freeze-thaw balance which controls many aspects of the polar environment (light, albedo, temperature, meltwater supply), and the general responsiveness of desert systems to variations in climate. These ecosystems, however, contain broad-niche genotypes that are highly tolerant of severe environmental variability.

In the desert environments of both polar regions, lakes are important foci of biological activity and can be considered downstream, multi-year integrators of landscape processes. Certain properties of these polar desert lakes such as their underwater UV optical properties offer exciting potential as integrative measures of catchment vegetation and hydrology, and as sensitive indicators of global change.

8 ACKNOWLEDGEMENTS

Financial support for my Arctic research group is provided by the Natural Sciences and Engineering Research Council of Canada, with logistic support from Centre d'études nordiques and Polar Continental Shelf Project (this is publication no. PCSP/ÉPCP/28). I also thank Mr. Barry Troke and the staff of Parks Canada (Nunavut District) for kindly providing access to their facilities in Ellesmere Island National Park Reserve for our research on Lake Hazen. The measurements in Fig. 4 were obtained in collaboration with C. Howard-Williams, I. Laurion, D. Lean, S. Markager, J. Priscu and R. Rae. I thank Dr. Reinhard Pienitz and two anonymous reviewers for helpful comments on the manuscript.

REFERENCES

Aleksandrova, V.D., 1988. *Vegetation of the Soviet polar deserts.* Cambridge University Press, Cambridge.

Bigl, S.R., 1984. Permafrost, seasonally covered ground, snow cover and vegetation in the USSR. *Cold Regions Research and Engineering Laboratory Special Report* 84-36. 128 p.

Bliss, L.C., 1979. Vascular plant vegetation of the Southern circumpolar region in relation to antarctic alpine and arctic vegetation. *Canadian Journal of Botany* 57:2167-78.

Bliss, L.C. & N.V. Matveyeva, 1992. Circumpolar arctic vegetation. In: Chapin, F.S., R. Jefferies, J. Reynolds, G. Shaver & J. Svoboda (Eds.), *Arctic ecosystems in a changing climate.* Academic Press, New York. pp. 59-89.

Bliss, L.C., J. Svoboda & D.I. Bliss, 1984. Polar deserts, their plant cover and plant production in the Canadian High Arctic. *Holarctic Ecology* 7:305-324.

Broady, P.A., 1981. The ecology of chasmolithic algae at coastal locations of Antarctica. *Phycologia* 20:259-72.

Bromley, A.M., 1985. Precipitation in the Wright Valley. *New Zealand Antarctic Record* 6:60-68.

Chapman, W.E. & J.E. Walsh, 1993. Recent variations in sea ice and air temperature in high latitudes. *Bulletin of the American Meterological Society* 74:33-47.

Chernov, Y. I., 1995. Diversity of the Arctic terrestrial fauna. In: Chapin, F.S. & C. Körner (Eds.). *Arctic and Alpine Diversity.* Springer-Verlag, Berlin. pp. 81-95.

Callaghan, T.V. & S. Jonasson, 1995. Implications for changes in Arctic plant biodiversity from environmental manipulation experiments. In Chapin, F.S. & C. Körner (Eds.). *Arctic and alpine diversity.* Springer-Verlag, Berlin. pp. 151-166.

Drewry, D.J., S.R. Jordan & E. Jankowski, 1982. Measured properties of the Antarctic Ice Sheet: surface configuration, ice thickness, volume and bedrock characteristics. *Annales of Glaciology* 3:83-91.

Henry, G.H.R. & J. Svoboda, 1986. Dinitrogen fixation (acetylene reduction) in high arctic sedge meadow communities. *Arctic Alpine Research* 18:181-7.

International Arctic Science Committee, 1995. Effects of increased ultraviolet radiation in the Arctic. *IASC Report No. 2*, Norway. 56 p.

Johnson, B.J. & T. Deshler, 1994. Record low ozone measured at McMurdo Station, Antarctica, during the austral spring of 1993. *Antarctic Journal of the United States* 29:249-51.

Kennedy, A.D., 1993. Water as a limiting factor in the Antarctic terrestrial environment: a biogeographical synthesis. *Arctic Alpine Research* 25:308-315.

Laurion, I., W.F. Vincent & D.R. Lean, in press. Underwater ultraviolet radiation: development of spectral models for northern high latitude lakes. *Photochemisty and Photobiology.*

Longton, R.E., 1988. *Biology of polar bryophytes and lichens*. Cambridge: Cambridge University Press.

McBean, G., 1994. Global change models - a physical perspective. *Ambio* 23:13-18.

Madronich, S., R.L. McKenzie, M.M. Caldwell & L.O. Björn, 1995. Changes in ultraviolet radiation reaching the Earth's surface. *Ambio* 24:143-52.

Margalef, R., 1968. *Perspectives in ecological theory*. University of Chicago Press, Chicago.

Nienow, J.A. & E.I. Friedmann, 1993. Terrestrial lithophytic (rock) communities. In Friedmann, E.I. (Ed.), *Antarctic microbiology*. Wiley Press, New York. pp. 343-412.

Powers, L.E., D.W. Freckman. & R.A. Virginia, 1995. Spatial distribution of nematodes in polar desert soils of Antarctica. *Polar Biology* 15:325-333.

Putnins, P., 1970. The climate of Greenland. In: Orvig, S. (Ed.). *Climates of the polar regions: world survey of climatology*, Volume 14:3-128. Elsevier, Amsterdam. pp. 3-128.

Remmert, H., 1980. *Arctic animal ecology*. Springer-Verlag, New York.

Schindler, D.W., P.J. Curtis, B.R. Parker & M.P. Stainton, 1996. Consequences of climate warming and lake acidification for UV-B penetration in North American boreal lakes. *Nature* 379: 05-708.

Schwarz, A-M.J., T.G.A. Green, & R.D. Seppelt, 1992. Terrestrial vegetation at Canada Glacier, southern Victoria Land, Antarctica. *Polar Biology* 12:397-404.

Schwerdtfeger, W. 1970. The climate of the Antarctic. In: S. Orvig (Ed.), *Climates of the polar regions: world survey of climatology*, 14. pp. 253-355.

Smith, R.I. Lewis, 1994. Vascular plants as bioindicators of regional warming in Antarctica. *Oecologia* 99:322-28.

Soper, J.H. & J.M. Powell, 1985. Botanical studies in the Lake Hazen region, northern Ellesmere Island. *National Museum of Natural Sciences Publications in the Natural Sciences, Ottawa* 5. 67 p.

Svoboda, J., 1994. The Canadian Arctic realm and global change. In: Riewe, R. & J. Oakes (Eds.), *Biological implications of global change: northern perspectives*. The Canadian Circumpolar Research Institute, Edmonton. pp. 37-47.

Van der Leeden, F., Troise, F.L. & D.K. Todd, 1990. *The water encyclopedia*. Lewis Publishers, Chelsea.

Vézina, S. & W.F. Vincent, in press. Arctic cyanobacteria and limnological properties of their environment: Bylot Island, Northwest Territories, Canada. (73°N, 80°W). *Polar Biology*.

Vincent, W.F. 1988. *Microbial ecosystems of Antarctica*. Cambridge University Press, Cambridge. 304 p.

Vincent, W.F. (Ed.), 1996. *Environmental management of a cold desert ecosystem*. University of Nevada Special Publication. 56 p.

Vincent, W.F & R. Pienitz, in press. Sensitivity of high latitude freshwater ecosystems to global change: temperature and solar ultraviolet radiation. *Geoscience*.

Vincent, W.F. & S. Roy, 1993. Solar ultraviolet-B radiation and aquatic primary production: damage, protection and recovery. *Environmental Reviews* 1:1-12.

Vowinckel, E. & S. Orvig, 1970. The climate of the north polar basin. In: Orvig, S. (Ed.), *Climates of the polar regions: world survey of climatology*, 14. Elsevier, Amsterdam. pp. 129-252.

Wilson, C., 1967. *Climatology of the cold regions*. Cold Regions Research and Engineering Laboratory Special Report. 147 p.

Wynn-Williams, D.D., 1990. Ecological aspects of Antarctic microbiology. *Advances in Microbial Ecology* 11:71-146.

Ecosystem Processes in Antarctic Ice-free Landscapes, Lyons, Howard-Williams & Hawes (eds)
© *1997 Balkema, Rotterdam, ISBN 90 5410 925 4*

Climate history of the McMurdo Dry Valleys since the last glacial maximum: A synthesis

W. Berry Lyons & Louis R. Bartek
Department of Geology, University of Alabama, Tuscaloosa, Ala., USA

Paul A. Mayewski
Institute for the Study of Earth, Oceans and Space, University of New Hampshire, Durham, N.H., USA

Peter T. Doran
Desert Research Institute, Reno, Nev., USA

ABSTRACT: The biological development of the McMurdo Dry Valleys (MDVs) terrestrial and aquatic ecosystems are closely tied to the climatic history of the region. Although much interest and recent debate has occurred about the longer term (i.e. millions of years) climatic variation of the MDVs, there has been little attempt at a synthesis of information relating the shorter term climatic history. We have utilized recent data from the ice-covered lakes in the MDVs and from ice cores within the region (Newell Glacier and Taylor Dome) along with marine sediment, paleo-biological studies, as well as glacial and geological information to develop a detailed picture of the climate in the MDV region in the Holocene. The data indicate that the region has been subjected to a number of significant climatic variations over this period that reflect responses to regional and global climate changes. These include a mid-Holocene warming, then cooling, and, what appears to be a Medieval Warm Period and a Little Ice Age signal. The climatic changes will be discussed in terms of the potential biological responses to these changes.

1 INTRODUCTION

The importance of climate and climatic change on ecosystem development has received a great deal of attention in the past decade as ecologists and paleoclimatologists have thought about future anthropogenically induced changes in atmospheric chemistry and landscape development. Even though the dry valleys of Antarctica are among the coldest and driest places on the planet, ecosystems exist there, as evidenced by the papers presented in this volume. These ecosystems, like their more temperate counterparts, must also be affected by climatic change. In fact, it is now clear that what would be considered very small variations in temperature, and perhaps, humidity and precipitation in more temperate regions, have potentially great impact on the environment in these polar desert areas. This is due, in part, to the fact that the change of state of water is delicately poised in these environments and small changes in temperature have profound impacts on the hydrologic budgets. For example, if there are many summer days when the temperatures rise above freezing, the liquid water produced results in increasing streamflow and lake levels, while fewer days above freezing creates the opposite effect.

The purpose of this paper is to summarise what is known about climatic change in the McMurdo region of Victoria Land since the last glacial maximum. This work is not meant to overlap with other recent reviews on the subject such as Prentice *et al.* (1993) and Doran *et al.* (1994), but rather draw on them as well as other types of data from the glaciochemical and oceanographic literature, in order to establish a more comprehensive picture of climate change in the region. Denton *et al.* (1989) have demonstrated that the Ross Sea ice lobe began its retreat out of the McMurdo region (Taylor Valley) by ~13 000 BP and it was completed by 6600 to 6000 y BP. So, by this time, the valleys were indeed valleys and since the anthropogenic period (last 100 years or so) there is evidence of increased glacier melting and lake ice thinning. Research covering the most recent time period indicates lake level rising and lake ice thinning over the past 30 to 90 years (Chinn 1993; Wharton *et al.* 1993; Webster *et al.* 1996). This lake rise was apparently associated with a decrease in annual snowpatch extent (Mayewski & Lyons 1982). What happened in between these two time periods?

2 PREVIOUS WORK

2.1 *Lakes*

The work of Wilson (1964), Hendy *et al.* (1977), Matsubaya *et al.* (1979) and Chinn (1993) and the review of Doran *et al.* (1994) all have discussed the paleoclimatic history of the lakes in Taylor and Wright Valleys (Fig. 1). The $\delta^{18}O$ and δD of water and Cl⁻ profiles from Lake Bonney (in Taylor Valley) and Lake Vanda (in Wright Valley) indicate a dramatic drawdown of these lakes, evaporation of water and evapoconcentration of solutes in both the east lobe of Lake Bonney and Lake Vanda that ended to ~1000 to 1500 y BP. This event has been interpreted as a prolonged cold/dry period. Matsubaya *et al.* (1979) have argued further that the west lobe of Lake Bonney maintained its ice cover because its isotopic chemistry has been unaffected by evaporation. In fact, the water column of Lake Bonney has $\delta^{18}O$ throughout that is similar to the waters entering today from streams emanating from the Taylor Glacier (Fig. 1). Those streams had $\delta^{18}O$ of -38 to -42‰ in 1993-94 field season, while the west lobe of Lake Bonney values reported by Matsubaya *et al.* (1979) ranged from -39.5 to -42.4‰. This suggests that although the climate was colder and drier than present, Taylor Glacier runoff continued to flow into the west lobe of Lake Bonney. This took place while sources of water to Lake Vanda and the east lobe of Lake Bonney decreased or stopped all together.

More recent work in Lake Hoare and Lake Fryxell indicates that Lake Fryxell probably went through a similar drawdown phase that ended ~1000 y BP, while Lake Hoare did not (Lyons *et al.*, unpubl.). The Lake Hoare data has been interpreted as suggesting that Lake Hoare itself is a relatively new lake having only been in existence since the climatic amelioration after ~1000 to 1500 y BP (Lyons *et al.* unpubl.). An alternative is that Lake Hoare evaporated to complete dryness at this time and the salts ablated prior to refilling. On the other hand, Doran (1996) sees no evidence of this drawdown event in Lake Hoare sediments, but Spaulding *et al.* (in press) demonstrate a dramatic change in diatom species make-up at ~1200 y BP. Evidence from Lake Wilson, 320 km south of the McMurdo Dry Valleys shows a similar pattern of a small brine lake at *ca.* 1000 y BP and a subsequent refilling (Webster *et al.* 1996).

2.2 *Support for drawdown/refill event from ice record*

There are other data sets available from both Northern and Southern Victoria Land highlighting this period of change at ~1000 to 1500 y BP. Mayewski *et al.* (1979) have documented alpine glacier retreat (i.e. warming) at ~1200 y BP in the Rennick Glacier area of Northern Victoria Land. Ice core records form the Newell Glacier in the Asgard Range (~2000 m) between Taylor and Wright Valleys (Fig. 1) indicate increased salt deposition, interpreted as drier and windier conditions, between ~3000 and 4000 y to 1500 y BP. In addition, there is an enrichment of $\delta^{18}O$ in the precipitation (i.e. warming), as well as an increase in methylsulfonic acid (MSA) at ~1500 y BP (Mayewski *et al.* 1995). The increase in MSA has been interpreted as a decrease in sea ice due to warming and a subsequent increase in biogenically produced MSA in the nearby Ross See environs (Welch *et al.* 1993). Ice core records from the Dominion Range, much further south (>85°S latitude), also indicate cooler temperatures between 1500 to 2000 y BP (Mayewski *et al.* 1995). Meserve Glacier (Wright Valley) records indicate a drier, windier period from 230 to 325 y BP, potentially associated with the Little Ice Age.

2.3   Support from coastal/marine record

Baroni & Orombelli (1994) have ^{14}C-dated abandoned penguin rookeries throughout the Ross Sea region. Their findings can be summarised as follows. The earliest Adelie rookeries date at between 11 000 to 13 000 y BP and there was a continuous presence of Adelies from 7000 y BP, or about the time of the complete retreat of the Ross Sea ice lobe from the region. They found that the greatest abundance of rookeries in the area occurred between 3000 to 4000 y BP. They termed this time the "Penguin Optimum"! From ~3000 to 1200 y BP there was a decrease in rookery numbers, which they attributed to an increase in sea ice extent. Finally, they observed an increase in rookeries from ~1000 to 600 y BP. The decrease in rookeries and increase in sea ice extent from ~3000 until 1200 y BP and their subsequent reversal coincides with both the ice core and the dry valley lake records. These data indicate that the change from cold/dry to warm/wet (?) conditions at ~1500 y BP was a regional phenomenon and was forced by a change in oceanic conditions. This is evidenced by the fact that there are no data showing this change from more inland locations such as the detailed ice core records from Taylor Dome ~100 km inland from Taylor Valley (Mayewski *et al.*, in press). The Taylor Dome records do, however, provide some additional information about this time period in the form of increasing/ decreasing Cl⁻ concentrations with time. Increased Cl⁻ values in the ice have been interpreted as times of increased marine storminess (Mayewski *et al.*, in press). Highs in Cl⁻ occurred at ~640 y, 1000 y, 2400 y and 3000 y BP, while lows occurred at 1200 y and 3550 y BP. Prior to 6000 y BP, the concentrations were very low, again supporting the presence of extensive sea ice which

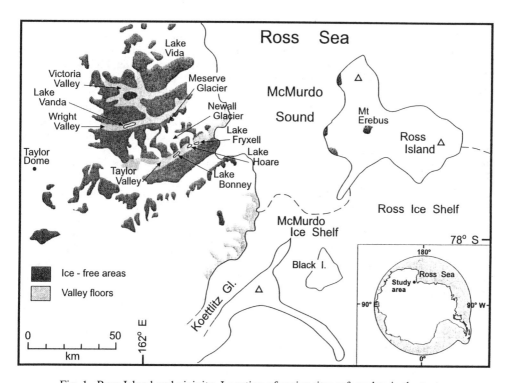

Fig. 1. Ross Island and vicinity. Location of major sites referred to in the text.

minimises the flux of Cl⁻ into the regional atmosphere. These low values occurred during the Penguin Optimum and at about the same time as the climate turned warmer/wetter (?).

Licht *et al.* (1996) have analysed a series of sediment cores from the Ross Sea. They observed that the Ross Ice Shelf reached its present position at ~7000 y BP, which coincides with the original hypothesis of Denton *et al.* (1989) as well as the work cited above. The area around Drygalski Ice Tongue (<75°S) was deglaciated at ~11 500 y BP, indicating a rather slow retreat over that period of 4500 y.

Goodwin (1995) has agreed that the Law Dome ice margin to the north and west of the McMurdo region advanced due to a positive surface mass balance between 4000 to 2500 y BP, coinciding with the Penguin Optimum. This was no doubt a period of warmer and more moist conditions along the East Antarctic coastline.

The scenario presented above, that is a warm/wet period from ~4000 y to ~3000 y BP followed by a cold/dry period from ~3000 y to between 1500 to 1000 y BP, followed by another warm/wet period is similar to the one proposed for James Ross Island, off the Antarctic Peninsula by Björck *et al.* (1996). These authors, using lake sediment data, propose a humid/warmer period starting at 4200 y BP and terminating at 3000 y BP. During this time, glaciers became absent in the region. The arid/colder period lasted until ~1200 y BP, but the current climatic optimum was not as warm and wet as the Penguin Optimum period. These authors regard the ~4000 to 3000 y BP period as a circumpolar event and relate it, in part, to the strength of the high pressure cell over the ice sheet. Indeed, recent work as far north as the Tasman Plateau in the Southern Ocean indicates a maximum Holocene ocean warming at ~3000 to 4000 y BP as evidenced by both alkenone and $\delta^{18}O$ of foramaniferan data (Ikehara *et al.* 1997). It is probably not a coincidence that the McMurdo Dry Valley data show synchronous climatic variation with these other coastal locations in both West and East Antarctica.

2.4   Most recent, pre-anthropogenic climate change

Leventer & Dunbar (1988) analysed the diatom flora from surface samples within McMurdo Sound and determined that there are floral assemblages that are indicative of both open water (*Thalassiosira* spp.) and sea ice covered conditions (*Nitzschia curta*) within McMurdo Sound. Analysis of these assemblages in cores from McMurdo Sound revealed a downcore increase in the percentage of open water species (Leventer & Dunbar 1988). Extrapolation of sedimentation rates determined through ^{210}Pb dating of the cores indicates that the increase in deposition of the open water assemblage occurred between 1600 to 1875 AD. The timing of the increased flux of open water species coincides with the Maunder Minimum ("Little Ice Age"). Leventer & Dunbar argue that katabatic winds were more persistent and more intense during the colder climatic conditions of the Little Ice Age, and that these winds caused extensive polynya formation along the coastline of the western Ross Sea. Therefore, deposition of primarily open water flora during the Little Ice Age is presumed to be a manifestation of extensive polynya formation (Leventer & Dunbar 1988). Cooler coastal water temperatures ~500 y BP observable in fossil shells at Terra Nova Bay, north of the McMurdo Dry Valleys (Bartoni & Orombelli 1991; Berkman 1994), along with these offshore data strongly suggest a Little Ice Age signature. The Wright Valley ice core data of Mayewski & Lyons (1982) support this notion. Apparently, this event was also a circum-coastal Antarctic phenomenon as lake sediment data from the Vestfold Hills, East Antarctica also demonstrate cold periods there from 400 to 250 y BP (Bronge 1992).

Bartek *et al.* (1991, 1996) have described a series of lithofacies in the McMurdo Sound region. Their "facies H" is composed of quartzose material, it is confined to the region that lies adjacent to Granite Harbor, MacKay Glacier, and its stratigraphic position correlates to the increase in open water flora in box cores and to the hard sand/gravelly layer that was described by Leventer & Dunbar (1988). Therefore, it is concluded that deposition of Facies H is partially related to the advance of the MacKay Glacier during the Little Ice Age.

A mechanism that may explain the expansion of the MacKay Glacier during the Little Ice Age has been proposed by Denton *et al.* (1970) and Drewry (1980). It has been proposed that the advances and retreats of valley glaciers in the McMurdo Sound region are out of phase with the Antarctic ice sheet fluctuations. It is presumed that the moisture supply (Ross Sea) for the valley glacier system is cut-off during glacial maxima and that the valley glaciers wane. When the Antarctic ice sheet wanes, the Ross Sea is not as extensively covered with ice, the moisture supply to the valley glacier system increases, and the valley glaciers wax. The intense katabatic winds of the Little Ice Age created an extensive polynya in the McMurdo Sound region which supplied moisture to the valley glacier system. Presumably, the increased moisture supply and cold conditions nourished MacKay Glacier and it advanced into McMurdo Sound. Thus, the Little Ice Age created environmental conditions for the valley glacier system in the McMurdo Sound region that mimicked and ice sheet waning episode.

3   CHRONOLOGY OF CLIMATE EVENTS

A summary of the climatic data for the McMurdo Dry Valley region is tabulated below:

- Ross Sea open to present stage by ~7000 y BP.
- Warming period between ~7000 and ~3000 y BP as evidenced by glacier advances (Domack *et al.* 1991).
- "Penguin Optimum" - 4000 to 3000 y BP - as evidenced by data from the McMurdo region as well as other coastal areas of Antarctica, including the Peninsula (Björck *et al.* 1996).
- Lake Vanda high water level at ≤3000 y BP (Friedman *et al.* 1995) forming "Great Lake Vanda".
- Cold/dry period begins at ~3000 y to ≤1500 y (~1000 y) BP, with extreme low stands of lakes in the McMurdo Dry Valleys (Lyons *et al.*, unpubl.) and further south (Webster *et al.* 1996).
- Wetter/warmer period since ~1000 y BP - coincident with Medieval Warm Period in the Northern Hemisphere.
- Colder beginning between ~1350 AD (ice core data) and 1600 AD (marine sediment data).
- Warming since at least 1975 AD, perhaps as early as ~1900 AD as evidenced by McMurdo Dry Valley lake levels and lake ice. Recent lake level increases have also been observed in Vestfold Hills, East Antarctica lakes since 1978 AD (Fulford-Smith & Sikes 1996).

4   BIOLOGICAL IMPLICATIONS OF CLIMATE CHANGE WITHIN THE DRY VALLEY ECOSYSTEMS

Past climates in polar desert environments such as the McMurdo Dry Valleys strongly overprint the present ecological conditions in these systems. This concept, referred to as legacy, implies that carryover or memory can be an extremely significant force in ecosystem development and evolution (Vogt *et al.* 1996). Although we are just beginning to understand the significance of legacy within the McMurdo Dry Valley ecosystem, examples within the lakes have been suggested. Priscu (1995) has demonstrated using both the dissolved inorganic nitrogen to soluble reactive phosphorus ratios within the lakes and nutrient enrichment experiments that Lakes Hoare and Fryxell are probably *N* deficient, while Lakes Bonney and Vanda are *P* deficient. He has shown that the primary reason for this is not any biochemical or physical process occurring today, but instead the geochemical variations of these nutrients in the deep waters (below the chemoclines) of these lakes. Because the nutrients in the deep waters of Lake Bonney (Matsubaya *et al.* 1979) and Lake Vanda (Wilson 1964) are remnants of previous, smaller, more hypersaline lakes (as described above), the present nutrient dynamics in the photic zones are a legacy of the past. This "nutrient memory" represents a major influence on the current biology of the lakes. The relationship between disturbance and recovery in these systems are probably greatly influenced by these climatic legacies that, in turn, may be induced by relatively small changes (a few degrees of temperature or cms of annual precipitation) in climatic variables. Because a change of only a few degree days in either direction above or below 0°C may have pronounced affects on the hydrologic cycle within the McMurdo Dry Valleys (Clow *et al.* 1988), these polar deserts may be much more influenced by small climatic changes and their legacies than either temperate or tropic aquatic environments.

What impacts did the above described climatic changes have on the terrestrial ecosystem in these valleys? Little can be substantiated at this time. However, it is clear that the extent of the terrestrial ecosystem has waxed and waned as the lakes have receded and grown. Certainly soil moisture, carbon flux and salt accumulation have varied as the hydrologic conditions changed over the past 8000 to 10 000 y. If liquid water is the key ingredient (Kennedy 1993), times of greater water fluxes from the glaciers must lead to important changes in the stream ecosystems and perhaps the soils as well. Much more information is needed on the understanding of how the aquatic and terrestrial ecosystems relate to and influence each other, before these questions can be answered in any definitive way.

5   CONCLUSIONS

In our review of the variations in climate in the McMurdo Dry Valleys over the Holocene, it is clear that there have been dramatic changes in both lacustrine and coastal oceanographic environments. These relatively small changes have undoubtedly had very significant impacts on the aquatic ecosystems within the valleys. The lakes within Taylor Valley have changed from very large, glacial lakes that occupied the majority of the valley floor to small, non-ice covered playas. Legacy effects on the ecosystems from post climates probably have been extremely significant in their evolution and development. To date, the impact of legacy has only been discussed in regard to the lacustrine ecosystems (Priscu 1995), but it has undoubtedly been important in all aspects of the ecology of the McMurdo Dry Valleys. Only through an integrated, interdisciplinary approach will future research be able to discern both better paleoclimate information and its role in understanding its influence on ecological change within the valleys.

6   ACKNOWLEDGMENTS

This work was supported in part by NSF grant OPP-9211773. The authors thank Clive Howard-Williams and Ian Hawes for inviting them to present this talk at the workshop. The senior author especially thanks his MDV-LTER (Long Term Ecosystem Research) site colleagues for an animated discussion of legacy and its importance on the ecology of the MDVs. They include Diana Freckman, Diane McKnight, Daryl Moorhead, Andrew Fountain, Cathy Tate and Bob Wharton. Thanks also to John Priscu for many discussions with regard to climate change and MDV lake evolution. We thank Pat Smith for typing the manuscript and Diane Norris for the production of Fig. 1.

REFERENCES

Baroni, C. & G. Orombelli, 1991. Holocene raised beaches at Terra Nova Bay, Victoria Land, Antarctica. *Quaterary Research* 36:157-177.
Baroni, C. & G. Orombelli, 1994. Abandoned penguin rookeries as Holocene paleoclimatic indicators in Antarctica. *Geology* 22:23-26.

Bartek, L.R., S.A. Henrys, J.B. Anderson & P.J. Barrett, 1996. Seismic stratigraphy of McMurdo Sound, Antarctica: implications for glacially influenced early Cenozoic eustatic change? *Marine Geology* 130:79-98.

Bartek, L.R., P.R. Vail, J.B. Anderson, P.A. Emmett & S. Wu, 1991. Effect of Cenozoic ice sheet fluctuations in Antarctica on the stratigraphic signature of the Neogene. *Journal of Geophysiology Research* 96:6753-6778.

Berkman, P.A., 1994. Geochemical signatures of meltwater in mollusc shells from Antarctic coastal areas during the Holocene. *Memoirs of the National Institute of Polar Research* 50:11-33.

Björck, S., S. Olsson, C. Ellis-Evans, H. Håkansson, O. Humlum & J.M. deLirio, 1996. Late Holocene palaeoclimatic records from lake sediments on James Ross Island, Antarctica. *Paleo* 121:195-220.

Bronge, C., 1992. Holocene climatic record from lacustrine sediments in a freshwater lake in the Vestfold Hills, Antarctica. *Geografiska annaler* 74A:47-58.

Chinn, T.J., 1993. Physical hydrology of the Dry Valley lakes. In: Green, W. & E.I. Friedmann (Eds.), *Physical and Biogeochemical Processes in Antarctic Lakes. Antarctic Research Series* 59. American Geophysical Union, Washington D.C. pp. 1-52.

Clow, G.D., C.P. McKay, G.M. Simmons, Jr. & R.A. Wharton, Jr. 1988. Climatological observations and predicted sublimation rates at Lake Hoare, Antarctica. *Journal of Climate* 7:715-728.

Denton, G.H., R.C. Armstrong & M. Stuiver, 1970. Late Cenozoic glaciation in Antarctica: the record in the McMurdo Sound region. *Antarctic Journal of the United States* 5(1):15-21.

Denton, G.H., J.G. Bockheim, S.C. Wilson & M. Stuiver, 1989. Late Wisconsin and early Holocene glacial history, inner Ross embayment, Antarctica. *Quaternary Research* 31:151-182.

Domack, E.W., A.J.T. Jull & S. Nakao, 1991. Advance of East Antarctic outlet glaciers during the Hypsithermal: implications for the volume state of the Antarctic ice sheet under global warming. *Geology* 19:1059-1062.

Doran, P.T. 1996. Paleolimnology of perennially ice-covered Antarctic oasis lakes. Ph.D. Dissertation. University of Nevada, Reno. 195 pp.

Doran, P.T., R.A. Wharton, Jr. & W.B. Lyons, 1994. Paleolimnology of the McMurdo Dry Valleys, Antarctica. *Journal of Paleolimnology* 10:85-114.

Drewry, D.J., 1980. Pleistocene bimodal response of Antarctic ice. *Nature* 287:214-216.

Friedman, I., A. Rafter & G.I. Smith, 1995. A thermal, isotopic and chemical study of Lake Vanda and Don Juan Pond. In: *Contributions to Antarctic Research IV. Antarctic Research Series* 67. American Geophysical Union, Washington D.C. pp. 47-74.

Fulford-Smith, S.P. & E.L. Sikes, 1996. The evolution of Ace Lake, Antarctica, determined from sedimentary diatom assemblages. *Paleo* 124:73-86.

Goodwin, I.D., 1995. On the Antarctic contribution to Holocene sea-level. PhD Dissertation. University of Tasmania. 315 pp.

Hendy. D.H., A.T. Wilson, K.B. Popplewell & D.A. House, 1977. Dating of geochemical events in Lake Bonney, Antarctica and their relation to glacial and climatic changes. *New Zealand Journal Geology and Geophysiology* 20:1103-1122.

Ikehara, M., K. Kawamura, N. Ohkouchi, K. Kimoto, M. Murayama, T. Nakamura, T. Oba & A. Taira, 1997. Alkenone sea surface temperature in the Southern Ocean for the last two deglaciations. *Geophysical Research Letters* 24:679-682.

Kennedy, A.D., 1993. Water as a limiting factor in the Antarctic terrestrial environment: a biogeographical synthesis. *Arctic Alpine Research* 25:308-315.

Leventer, A.R. & E.R.B. Dunbar, 1988. Diatom flux in McMurdo Sound, Antarctica. *Marine Micropaleology* 12:49-64.

Licht, K.J., A.E. Jennings, J.T. Andrews & K.M. Williams, 1996. Chronology of late Wisconsin ice retreat from the western Ross Sea, Antarctica. *Geology* 24:223-226.

Lyons, W.B., S.W. Tyler, R.A. Wharton, Jr., D.M. McKnight & B.H. Vaughn, unpublished. The late Holocene paleoclimate history of the McMurdo Dry Valleys, Antarctica as derived from lacustrine isotope data.

Matsubaya, O., H. Sakai, T. Torii, H. Burton & K. Kerry, 1979. Antarctic saline lakes—stable isotopic ratios, chemical compositions and evolution. *Geochimica et Cosmochimica Acta* 43:7-25.

Mayewski, P.A., J.W. Attig, Jr. & D.J. Drewry, 1979. Pattern of ice surface lowering for Rennick Glacier, northern Victoria Land, Antarctica. *Journal of Glaciology* 22:53-65.

Mayewski, P.A. & W.B. Lyons, 1982. Source and climatic implication of the reactive iron and reactive silicate concentration found in a core from Meserve Glacier, Antarctica. *Geophysiology Research Letters* 9:190-192.

Mayewski, P.A., W.B. Lyons, G. Zielinski, M. Twickler, S. Whitlow, J. Dibb. P. Grootes, L. Forsberry, C.

Wake & K. Welch, 1995. An ice core based late Holocene history for the Transantarctic Mountains., Antarctica. In: *Contributions to Antarctic Research* IV. *Antarctic Research Series* 67. American Geophysical Union, Washington D.C. pp.33-45.

Mayewski, P.A., M.S. Twickler, S.I. Whitlow, L.D. Meeker, Q. Yang, J. Thomas, K. Krentz, P. Grootes, D. Morse, E. Steig & E.D. Waddington, in press. Climate change during the last deglaciation in Antarctica. *Science.*

Prentice, M.L., J.G. Bockheim, S.C. Wilson, L.H. Burckle, D.A. Hodell, C. Schlüchter & D.E. Kellogg, 1993. Late Neogene Antarctic glacial history: evidence from central Wright Valley. In: *The Antarctic Paleoenvironment: A Perspective on Global Change. Antarctic Research Series* 60. pp. 207-250.

Priscu. J.C., 1995. Phytoplankton nutrient deficiency in lakes of the McMurdo dry valleys, Antarctica. *Freshwater Biology* 34:215-227.

Spaulding, S.A., D.M. McKnight, E.F. Stoermer & P.T. Doran, in press. Diatoms in sediments of Lake Hoare, Antarctica. *Journal of Paleolimnology.*

Vogt, K.A. & ten others, 1996. Eco-systems: balancing science with management. Springer. New York.

Webster, J., I. Hawes, M.T. Downes, M. Timperley & C. Howard-Williams, 1996. Evidence for regional climate change in the recent evolution of a high latitude, pro-glacial lake. *Antarctic Science* 8:49-59.

Welch, K.A., P.A. Mayewski & S.I. Whitlow, 1993. Methane sulfonic acid in coastal Antarctic snow related to sea-ice extent. *Geophysical Research Letters* 20:443-446.

Wharton, Jr., R.A., C.P. McKay, C.D. Clow & D.T. Anderson, 1993. Perennial ice covers and their influence on Antarctic lake ecosystems. In: Green, W/ & E.I. Friedmann (Eds,), *Physical and Biogeochemical Processes in Antarctic Lakes. Antarctic Research Series* 59. American Geophysical Union, Washington D.C. pp. 53-70.

Wilson, A.T., 1964. Evidence from chemical diffusions of a climate change in the McMurdo Dry Valley 1200 years ago. *Nature* 201:176-177.

Ecosystem Processes in Antarctic Ice-free Landscapes, Lyons, Howard-Williams & Hawes (eds)
© *1997 Balkema, Rotterdam, ISBN 90 5410 925 4*

UV radiation in polar regions

Greg Bodeker
National Institute of Water and Atmospheric Research Ltd, Lauder, Central Otago, New Zealand

ABSTRACT: Recent polar ozone depletion and the concomitant increase in surface ultraviolet (UV) radiation may be expected to adversely influence the polar biosphere. To investigate the effect of long-term changes in total column ozone on the spatial and temporal variability of biologically harmful radiation, daily global ozone maps have been obtained from two satellite experiments and used as input to a simple radiative transfer model. Four different data products of erythemally weighted irradiance, at monthly resolution, and in 1° latitude zones, were produced for the period 1979 to 1994; the time period spanned by the satellite experiments. The results show that significant increases in erythemal UV can be expected over Antarctica and the surrounding oceans over this period. In contrast, increases over the high latitudes of the Northern Hemisphere are significantly smaller. The magnitude of the hemispheric differences in the long-term change in surface UV irradiance will be sensitive to the action spectrum (biological weighting function) applied. For biological processes responding to UV in a similar way to the erythemal action spectrum, this study suggests that the marginal ice zone of the Antarctic continent provides the best site for investigations of the effects of enhanced UV-B as a result of ozone depletion.

1   INTRODUCTION

Anthropogenic production and release of chlorofluorocarbons (CFCs) has resulted in significant depletion of stratospheric ozone, particularly in the polar regions, since the early 1980s (Farman *et al.* 1985). While this depletion has been constrained primarily to the high latitudes of the Southern Hemisphere, significant Northern Hemisphere depletion and weaker extratropical decreases have recently become apparent (Herman *et al.* 1993; Gleason *et al.* 1993; Herman & Larko 1994). These ozone trends and the concomitant increase in ultraviolet (UV) radiation may be expected to adversely influence the biosphere. Increased biologically damaging UV-B radiation (280 to 315 nm) has been measured in the Antarctic, both at the surface (Lubin *et al.* 1989; Stamnes *et al.* 1992; Roy *et al.* 1994) and at ecologically significant depths in the marginal ice zone (Smith *et al.* 1992; Herndl *et al.* 1993; Prezelin *et al.* 1994).

For a given decrease in total column ozone, the increase in UV-B is greater than the increase in UV-A (315 to 400 nm) which in turn is greater than the increase in photosynthetically active radiation, PAR (400 to 700 nm) as a result of the wavelength dependence of the ozone absorption cross-section. This has important implications for organisms which rely on UV-A or PAR to trigger defense mechanisms.

At the time of the discovery of the Antarctic ozone hole no routine UV-B monitoring programme was in place in the Antarctic. In 1988 the US National Science Foundation (NSF) established a network of scanning UV-B spectroradiometers at three Antarctic stations (McMurdo (166.40°E, 77.51°S), Palmer (64.03°E, 64.46S) and South Pole (0°E, 90°S)) and one station in Argentina (Ushuaia (68.19°W, 54.59°S)). Later, additional sites at San Diego and Barrow (Alaska) were added. A description of these sites and their data products is given by Booth *et al.* (1994). Data obtained from this monitoring programme shows that while the maximum monthly and yearly erythemal doses (sunburning irradiance; see later) among the six sites occurs at San Diego, the maximum weekly erythemal dose occurs at the South Pole and the maximum daily and hourly doses occur at Palmer station.

Calculations suggest that for equivalent latitudes, locations in the Southern Hemisphere should receive approximately 15% more UV than in the Northern Hemisphere. However, measurements have shown that erythemal UV in the Southern Hemisphere can exceed that at comparable latitudes in Europe by more than 50% (Seckmeyer *et al.* 1995). Over the summer months, the UV received can be relatively large at all southern latitudes. Summertime monthly means of spectroradiometer measurements of daily erythemal doses show only a small latitudinal gradient in the Southern Hemisphere and a large gradient in the Northern Hemisphere (Seckmeyer *et al.* 1995). The Southern Hemisphere latitudinal gradient in the monthly maximum daily integrals is almost negligible.

Measurements have shown that exposure to enhanced levels of UV-B radiation decreases algal productivity and causes damage to various forms of aquatic larvae and other organisms (Smith *et al.* 1992 and references therein). Ocean experiments in the Bellingshausen Sea in the austral spring of 1990 ("Icecolors") showed that decreases in ozone increased UV-B inhibition of photosynthesis. A minimum 6 to 12% reduction in primary production associated with ozone depletion was estimated for the duration of the cruise (Smith *et al.* 1992). Further results from the "Icecolors" campaign showed that strong vertical stratification in the ocean concentrates and restricts algal blooms to near surface waters of the marginal ice zone. These blooms proceed southward as the ice edge recedes. Coincident measurements of spectral irradiance at and below the ocean surface showed evidence of ozone related UV-B inhibition of photosynthesis to depths of 25 m. The presence of the Antarctic ozone hole increases the effective ocean penetration depth, the additional depth required to ensure the same UV-B levels, by about 7m. Since phytoplankton are cued to short-term changes in UV-A flux as an index of changing total UV flux, the ratio of UV to UV-A or visible (vis) radiation is important when investigating the effects of changes in UV on marine organisms. The UV/vis ratio in the presence of the ozone hole is larger than would ever exist, even at summer solstice, under normal ozone conditions. However, care must be taken when interpreting these results. As concluded during the "Icecolors" campaign, the estimated 2 to 4% UV-B induced loss to marginal ice zone productivity should be viewed in the context of a presumed natural variability of ±25%. Analyses of diatom assemblages from high-resolution stratigraphic sequences from anoxic basins in fjords of the Vestfold Hills, Antarctica, showed that the compositional changes in the diatom component of the phytoplankton community over the past 20 years cannot be distinguished from long-term natural variability, although there is some indication of a decline in the production of some sea-ice diatoms (McMinn *et al.* 1994).

2   THE POLAR UV ENVIRONMENT

Within the bounds of currently observed ozone depletion, the factors influencing surface UV irradiance received on a horizontal surface, in approximate order of decreasing importance for polar regions, are:

- Solar zenith angle (SZA) which determines the optical path length.
- Ozone; both the integrated column amount and its vertical distribution.
- Cloud cover, and in particular the geometry of the cloud cover (Lubin & Jensen 1995).
- Surface albedo.
- Aerosol loading of the atmosphere.
- Earth-Sun separation which modulates the extraterrestrial solar irradiance.
- Altitude (air pressure): Model calculations suggest an approximate 5% increase in UV-B per 1 km increase in altitude but may vary depending on other atmospheric parameters.

An objective of current research is to investigate how the relative importance of these factors depends on wavelength and geographic location. Each of these factors plays a unique role in determining the UV environment of the polar regions.

Solar zenith angles (SZAs) are generally large at high latitudes which result in long atmospheric light paths, increased absorption of solar UV radiation, and larger diffuse/direct ratios (the ratio of scattered to unscattered solar irradiance, incident at the surface). During polar summer, when the sun is continuously above the horizon, daily doses of UV can be large. Furthermore, it has been shown theoretically that at large solar zenith angles and high surface albedos (as found in polar regions), scattering by stratospheric aerosols can increase the surface UV flux (Tsitas & Yung 1996). In the UV, albedos for typical surfaces such as grass or sand seldom exceed 5%. However, for snow or ice covered surfaces, the albedo can vary between 70% and 95% (Beaglehole & Carter 1992).

There are large geographic differences in cloud cover over the Antarctic continent. More than half the days at Australian stations (Mawson (67.6°S, 62.9°E), Casey (66.3°S, 110.5°E) and Davis (68.6°S, 77.97°E)) experience solar radiation equal to or greater than 90% of the maximum expected. At Palmer station on the Antarctic Peninsula, more cloudy conditions result in the average surface UV irradiance of just 50 to 60% of that expected under clear skies (Lubin & Frederick 1991). There is very little cloud cover at the South Pole station throughout the year (Booth *et al.* 1994). The high surface albedos of the polar regions reduce the effective optical depths of clouds through multiple scattering of irradiance between the cloud base and the surface.

Clouds increase the tropospheric light path, and thereby absorption by tropospheric ozone (Erle *et al.* 1995). However, the effect plays a greater role at smaller SZAs and for large tropospheric ozone concentrations, more typical of the midlatitudes of the Northern Hemisphere. Enhanced backscattering by clouds increases the fraction of radiation absorbed in the atmosphere.

Transmittance of UV and visible radiation through clouds is generally independent of wavelength. However, different cloud geometries can cause spectrally dependent effects (Bodeker & McKenzie 1996) viz.:

- Partially cloudy skies with the sun not obscured can increase surface UV-B by up to 25% above clear-sky values. Figure 1 shows two pairs of daily erythermal UV irradiance time series measured at Lauder (45.04°S, 169.68°E) using a broadband instrument (see below). The data plotted in each panel are from nearly consecutive days. Clear enhancements of the UV can be seen during periods when the sun is not obscured by cloud. Visible radiation is even further enhanced allowing UV-A triggered defense and repair mechanisms in biological organisms to be activated.
- Partially cloudy skies with the sun obscured result in decreases in both the UV and visible radiation. However, since the contribution of sky radiance is higher in the UV than in the visible, the effect of a cloud obscuring the sun is greater in the visible than in the UV. The relative increase of UV radiation over visible radiation has important implications for lifeforms relying on UVA or PAR stimulated photorepair or defense.

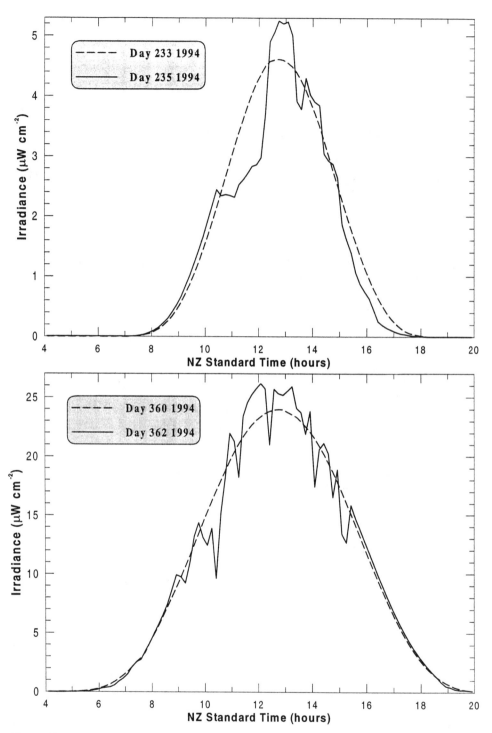

Fig. 1. Diurnal variations of surface erythermal irradiance measured by a broadband instrument at Lauder, New Zealand. Surface erythermal irradiances under partly cloudy conditions (days 235 and 362 of 1994) show enhancements over the clear-sky values expected (days 233 and 360 of 1994) when the sun is not obscured by cloud.

- Uniform cloud cover results in approximately equal reduction of irradiance at all wavelengths in the UV and visible.

The inclusion of clouds in radiative transfer models is particularly problematic. A semi-empirical algorithm has recently been developed which takes advantage of the fact that plots of erythemal UV (defined later) against broadband (visible) irradiance form compact curves for fixed SZA, allowing estimation of surface UV in the presence of clouds (Bodeker & McKenzie 1996). Examples of the performance of this algorithm are shown in Fig. 2. The model is seen to be capable of tracking measurements of erythemal irradiance under a range of different cloud conditions from clear-sky days (day 305) to partially cloudy days (day 315 and day 323) to completely overcast days (day 325).

Recent springtime Antarctic ozone depletion, resulting from the increase in stratospheric chlorine loading, has reduced October total column ozone levels to approximately one third of their pre-1980 values. A Southern Hemisphere map of total column ozone from the Total Ozone Mapping Spectrometer (TOMS) instrument, flown onboard the Earth Probe satellite, is shown for 5 October 1996 in Fig. 3. The Antarctic ozone hole is clearly visible with ozone levels close to 110 Dobson Units (DU) in places.

Using total column ozone data from the Nimbus-7 satellite TOMS experiment, a measure of the Antarctic ozone hole size and depth has been derived for each day from 1979 to 1992 (Bodeker & Scourfield 1995). Failure of Nimbus-7 in May 1993, and gaps in the Meteor-3 TOMS data prevented calculation of values after 1992. For each TOMS data cell south of 40°S, an equivalent mass of total ozone is calculated. The difference between this value and that equivalent to a 220 DU value (the generally used threshold for identifying the ozone hole) gives the "mass of depleted ozone" per cell. These values are summed to produce one value per day and then averaged over the Antarctic vortex period, AVP (19 July to 1 December) to produce one value per year. Fig. 4 shows these averaged depleted mass values together with stratospheric chlorine loading over the same period. The AVP averaged ozone depletion increases with the rise in stratospheric chlorine from close to zero in 1979 to about 9×10^9 kg per day in 1992, showing considerable annual variability induced by variability in planetary wave activity (Bodeker & Scourfield 1995).

The unique meteorological conditions of the Antarctic stratosphere result in a stable long-lived circumpolar vortex which isolates the stratosphere from lower latitudes, creating the cold conditions necessary for the formation of polar stratospheric clouds (PSCs) and preventing transport of lower latitude air into the Antarctic stratosphere. PSCs accelerate ozone depletion through heterogeneous chemical processes. Ozone depletion in the Arctic has been less severe since the stratosphere is generally 10°C warmer than that of the Antarctic and the circumpolar vortex is weaker and shorter lived.

It is not so much the depth or areal extent of the ozone holes which is important for the respective UV environments of the polar regions, but more their seasonal longevity. For both polar regions, the period of ozone depletion commences in later winter, with the return of the sun. In the case of the Antarctic however, the circumpolar vortex remains intact into late spring and even early summer, extending the period of ozone depletion to when smaller SZAs (greater solar elevations) are encountered. The minimum in Antarctic total column ozone levels is usually reached in the first week of October (Herman *et al.* 1995). As shown later, it is this combination of low ozone and small SZA which produces the anomalously high UV measured recently in the Antarctic. The Arctic vortex breaks down in late winter or early spring when SZAs are still large. Furthermore, Antarctic vortex breakdowns have occurred later in recent years, inducing strong upward trends in the surface UV irradiance later in the season. Booth *et al.* (1994) concluded that while ozone depletion occurring early in the season causes elevated levels of UVB, and while the lowest levels of ozone are found in late September or early October, ozone depletion persisting toward summer solstice is more significant in both instantaneous maximum readings and weekly exposures.

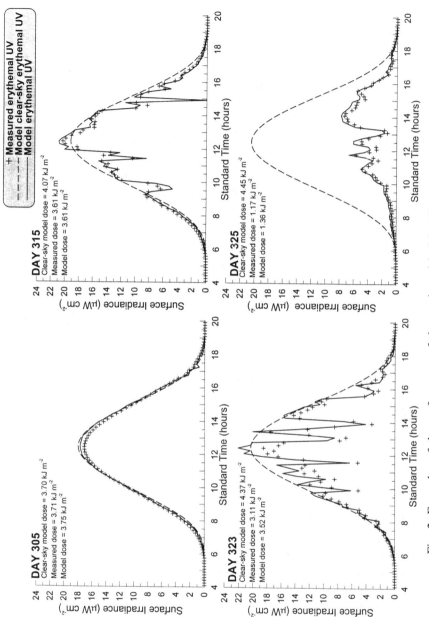

Fig. 2. Examples of the performance of the semi-empirical irradiance model of Bodeker & McKenzie (1996). The predictions of the model (thick solid line) are shown together with the measured erythemal UV (crosses) and the clear-sky spectral model erythemal UV (dashed heavy line) for four case study days in November 1994.

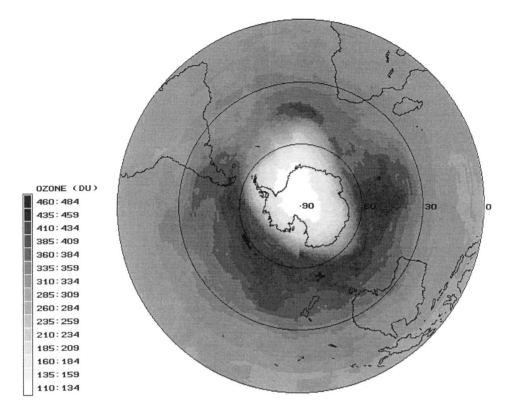

OZONE (DU)

460 : 484
435 : 459
410 : 434
385 : 409
360 : 384
335 : 359
310 : 334
285 : 309
260 : 284
235 : 259
210 : 234
185 : 209
160 : 184
135 : 159
110 : 134

Fig. 3. Southern Hemisphere polar projection of the total column ozone distribution on 5 October 1996 as measured by the TOMS experiment flown onboard the Earth Probe satellite.

Short-term process studies have conclusively demonstrated the strong anti-correlation between ozone and UV (McKenzie *et al.* 1991), in agreement with predictions by models. Thus there is no doubt that, in the absence of other changes, reductions in stratospheric ozone will result in UV increases. In cases where surface UV has not increased with the long-term decrease in stratospheric ozone, this has resulted from a concomitant rise in tropospheric pollution, a portion of which is tropospheric ozone.

The effect of ozone depletion on the biosphere is worsened by the fact that it is primarily UV-A that plays a role in photorepair and is essentially unaffected (for wavelengths greater than 330 nm) by ozone depletion. Hence while stratospheric ozone depletion leads to a dramatic increase in damaging UV-B radiation, the energy required to drive photorepair processes is left unchanged (Smith *et al.* 1992; Roy *et al.* 1994). The inclusion of ozone in radiative transfer models is not difficult, as will be demonstrated below.

The pristine environment of the polar regions generally results in low aerosol loading of the local atmosphere. However, volcanic eruptions can inject large amounts of aerosol into the atmosphere, and, at large SZA, scattering by these aerosols can actually increase UV-B radiation reaching the Earth's surface as discussed above (Davies 1993; Tsay & Stamnes 1992). This phenomenon makes Antarctica during spring the most susceptible place on Earth to the scattering effect of volcanic aerosols, due to the combined effect of the spring ozone hole and the large SZAs characteristic of this time of year. As well as increasing the aerosol loading of the atmosphere, volcanic eruptions enhance ozone depletion.

In addition to these factors discussed above, differences in the Antarctic and Arctic UV environments result from differences in Earth-Sun separation and differences in surface

29

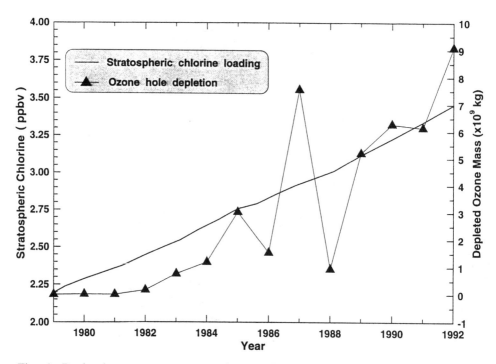

Fig. 4. Depleted ozone mass averaged over the Antarctic Vortex Period (AVP) and stratospheric chlorine loading. Chlorine data as cited in Solomon (1990) and Toohey (pers. comm.).

topography and albedo. As a result of the ellipticity of the Earth's orbit, the Earth is closer to the Sun during the Southern Hemisphere summer than during the Northern Hemisphere summer, resulting in a 7% peak-to-peak increase in erythemal UV compared to the Northern Hemisphere.

3   UV RADIATION

Surface irradiance spectra are steeply sloping in the UV portion of the spectrum as a result of ozone absorption at low wavelengths. To use measured or modeled irradiance spectra to investigate the impact of the irradiance on any process (biological or chemical), it is necessary to weight the spectra with a function which specifies the relative important of the energy at each wavelength in the process under consideration. Such weighting functions are called action spectra or biological weighting functions (BWFs). Typical noon spectra of UV irradiance measured at a midlatitude site for summer and winter are shown in Fig. 5a. Also shown is an erythemal weighting function appropriate for sunburn of human skin (McKinlay & Diffey 1987). When the measured spectra are multiplied with this action spectrum, erythemally weighted UV irradiance spectra are produced as shown in Fig. 5b. Integration over wavelength allows calculation of erythemal UV dose (usually in $\mu W \ cm^{-2}$) while integration over the day allows calculation of the daily erythemal dose (usually in kJ m^{-2}). There are a range of different action spectra which can be applied to measured or modeled irradiance spectra to estimate the impact on the biological process under consideration (Madronich 1992; Madronich 1994) and their choice has significant implications for the conclusions reached in this study. Processes whose action spectra are weighted more strongly towards the lower wavelengths can be

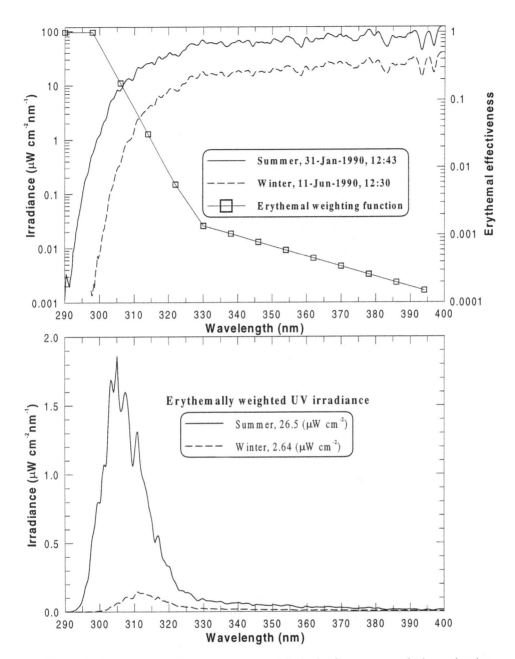

Fig. 5 a - typical noon spectra of UV irradiance at midlatitudes for summer and winter, showing the erythemal action spectrum used in the calculations that follow, b - corresponding erythemally weighted irradiances; from McKenzie *et al.* 1991.

expected to show greater sensitivity to changes in column ozone. For example, application of plant or DNA damage action spectra would show larger responses to changes in ozone compared with the erythemal damage action spectrum assumed here. While the erythemal action spectrum provides an indication of the damage to biological systems, it is specific to erythemal damage, and the absolute values of the irradiances and trends derived in this study can be expected to change to some extent if different action spectra are used. It should also be noted that measurement of an action spectra is difficult (Vincent & Roy 1993) and the derived sensitivity of a biological processes to changes in UV must be interpreted within the accuracy of the action spectrum.

4   THE MEASUREMENT OF UV RADIATION IN THE POLAR REGIONS

Measurements of surface UV spectra are difficult since the irradiance changes by 5 to 6 orders of magnitude across the UV-B portion of the spectrum (see Fig. 5) and because calibration standards in this spectral region are imperfect. In additional to coping with the wide dynamic range, instruments need to reject stray light adequately and have accurate wavelength alignment. Clearly, in the region where the irradiance is changing rapidly with wavelength, such as the UV-B, a small wavelength error can result in a large error in the measured UV irradiance. An additional problem concerns tracing the absolute calibration to a common standard. National standards laboratories themselves disagree by more than ±2% in the UV-B region (Walker *et al.* 1991).

Excellent radiometric stability is required to measure UV trends or geographic differences. However, recent intercomparisons have revealed large calibration differences between some spectro-radiometers. Major sources of uncertainty are instability of sensitivity and cosine errors. Agreement at the ±5% level is a good as can be expected at present.

There are two approaches to measuring UV radiation in the polar regions, viz.:

Broadband instruments: These instruments measure the total energy over a broad range of UV wavelengths. The response of the instrument is designed to mimic the response of human skin to UV radiation. The advantages of such instruments are:

- They are capable of making instantaneous measurements of erythemal UV and can therefore make continuous (every 5 or 10 minutes) readings throughout the day.
- They have no moving parts and therefore require little or no maintenance.
- They are less expensive than spectroradiometers.

The disadvantages of these instruments are:

- They are difficult to calibrate.
- They require accurate temperature stabilisation if they are to be used in long-term trend studies.
- Different biological processes have different wavelength sensitivities and the response of these instruments cannot be changed accordingly.

Spectroradiometers: These instruments scan the whole UV band at fixed wavelength intervals. The advantages of such instruments are:

- They produce high quality results when accurately calibrated. A description of an example of the calibration process required to maintain accurate results is given by Booth *et al.* (1994).
- Any action spectrum weighting can be used on the spectral output to simulate the effect of the irradiance on a given process.

The disadvantages of these instruments are:

- Since the instrument takes some time to perform a scan (e.g. 5 or 6 minutes to create a scan from 280 to 450 nm in 0.2 nm steps) continuous time series recording is not practical.
- Interference from clouds during scans can create inaccurate spectra.
- They require accurate temperature stabilisation to reduce photomultiplier dark current and noise. Generally the temperature of the monochromator must be maintained constant to ±0.5°C.
- They are more expensive than broadband meters.

5   MODELLING THE TEMPORAL AND SPATIAL VARIABILITY OF GLOBAL UV

To show the temporal and spatial variability of global erythemal UV, satellite total column ozone data were used as input to a single layer radiative transfer model. Daily maps of global total column ozone from TOMS instruments flown onboard Nimbus-7 (1979 to 1992) and Meteor-3 (1993 and 1994) were obtained from NASA (Jay Herman, pers. comm.). These were the most recent, version 7, TOMS data products. Each global map consists of a grid of 288×180 cells, each containing a total column ozone reading in DU. The radiative transfer model used was similar to that described by Bird & Riordan (1986). Model parameters were as follows:

- 280 to 450 nm in 0.2 nm steps.
- No topography. The surface pressure was fixed at 1013.25 hPa.
- The altitude of the ozone layer was assumed constant at 20 km.
- The temperature of the ozone layer was assumed constant at 213 K.
- Constant atmospheric aerosol loading appropriate to clean polar air was used.
- A constant ground albedo of 2% was used since this should apply over most regions of the globe. This value is too low for polar regions and calculated erythemal irradiances can be expected to be 20% too low as a result.
- Clear-sky conditions were assumed (no clouds).

6   MODEL RESULTS

The model runs were used to produce the following data sets:

- Monthly means of daily doses: erythemal doses were calculated using zonal mean total column ozone, in 180 latitude zones, at hourly resolution, for each day from 1979 to 1994. Monthly means for each zone were then derived.
- Monthly means of daily maxima: noon erythemal doses were calculated using zonal mean total column ozone, in 180 latitude zones, at hourly resolution, for each day from 1979 to 1994, and monthly means were calculated.
- Maximum daily dose in the month: erythemal doses were calculated using the minimum total ozone in each of the 180 zones, at hourly resolution, for each day. The maximum value in each month was then extracted.
- Maximum noon irradiance in the month: noon erythemal irradiances were calculated using the minimum total ozone in each of the 180 zones, at hourly resolution, for each day, and the maximum value in each month was calculated.

Monthly means of the daily doses, for each of the 180 latitude zones, and for the first four years of the analysis period (1979 to 1982) are shown in Fig. 6a. The blanked areas during the winter of the high latitudes of both hemispheres result from the TOMS instrument being unable to make measurements during the polar night. Fortunately, the surface erythemal

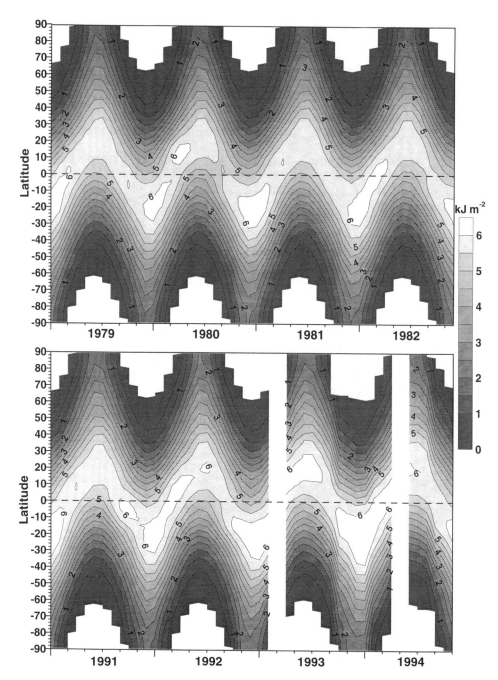

Fig. 6. Monthly means of modeled daily doses of erythemal UV for a - 1979 to 1982, and b - 1991 to 1994.

irradiances during these periods may be assumed to be zero. A similar data set for the last four years of the analysis period (1991 to 1994) are shown in Fig. 6b. In this case there are additional missing data as a result of data gaps from the TOMS instrument flown onboard the Meteor-3 satellite. The general temporal and spatial morphology of the erythemal UV

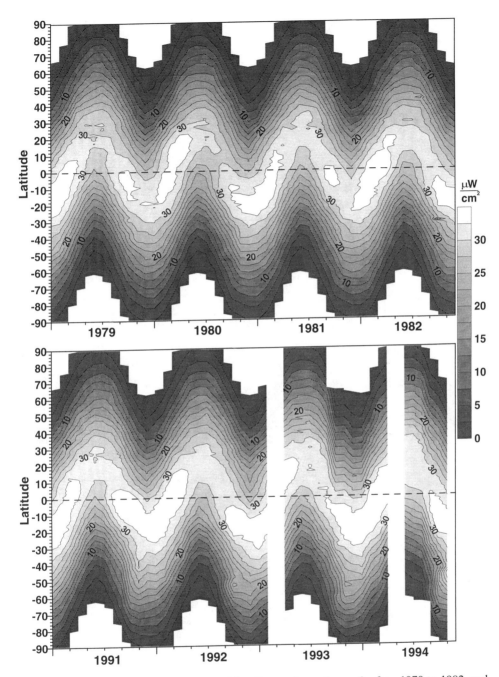

Fig. 7. Modeled maximum noon erythemal irradiances for each month of a - 1979 to 1982, and b - 1991 to 1994.

distribution is dominated by the SZA dependence. The effects of the Antarctic ozone hole are visible where monthly mean doses in excess of 2 kJ m^{-2} have encroached southward to the South Pole in recent years. No such general trend is visible in the Northern Hemisphere data. The anomalously high values occurring at the high latitudes of the Northern Hemisphere in

35

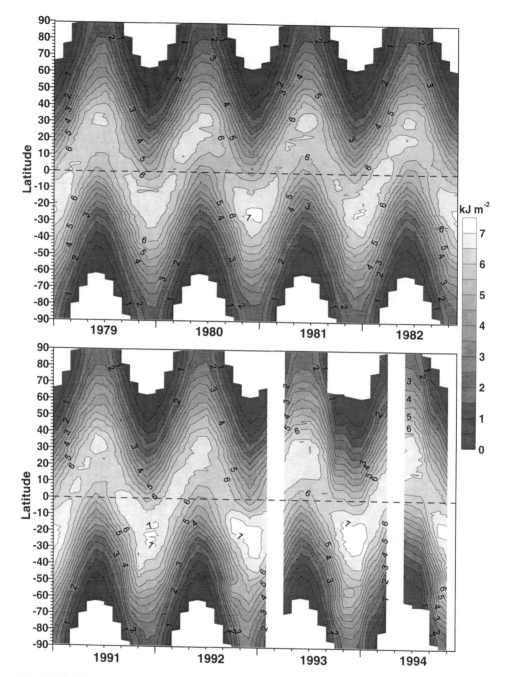

Fig. 8. Modeled maximum daily integrated erythemal doses for each month of a - 1979 to 1982, and b - 1991 to 1994.

1992 may result from enhanced ozone depletion following the Mount Pinatubo volcanic eruption in June 1991. Higher monthly means of daily erythemal doses are seen in the Southern Hemisphere compared to the Northern Hemisphere as a result of the eccentricity of the Earth's orbit.

The modeled maximum noon erythemal irradiance at each latitude for each month of the first four years of the analysis period are shown in Fig. 7a. A similar data set for the last four years of the analysis period are shown in Fig. 7b. The poleward encroachment of higher erythemal UV doses is not as clear in this plot since the enhancement due to longer summer days at high latitudes is not accounted for when examining only the noon irradiances. During the first four years of the period, the latitudinal gradient in the noon erythemal irradiances is generally monotonic, decreasing from the equator to the pole in both hemispheres. This results from the SZA dominance of the surface irradiance. In more recent years however (1991 to 1994), there are times when the latitudinal gradient is either zero or even reversed, though primarily in the Southern Hemisphere. The severe ozone depletion over the Antarctic continent produces sufficiently low ozone levels to dominate the SZA dependence. For example, in September 1992 at 55°S, a movement either north or south results in a decrease in monthly maximum noon erythemal irradiances.

The modeled maximum daily integrated erythemal doses at each latitude for each month of the first four years, and last four years of the data analysis period are shown in Figs. 8a and 8b respectively. Note that the latitudinal gradients are less steep in this plot than in Fig. 7 as a result of the implicit inclusion of the length of day in the calculation of the daily doses. Again the increase in Antarctic UV levels, and the lack of any clear trend in the Arctic, is evident. As in Fig. 7, the occasional reversal of the latitudinal gradient in erythemal UV in recent years can be seen. Note that the global maximum daily doses of erythemal UV occur near 22°S and not at the equator as may have been expected.

To highlight the hemispheric differences in these data sets, monthly means of the daily doses in both hemispheres have been ratioed after applying a 6 month shift to the Northern Hemisphere data. The results, in the form of percentages (positive percentages indicate Southern Hemisphere values higher than Northern Hemisphere values by this amount), are shown in Fig. 9. Note that the scale on the ordinate ranges from the equator to the South Pole. During the first four years of the analysis period (Fig. 9a), the effects of weak ozone holes in 1980 and 1981 can be seen. Southern Hemisphere values were 35% higher than Northern Hemisphere values in spring (Southern Hemisphere) near 80° latitude. At low latitudes the eccentricity of the Earth's orbit tends to create higher Northern Hemisphere irradiances in the winter compared to the Southern Hemisphere while the opposite is true in summer. These features are slightly obscured by the fact that there is an additional asymmetry in the hemispheric ozone distribution where ozone levels at 60°S tend to be consistently higher than levels at 60°N (Gleason *et al.* 1993). Hemispheric differences in more recent years are shown in Fig. 9b (1993 and 1994 were excluded as a result of an excess of missing data). The effects of the Antarctic ozone hole are far more dramatic with hemispheric differences in the spring at 80° latitude exceeding 50% and are at times as high as 75%. The Antarctic ozone hole of 1991 was less severe than in neighbouring years and this is also evident in Fig. 9b.

To highlight the temporal change in erythemal UV, global distributions of the monthly means of the noon values have been ratioed for 1979 and 1992. The results are shown in Fig. 10a where positive percentages indicate increases in the monthly mean noon erythemal UV from 1979 to 1992. The development of the Antarctic ozone hole has resulted in a increase of more than 60% in the values from 1979 to 1992. Changes in the Northern Hemisphere are smaller. Madronich (1994) reported spring time increases in DNA daily dose, from 1979 to 1992, of 40% in the high latitudes of the Northern Hemisphere and 150% in the high latitudes of the Southern Hemisphere. The most important cause of the difference in results between those reported here and those of Madronich (1994) is that processes applicable to DNA action spectra will be more sensitive to changes in ozone that those applicable to erythemal action spectra since that for DNA is more strongly weighted towards lower wavelengths. The second reason is the study of Madronich (1994) made use of version 6 TOMS data while version 7 TOMS data were used here. Corrections to the data reduction algorithm result in smaller long-term trends using the version 7 data compared with version 6. Finally it must be noted that

37

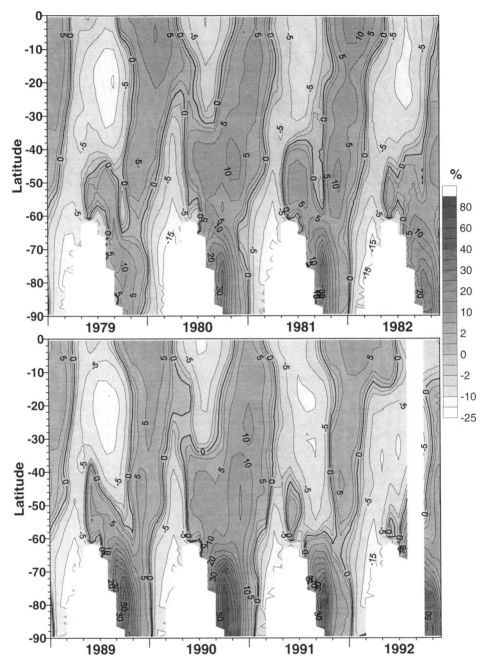

Fig. 9. Ratios between Southern Hemisphere and Northern Hemisphere (6 month shifted) monthly means of modeled daily doses of erythemal UV for a - 1979 to 1982, and b - 1989 to 1992.

Figure 10a does not show a trend analysis; it only shows the difference between 1979 and 1992. A linear trend analysis making use of data from all intermediate years might be expected to show different results, but is beyond the scope of this paper.

A similar data set for the monthly maximum noon erythemal irradiances is shown in Fig.

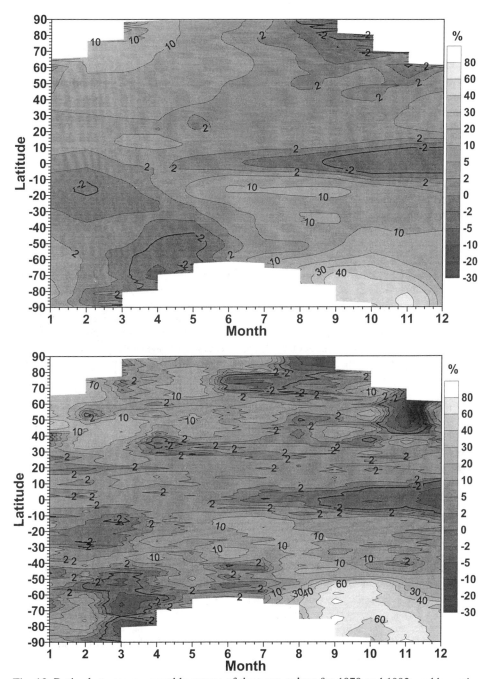

Fig. 10. Ratios between a - monthly means of the noon values for 1979 and 1992, and b - ratios of the monthly maximum noon erythemal irradiances for 1979 and 1992, expressed as percentages.

10b. The effects of the Antarctic ozone hole are even more pronounced than in Fig. 10a with increases of monthly maximum noon erythemal UV exceeding 80% from 1979 to 1992. In addition to the high latitude increases seen in both Figs. 10a and 10b, there is a second region of increase centered on the month of September between 50°S and 80°S. Its source is occasional elongation of the Antarctic vortex by midlatitude planetary wave activity which displaces the vortex center from the South Pole towards South America. This results in significantly lower total column ozone over South America during these periods, and together with the smaller SZAs experienced there, produces anomalously high noon erythemal irradiances. Such effects are not evident in Fig. 10a since zonal mean total column ozone fields were used as input to the model in this case and any zonal asymmetry introduced by the displacement of the vortex from the pole is smoothed.

7   DISCUSSION AND CONCLUSION

Note that in the three data sets shown in Figs. 6, 7 and 8, the modeled high latitude values are significantly less than those measured (Seckmeyer et al. 1995). These differences result partially from the fact that the model calculations have made no provision for higher polar ground albedos which can be expected to increase erythemal irradiances by 20%. Furthermore, higher aerosol loading over the tropical regions will decrease irradiances there below the predictions of the model. Higher aerosol loading and tropospheric pollution, of which ozone is a part, also decreases the surface erythemal irradiance at the midlatitudes of the Northern Hemisphere.

The analysis presented above assumes cloud free conditions. In practice, cloud variability causes large year-to-year changes in UV. The theory of radiative transfer through clouds is well developed, and algorithms for its numerical implementation are available (e.g. Stamnes et al. 1988). However, the practical application of the theory to the atmosphere is still limited because of incomplete cloud characterization. Cloud cover at most surface sites is specified only as the fraction of sky covered by cloud, with little or no information about the optical depth or layering. The importance of different cloud geometries on the surface erythemal irradiance was discussed above. Further, although cloud optical depth is not a strong function of wavelength, there is a nonlinear relationship between observed cloud cover and its effect in the UV-B region where a much larger fraction of the energy is diffuse. Measured reductions in UV-B are relatively small even for large fractional cloud covers (Bais et al. 1993).

Care must be taken when using zonal mean total column ozone fields as input to models to predict the spatial and temporal variability of surface UV radiation. While this practice reduces the dimensionality of the problem from 4 to 3, it masks the effects of zonal asymmetry which can be very important as seen in Fig. 10b.

While changes in surface erythemal UV have been far greater in the Antarctic than in the Arctic, the impact on human life may be expected to be greater in the Arctic where the population density is far higher. This may have significance for native Arctic populations such as the Inuit who live for long periods of time out on the sea ice. Furthermore, there is a great deal of interest at present in the effects of global change on water column attenuation of UV radiation (Laurion et al., in press), which could be more substantial in the North Polar region (and much greater than any ozone-depletion effect) because of the greater importance of freshwater ecosystems in the North and the greater influence of freshwater discharge (with its UV-screening constituents) on the Arctic Ocean relative to the Southern Ocean (Vincent, pers. comm.).

The increases in surface UV-B radiation observed in the Antarctic over the past 20 years are larger than will ever be observed in the Northern Hemisphere given our current understanding of the characteristics of polar ozone depletion. For this reason, the marginal ice zone of the Antarctic continent provides the best site for investigations of the effects of enhanced UV-B on marine organisms.

REFERENCES

Bais, B.A., C.S. Zerefos, C. Meleti, I.C. Ziomas & K. Tourpali, 1993. Spectral measurements of solar UVB radiation and its relations to total ozone, SO_2, and clouds. *Journal of Geophysical Research* 98(D3):5199-5204.

Beaglehole, D. & G.G. Carter, 1992. Antarctic skies 1. Diurnal variations of the sky irradiance, and UV effects of the ozone hole, spring 1990. *Journal of Geophysical Research* 97(D2):2589-2596.

Bird, R.E. & C. Riordan, 1986. Simple solar spectral model for direct and diffuse irradiance on horizontal and tilted planes at the Earth's surface for cloudless atmospheres. *Journal of Climate and Applied Meteorology* 25(1):87-97.

Bodeker, G.E. & R.L. McKenzie, 1996. An algorithm for inferring surface UV irradiance including cloud effects. *Journal of Applied Meteorology* 35(10):1860-1877.

Bodeker, G.E. & M.W.J. Scourfield, 1995. Planetary waves in total ozone and their relation to Antarctic ozone depletion. *Geophysical Research Letters* 22(21):2949-2952.

Booth, C.R., T.B. Lucas, J.H. Morrow, C.S. Weiler & P.A. Penhale, 1994. The United States National Science Foundation's polar network for monitoring ultraviolet radiation. In: Weiler, C.S. & P.A. Penhale (Eds.), *Ultraviolet Radiation in Antarctica: measurements and biological effects. Antarctic Research Series* 62. American Geophysical Union, Washington D.C. pp. 17-37.

Davies, R.. 1993. Increased transmission of ultraviolet radiation to the surface due to stratospheric scattering. *Journal of Geophysical Research* 98(D4):7251-7253.

Erle, F., K. Pfeilsticker & U. Platt, 1995. On the influence of tropospheric clouds on zenith-scattered-light measurements of stratospheric species. *Geophysical Research Letters* 22(20):2725-2728.

Farman, J.C., B.G. Gardiner & J.D. Shanklin, 1985. Large losses of total ozone in Antarctica reveal seasonal ClO_x/NO_x interaction. *Nature* 315:207-210.

Gleason, J., P.K. Bhartia, J.R. Herman, R. McPeters, P. Newman, R.S. Stolarski, L. Flynn, G. Labow, D. Larko, C. Seftor, C. Wellemeyer, W.D. Komhyr, A.J. Miller & W. Planet, 1993. Record low global ozone in 1992. *Science* 260: 523-526.

Herman, J.R., R.D. McPeters & D. Larko, 1993. Ozone depletion at Northern and Southern latitudes derived from January 1979 to December 1991 Total Ozone Mapping Spectrometer Data. *Journal of Geophysical Research* 98(D7):12783-12793.

Herman, J.R. & D. Larko, 1994. Low ozone amounts during 1992-1993 from Nimbus 7 and Meteor 3 total ozone mapping spectrometers. *Journal of Geophysical Research* 99(D2):3483-3496.

Herman, J.R., P.A. Newman & D. Larko, 1995. Meteor-3/TOMS observations of the 1994 ozone hole, *Geophysical Research Letters* 22(23):3227-3229.

Herndl, G.J., G. Muller-Niklas & J. Frick, 1993. Major role of ultraviolet-B in controlling bacterioplankton growth in the surface layer of the ocean. *Nature* 361:717-719.

Laurion, I., W.F. Vincent & D.R.S. Lean, in press. Underwater ultraviolet radiation: development of spectral models for Northern high latitude lakes. *Photochemistry and Photobiology*.

Lubin, D. & J.E. Frederick, 1991. The ultraviolet radiation environment of the Antarctic Peninsula: the roles of ozone and cloud cover. *Journal of Applied Meteorology* 30:478-492.

Lubin, D. & J.H. Jensen, 1995. Effect of clouds and stratospheric ozone depletion on ultraviolet radiation trends. *Nature* 377:710-713.

Lubin, D., J. Frederick, R. Booth, T. Lucas & D. Neuschuler, 1989. Measurements of enhanced spring time ultraviolet radiation at Palmer station, Antarctica. *Geophysical Research Letters* 16(8):783-785.

Madronich, S., 1992. Implications of recent total atmosperic ozone measurements for biologically active ultraviolet radiation reaching the Earth's surface. *Geophysical Research Letters* 19:37-40.

Madronich, S., 1994. Increases in biologically damaging UV-B radiation due to stratospheric ozone reductions: A brief review. *Archiv fuer Hydrobiologie. Beihefte* 43:17-30.

McKenzie, R.L., W.A. Matthews & P.J. Johnston, 1991. The relationship between erythemal UV and ozone, derived from spectral irradiance measurements. *Geophysical Research Letters* 18:2269-2272.

McKinlay, A.F. & B.L. Diffey, 1987. A reference spectrum for ultraviolet induced erythema in human skin. In: Passchler, W.R. & B.F.M. Bosnajokovic, (Eds.), *Human Exposure to Ultraviolet Radiation: risks and regulations.* Elsevier, Amsterdam.

McMinn, A., H. Heijnis & D. Hodgson, 1994. Minimal effects of UVB radiation on Antarctic diatoms over the past 20 years. *Nature* 370:547-549.

Prèzelin, B.B., N.P. Boucher & R.C. Smith, 1994. Marine primary production under the influence of the Antarctic ozone hole: Icecolors '90. In: Weiler, C.S. & P.A. Penhale (Eds.), *Ultraviolet Radiation in Antarctica: measurements and biological effects. Antarctic Research Series 62.* American Geophysical Union, Washington D.C. pp. 159-186.

Roy, C.R., H.P. Gies, D.W. Tomlinson, D.L. Lugg, 1994. Effects of ozone depletion on the ultraviolet radiation environment at the Australian stations in Antarctica. In: Weiler, C.S. & P.A. Penhale (Eds.), *Ultraviolet radiation in Antarctica: measurements and biological effects. Antarctic Research Series 62.* American Geophysical Union, Washington D.C. pp. 1-15.

Seckmeyer, G., B. Mayer, G. Bernhard, R.L. McKenzie, P.V. Johnston, M. Kotkamp, C.R. Booth, T. Lucas, T. Mestechkina, C.R. Roy, H.P. Gies & D. Tomlinson, 1995. Geographical differences in the UV measured by intercompared spectroradiometers. *Geophysical Research Letters* 22(14):1889-1892.

Smith, R.C., B.B. Prèzelin, K.S. Baker, R.R. Bidigare, N.P. Boucher, T. Coley, D. Karentz, S. McIntyre, H.A. Matlick, D. Menzies, M. Ondrusek, Z. Wan & K.J. Waters, 1992. Ozone Depletion: Ultraviolet Radiation and Phytoplankton Biology in Antarctic Water. *Science* 255:952-959.

Solomon, S., 1990. Progress towards a quantitative understanding of Antarctic ozone depletion. *Nature* 347:347-354.

Stamnes, K., S-C. Tsay, W. Wiscombe & K. Jayaweera, 1988. Numerically stable algorithm for discreet-ordinate-method radiative transfer in multiple scattering and emitting layered media. *Applied Optics* 27:2502-2509.

Stamnes, K., Z. Jin, J. Slusser, C. Booth & T. Lucas, 1992. Several-fold enhancement of biologically effective ultraviolet radiation levels at McMurdo station Antarctica during the 1990 ozone "hole", *Geophysical Research Letters* 19(10):1013-1016.

Tsay S-C. & K. Stamnes, 1992. Ultraviolet radiation in the Arctic: the impact of potential ozone depletions and cloud effects. *Journal of Geophysical Research* 97(D8):7829-7840.

Tsitas, S.R. & Y.L. Yung, 1996. The effect of volcanic aerosols on ultraviolet radiation in Antarctica. *Geophysical Research Letters* 23(2):157-160.

Vincent, W.F. & S. Roy, 1993. Solar ultraviolet-B radiation and aquatic primary production: damage, protection and recovery. *Environmental Reviews* 1:1-12.

Walker, J.H., R.D. Saunders, J.K. Jackson & K.D. Mielenz, 1991. Results of a CCPR intercomparison of spectral irradiance measurements by National Laboratories. *Journal of Research of the National Institute of Standards and Technology (NIST)* 96:647-669.

2 Terrestrial environments

Ecosystem Processes in Antarctic Ice-free Landscapes, Lyons, Howard-Williams & Hawes (eds)
© *1997 Balkema, Rotterdam, ISBN 90 5410 925 4*

Thermal regimes of some soils in the McMurdo Sound region, Antarctica

D.I.Campbell & R.J.L.MacCulloch
Department of Earth Sciences, University of Waikato, Hamilton, New Zealand
I.B.Campbell
Land and Soil Consultancy Services, Nelson, New Zealand

ABSTRACT: This paper presents the results of research carried out during the 1994/95 summer, which investigated the diurnal thermal regime, thermal properties, and radiation balance of soils at three locations in the McMurdo Sound region.

The major soil factor affecting the diurnal thermal regime of Antarctic soils in summer is the surface albedo, because this determines the proportion of incoming solar radiation that is available to heat the soil. Soil surface albedo was 0.06 at Scott Base, 0.07 at the Coombs Hills and 0.26 for the pale polished sandstone surface in the Northwind Valley. The high albedo of the latter site meant that the soil was colder than at the Coombs Hills, despite the Northwind Valley's lower altitude and warmer air temperature. All soils had thermal regimes typical of desert regions, with large diurnal temperature extremes near the surface. The Scott Base soil experienced less temperature variation with depth, probably because of latent heat effects associated with freezing and thawing of soil water at around 0°C, which did not occur in the other colder, drier and more saline soils.

1   INTRODUCTION

The Antarctic environment offers severe challenges for biological processes. Antarctica is the coldest, highest and driest continent on Earth, and only a few percent of the continent is ice free. Within these ice-free "oases", of which the McMurdo dry valley region is the largest, there exist, for relatively short periods of time, temperatures above freezing which allow liquid water to become freely available in the soils, rivers and lakes. Weyant (1966) noted that the contrasting albedos of snow-covered and snow-free surfaces is the main reason for the large differences in their climates, since snow surfaces typically absorb less that 20% of the solar radiation that soil and rock surfaces do.

The soil biological environment is particularly fragile because of the extreme range of temperatures, aridity, and the short summer season. Human impacts on the soil surface can have profound effects on soil physical processes and are slow to recover.

The thermal regimes of desert soils are extreme, partly because of the absence of surface vegetation cover, but mainly because of a lack of soil moisture (Weyant 1966; Oke 1987). The thermal properties of dry soils limit temperature changes to shallow depths, hence for a given energy input their near-surface thermal regimes are extreme compared to moist soils.

The soil environment is highly variable at both regional, local and micro scales. This paper reports the results of a study which examined the principal climate, surface and soil factors which affect the diurnal soil thermal regime at three sites across a range of climate zones during the 1994/95 summer.

2.1   Study sites

Three field sites were chosen across a range of climate zones and bare surface types in the McMurdo Sound region (Fig. 1; Table 1). The Scott Base (SB) site was a flat, relatively undisturbed bench 200 m north west of Scott Base, with soil formed on recent, unweathered

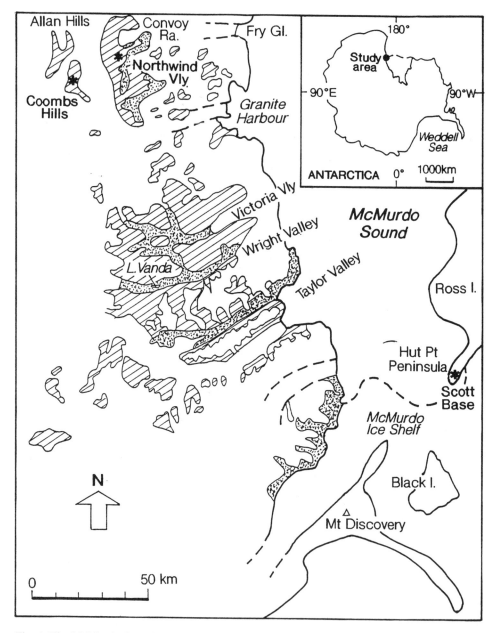

Fig. 1. The McMurdo Sound region, showing the locations of the three field sites (*) and extent of ice-free areas. Valley floors stippled.

basaltic deposits (≈0.4 ma, Kyle 1981). The Northwind Valley (NW) site was located in the Convoy Range, on a low moraine ridge composed of sandstone derived from a nearby outcrop of Weller Coal Measure Formation (Beacon Supergroup; Chinn *et al.* 1994). The soil had a sandy texture with noticeable salt deposits, and was armoured with a surface layer of creamy white polished sandstone cobbles. The Coombs Hills (CH) site was located on the flat floor of a valley to the north of Mt. Brooke, in the Coombs Hills. The soil was formed from dolerite, including locally-derived volcanic materials of the Mawson Formation (Grapes *et al.* 1972) and armoured with dark brown polished dolerite cobbles.

2.2   Measurements

Soil temperatures were measured with type-T thermocouples at closely spaced vertical intervals. At each site a small pit was carefully excavated in layers and the soil was bagged and insulated. Thermocouple probes were inserted in the undisturbed face of the pit and the soil carefully replaced to the original bulk density. The surface was restored until the pit site was indistinguishable from its surroundings. Soil surface temperature was measured using a 4-junction averaging thermocouple with local surface materials glued to the sensors which were then laid on the surface, and cables covered.

Incoming and reflected solar radiation were measured with two Kipp and Zonen pyranometers whose outputs had been matched by previous comparison. One of the pyranometers was inverted to measure reflected solar radiation. Net radiation was measured with a Radiation and Energy Balance Systems Inc. Q6 net radiometer. All radiation instruments were mounted on a horizontal pole extending from a surveying tripod at a height of approximately 0.7 m.

Air temperature was measured by a shielded probe (Skye 2031) at a height of 0.8 m, and windspeed was measured with a 3-cup anemometer (Vector A101M) at a height of 1.7 m. All measurements were recorded with Campbell Scientific Inc. CR10 data loggers.

Physical and thermal properties were measured for each soil. Soil samples were carefully excavated from shallow holes which were backfilled to determine their volume (Burke *et al.* 1986). Samples were bagged and then oven dried at Scott Base, to determine dry bulk density and volumetric moisture content. Measurements of soil volumetric heat capacity and thermal conductivity were made on soil samples repacked to their original bulk density using Soiltronics (Seattle, USA) probes (Bristow *et al.* 1994; Campbell *et al.* 1991).

Table 1. Details of the field sites.

Site (dates)	Abbreviation	Altitude (m asl)	Coordinates	Climate zone	Surface type
Scott Base 19-31 Decenber 1994	SB	44	75°50'S 166°45'E	coastal	vesicular basalt, dull black
Northwind 18-22 January 1995	NW	≈1500	76°46.3'S 160°45.1'E	central mountain	sandstone, creamy white
Coombs Hills 10-16 January 1995	CH	≈2200	76°48.3'S 159°54.1'E	central mountain	dolerite, polished dark brown

3   RESULTS

3.1   Soil temperature

Fig. 2a shows the diurnal soil temperature behaviour at Scott Base for a clear-sky day. There is strong heating at the soil surface, with the peak 30 minute mean temperature of 17.8°C experienced late morning, coinciding with minimum windspeed. At night the soil surface cools by contact with the overlying cold airstream to a minimum temperature of –4.7°C. Because of 24 hour insolation the soil surface always remains warmer than the overlying air. The large diurnal variation of surface temperature drives a wave of heating into the soil profile. The amplitude of the diurnal wave decreases with depth and displays a clear time lag, so that by 12 cm depth the peak temperature lags six hours behind the surface peak. The diurnal wave is almost completely damped out at the 24 cm depth, hence, the diurnal thermal regime is a relatively shallow phenomenon at this site.

Fig. 2b shows the effect of a shallow (≈3 cm) fresh snowpack covering the site. Radiation receipt is greatly reduced because of the high albedo of the snow and partly cloudy conditions, so there is less heating and the diurnal temperature wave is very small. At depths below 6 cm the soil is nearly isothermal. The increase in near-surface temperature in the afternoon coincided with snowpack melting and peak net radiation receipt at 1400 hrs.

The diurnal thermal regime is driven by the rate of heating and cooling at the soil surface. Table 2 lists representative soil surface and air temperature data for the three sites used in this study. On clear sky days there is a very large range in soil surface temperatures (>22°C) at all of the sites, but greatest at CH (28°C). At SB the higher windspeed probably acted to reduce the maximum temperature. Maximum surface temperature is greatest at SB and smallest at NW. At CH soil surface temperatures were significantly higher than NW, despite the colder

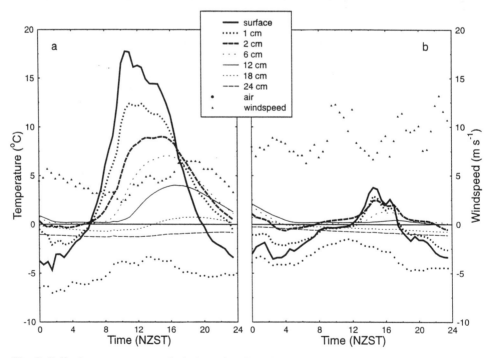

Fig. 2. Soil. air temperature, and windspeed at Scott Base for a - a clear-sky day, 23 December 1994; b - under a thin snow pack, 28 December 1994.

48

air. All sites attain soil surface temperatures well above 0°C when radiation inputs are large, and in these conditions a large surface–air temperature difference is maintained, exceeding 20°C at both SB and CH. Soil surface temperatures never fell below air temperature at any of the three sites during the measurement periods indicating that the soil is always acting as a sensible heat source and the air a sink during summer. The coldest soil surface was at NW, and this resulted in a smaller difference between surface and air temperatures at this site.

Fig. 3 shows vertical profiles of mean temperature at the three sites for the days presented in Table 2. SB has the warmest soil and both cloudy and clear-sky days have similar mean temperatures. At the inland sites the mean temperature of the soil drops dramatically when radiation inputs decline on cloudy days, with greatest cooling near the surface. The CH soil is warmer than the NW soil, despite colder air temperatures at CH (Table 2). Only SB had mean soil temperatures above 0°C. Under snow cover the SB soil quickly became almost isothermal at approximately 0°C, while the CH soil continually loses heat to the air.

Fig. 4 shows the patterns of temperature variation at each measurement depth for the clear-sky days, and this shows a clear difference between the thermal regimes of the warmer, coastal SB site and the colder inland sites. At SB there are large variations in soil temperature close to the surface, but these rapidly decline with depth, becoming almost negligible at 20 cm. The inland sites have large variations near the surface, with CH having the largest near-surface range, but significant (4 to 5°C) diurnal variations still occur deep in both profiles. Thus the full depth of diurnal temperature variation was not sampled at these sites. On 13 January 1995 a soil pit was dug at the CH site and temperatures measured throughout its depth. Below 40 cm depth the temperature declined in a linear fashion to –16°C at 130 cm, hence in the absence of surface heating, strong divergence occurs in both upward and downward directions, causing the upper part of the soil profile to cool rapidly.

3.2   The surface radiation balance

The diurnal soil thermal regime is largely driven by the gain and loss of energy by radiation at the soil surface. The quantity of incoming solar radiation is determined by solar altitude, cloudiness and site shading, etc., however the amount of solar energy absorbed by the surface ($K*$) is determined by the soil surface albedo (α). Albedo is defined as the ratio of reflected ($K\uparrow$) to incoming ($K\downarrow$) shortwave radiation, $\alpha = K\uparrow/K\downarrow$, and is largely determined by soil surface colour, roughness and moisture content (Oke 1987). The soil emits and receives longwave radiation to and from the atmosphere respectively. The balance of all incoming and

Table 2. Soil surface and air temperatures (T_{air}) and mean windspeed (u) for the three sites. "Snow" refers to the presence of a surface snowpack.

| Site | Conditions | Date | Soil surface temperature (°C) | | | | T_{air} | u |
			max.	min.	range	mean	°C	m s^{-1}
Scott Base	clear sky	23 December 1994	17.8	–4.7	22.5	4.5	–5.2	4.6
44 m asl.	cloudy	25 December 1994	11.0	–2.3	13.3	4.6	–4.8	2.9
	snow	28 December 1994	3.8	–3.6	7.4	–0.9	–3.4	8.9
Northwind	clear sky	21 January 1994	8.5	–13.5	22	–2.9	–9.6	2.2
≈1500 m asl.	cloudy	20 January 1994	–1.8	–14.9	13.1	–7.9	–10.8	2.7
Coombs Hills	clear sky	11 January 1995	14.3	–13.6	27.9	0.4	–12.9	3.0
≈2200 m asl.	cloudy	13 January 1995	2.5	–13.0	15.5	–6.7	–15.6	2.1
	snow	14 January 1995	–6.4	–14.3	7.9	–10.5	–17.1	4.7

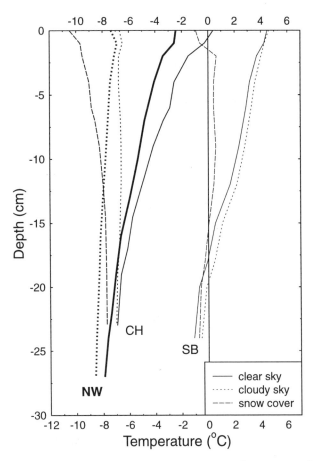

Fig. 3. Vertical profiles of 24 hour mean soil temperature for a range of conditions at Scott Base (SB), Northwind Valley (NW) and Coombs Hills (CH). For dates and surface temperature data refer to Table 1.

outgoing radiation fluxes is the net radiation, $Q*$, which is the energy available to heat the soil and overlying air, and evaporate moisture or melt ice.

Table 3 lists the measured and estimated total radiation flux densities for the days included in Table 2. On clear-sky days $K\downarrow$ was similar for all sites. SB received the greatest $K\downarrow$ because this day was close to the solstice. Longwave radiation losses were large for all surfaces. The major variation between sites is the surface albedo. SB had the lowest albedo (0.06) because of the rough, dark-coloured basalt surface. The polished, dark dolerite pavement at CH had a similarly low albedo (0.07), while the light-coloured sandstone pavement at NW is highly reflective, with $\alpha = 0.26$. The effect of the differing albedos can be seen most clearly by considering $Q*$. At NW $Q*$ is approximately half that at SB, in other words there is only half the energy available for heating the soil and air or driving latent heat processes. The effect of snow cover is to increase α and reduce $Q*$.

Snow covering the soil has a dramatic effect on the radiation balance because it increases the albedo. At CH on 14 January 1995 a thin snow pack with $\alpha = 0.7$ reduced $K*$ and large longwave radiation losses to the clear sky meant that $Q*$ was negligible, resulting in strong soil cooling (Fig. 3). In contrast, for SB on 28 December 1994, a thin snow pack with

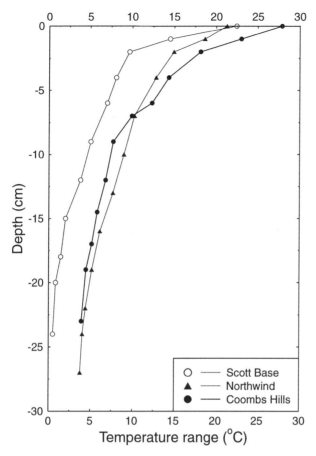

Fig. 4. Vertical profiles of daily temperature range (T_{MAX}–T_{MIN}) for clear-sky days: SB, 23 December 1994; NW, 21 January 1995; CH, 11 January 1995.

$\alpha = 0.57$ reduced K^*, however Q^* was still significant because large longwave losses were prevented by the presence of cloud cover.

Fig. 5 shows the diurnal pattern of albedo on clear-sky days at the three sites. At both NW and CH there was shading by topography for several hours at night, and the effect of this can be seen as a discontinuity in the data for these sites as incoming solar radiation switched from mainly direct to diffuse, and the albedo differs for each (Oke 1987). At CH snowfall began after 2000 hrs and the albedo increases dramatically. The general pattern during the daytime at all sites is for minimum α around midday. This is a commonly observed phenomenon for many surfaces, and results from differences in reflective behaviour at different solar angles. At high solar altitude natural surfaces scatter radiation in a diffuse manner and there are more opportunities for absorption, particularly for rough surfaces so α is lowest. At low solar altitude reflection is specular (mirror-like). The difference in albedo as solar altitude changes is generally greatest for smooth surfaces (Oke 1987). The magnitude and diurnal pattern of α at both SB and CH is remarkably similar.

At all three sites the albedo declines during the morning at a greater rate than it increases during the afternoon (Fig. 5). This phenomenon was also noted by Arnfield (1975),

Table 3. Radiation flux densities for representative days at the three field sites (MJ m^{-2} day^{-1}), albedo (1000 to 1400 h) and mean surface temperature, T_s (°C). Net longwave flux is calculated as $L*=Q*-K*$. Sign definition is that fluxes representing energy gains by the soil surface are positive, losses are negative (Oke 1987).

Site	Conditions	Date	$K\downarrow$	$K*$	$L*$	$Q*$	α	T_S
Scott Base	clear sky	23 December 1994	35.2	32.9	−14.3	18.6	0.061	4.5
44 m	cloudy	25 December 1994	22.6	21.0	−8.6	12.4	0.063	4.6
	snow, cloudy	28 December 1994	20.0	10.9	−3.5	7.4	0.57	−0.9
Northwind	clear sky	21 January 1994	31.3	22.8	−13.1	9.7	0.26	−2.9
≈1500 m	cloudy	20 January 1994	21.0	13.6	−10.5	3.1	0.34	−7.9
Coombs Hills	clear sky	11 January 1995	33.0	29.9	−14.4	15.5	0.071	0.4
≈2200 m	cloudy	13 January 1995	20.6	12.4	−7.1	5.3	0.078	−6.7
	snow, clear sky	14 January 1995	33.1	8.9	−8.7	0.2	0.7	−10.5

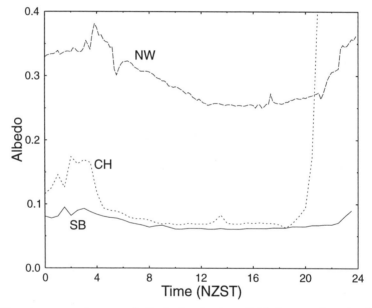

Fig. 5. Diurnal pattern of measured albedo for the three field sites on clear-sky days: SB, 23 December 1994; NW, 21 January 1995; CH, 11 January 1995.

for crop surfaces, and was attributed to forenoon-afternoon surface changes. In the present study we attribute this behaviour to variations in surface roughness caused by preferential polishing of the desert pavement by prevailing winds, so that the albedo is dependent, to a small degree, on solar azimuth.

3.3   Soil thermal properties

The thermal regime of a soil is dependent on its thermal properties as well as its energy balance. The temperature change of a soil volume depends on the change in stored energy and on the soil heat capacity, C_s. The bulk soil value of C_s is determined by the heat capacity of individual soil constituents (mineral, organic, air, water, ice) and their respective volume fractions. Most of these fractions are effectively constant with time, except for the soil water/ice content. Heat is conducted through the soil in response to temperature gradients and the ability of the soil to conduct heat energy is determined by the thermal conductivity, k_s. k_s depends on the thermal conductivity of the soil particles, the porosity (air is a poor conductor), and the moisture content (films of water linking soil grains improves thermal contact). The soil thermal diffusivity, $\kappa_s = k_s/C_s$, describes the ability of a soil to diffuse temperature changes. Soils with a high κ_s transport heat rapidly into the soil so that temperature changes are spread over a greater depth. These soils have a less extreme thermal regime than soils with low κ_s. Dry desert soils typically have low κ_s and display extreme thermal behaviour near the surface.

Table 4 lists physical and thermal properties of the three soils. Bulk density is highly variable both within a profile and spatially, largely because of variable grainsize distributions. Moisture content is a key variable that affects the thermal behaviour of a soil, and SB is the moister soil, however all soils are relatively dry. Balks *et al.* (1995) found that soil moisture content in the near-surface soil at a site near Scott Base was low ($\approx$10%) but increased rapidly with depth at the top of the ice cemented zone at 30 to 40 cm depth (30 to 60%). The SB soil in the present study was clearly drier than that studied by Balks *et al.* (1995), however ice cement was present at a depth of approximately 25 cm coinciding with a mean temperature for the study period of approximately $-0.7°C$. Ice cement was also noted close to the surface when the soil temperature fell below $0°C$ under a shallow snowpack (Fig. 3). At NW ice cement was not present at any depth, despite the much lower temperatures. At CH ice cement was occasionally found, but at irregular depths and was spatially variable. A probable reason for this is the higher soluble salt contents of these soils (Table 4). Anderson & Tice (1989) found that saline soils from the Beacon and Wright Valleys retained considerable unfrozen water at extremely low temperatures and that these soils did not gain strength on freezing. Latent heat release and uptake during phase changes are, therefore, spread over a wide temperature range. The SB soil has very low soluble salt content so phase change appears to occur close to $0°C$.

There are differences in the thermal properties of the soils. SB has the highest heat capacity and NW the lowest, hence temperature changes for a given change in stored energy will be less at SB and greatest at NW. Thermal conductivity values are similar for all soils when oven dry, and a laboratory experiment showed that k_s was relatively constant through the range of soil moisture content found in the field (MacCulloch 1996). Thermal diffusivity is relatively low for all soils, indicating that extreme temperature changes are confined to shallow layers, which is typical behaviour for dry desert soils (Oke 1987).

Table 4. Physical and thermal properties of soils at the study sites. ρ_b - dry bulk density; θ - volumetric moisture content; C_s - heat capacity (at field moisture content); k_s - thermal conductivity (oven dry); κ_s - thermal diffusivity. Standard errors are included for C_s and k_s.

Site	ρ_b kg m^{-3}	θ %	Soluble salt (%)	C_s MJ m^{-3} K^{-1}	k_s W m^{-1} K^{-1}	κ_s m^2 s^{-1}×10^{-6}
SB	1778	5.8	0.52	1.939 ±0.028	0.261 ±0.007	0.138
NW	1638	2.9	8.08	1.549 ±0.02	0.205 ±0.006	0.132
CH	1643	4.2	10.46	1.757 ±0.02	0.217 ±0.01	0.124

4   DISCUSSION AND CONCLUSIONS

This research has examined the diurnal thermal regimes of three soils for short periods during the 1994/95 summer. Antarctic soil temperatures rise above 0°C for short periods in summer each year. At low altitudes the mean temperature of the near-surface layers may exceed 0°C in summer, as measurements by Thompson *et al.* (1971), for Vanda Station in the Wright Valley, and Balks *et al.* (1995), for Marble Point, demonstrate. At higher altitudes, such as at the NW and CH sites, summertime near-surface soil temperatures may exceed 0°C for short periods however mean daily temperatures are probably only rarely above freezing.

Large variations in near-surface soil temperature occur when radiation inputs are high, which is typical behaviour for dry desert soils in all climates. With 24 hr insolation in summer, there is considerable energy available to heat the soil and the overlying air, and to melt ice and evaporate water, if the soils are moist. Balks *et al.* (1995) measured daily evaporation rates of up to 3 mm/day at Scott Base on a disturbed soil surface where icy permafrost had been exposed, but evaporation was negligible for an adjacent undisturbed site.

The low thermal diffusivity of the soils studied limits diurnal temperature changes to relatively shallow soil layers, hence the extreme variations in temperature (Fig. 4).

There are three main causes for the differences in diurnal thermal regimes of the soils studied. The major difference is due to location: the higher, colder NW and CH sites had mean soil temperatures below 0°C (Fig. 3), and these soils cooled dramatically when surface heating did not occur. The thermal regime of a soil at a particular location will be most affected by the albedo of the soil surface since this regulates the amount of absorbed solar radiation. The light-coloured sandstone soil at NW was the coldest soil measured, despite having warmer air temperatures than the higher CH site. The albedo differences between basalt/dolerite (α <0.1) and sandstone ($\alpha \approx 0.3$) are particularly dramatic, and would be a major ecological factor where these surface types are adjacent to one another.

Soil salinity is an important factor in the soil thermal regime. Despite broadly similar soil moisture content at all three sites, only the non-saline SB soil had a definite ice-cement zone which coincided with temperatures of approximately 0°C. Coincident with this was a thermal regime characterised by almost no daily temperature variations at depth (Fig. 4). Freezing and thawing are processes involving the release and uptake of latent heat which have a strong conservative influence on temperature change. At the saline NW and CH sites, where soil freezing and thawing occur over a wide temperature range, this influence is absent and so diurnal temperature changes continue throughout the profile in response to upward and downward heat conduction.

The soil "active layer" is usually defined as the near-surface soil layer down to the top of the permafrost, below which temperatures never rise above 0°C. At the sites examined the active layer depth for the brief study periods were approximately: SB, 24 cm; NW, 9 cm; CH, 9 cm. At SB this depth coincided with the occurrence of ice-cement so the active layer as usually defined is relevant to the seasonal occurrence of liquid soil water. At the saline NW and CH sites, however, where freezing of soil moisture apparently occurs at much lower temperatures, the term has much less relevance.

Further research is required to investigate the effect that thermal regimes of differing soil surfaces have on the soil ecology, preferably at adjacent sites so that the effects of differences in albedo can be evaluated. There is also a need to further investigate the role of salinity in determining the presence or absence of ice-cement within the soil, and its moderating influence on diurnal temperature variations.

5   ACKNOWLEDGMENTS

This research was funded by the New Zealand Foundation for Research, Science and Technology. The University of Waikato Research Committee provided funding for travel costs. Antarctica New Zealand provided accommodation and logistical support in Antarctica. The University of Otago Geography Department is thanked for the loan of its pyranometers. Dr. Graham Buchan, Soil Science Department, Lincoln University provided many helpful suggestions.

REFERENCES

Anderson, D.M. & A.R. Tice, 1989. Unfrozen water contents of six Antarctic soil materials. In: Michalowski, R.L. (Ed.), *Cold Regions Engineering. Proceedings of the Fifth International Conference, St Paul, Minnesota, February 1989*. American Society of Civil Engineers, New York. pp. 353-366.

Arnfield, A.J., 1975. A note on the diurnal, latitudinal and seasonal variation of the surface reflection coefficient. *Journal of Applied Meteorology* 14:1603-1608.

Balks, M.R., D.I. Campbell, I.B. Campbell & G.G.C. Claridge, 1995. Interim results of the 1993/94 soil climate, active layer and permafrost investigations at Scott Base, Vanda and Beacon Heights, Antarctica. *University of Waikato, Antarctic Research Unit Special Report* 1. 64 p.

Bristow, K.L., R.D. White & G.G. Kluitenberg, 1994. Comparison of single and dual probes for measuring soil thermal properties with transient heating. *Australian Journal of Soil Research* 32:447-464.

Burke, W., D. Gabriels & J. Bouma, 1986. *Soil structure assessment*. Balkema, Rotterdam.

Campbell, G.S., C. Calissendorff & J.H. Williams, 1991. Probe for measuring soil specific heat using a heat-pulse method. *Soil Science Society of America Journal* 55:291-293.

Chinn, T.J., D.T. Pocknall, D.N.B. Skinner & R. Sykes, 1994. *Geology of the Convoy Range Area, Southern Victoria Land, Antarctica*. Institute of Geological and Nuclear Sciences, New Zealand.

Grapes, R.H., D.L. Reid & J.G. McPherson, 1972. Shallow dolerite intrusion and phreatic eruption in the Allan Hills region, Antarctica. *New Zealand Journal of Geology and Geophysics* 17:564-577.

Kyle, P.R., 1981. Geologic history of Hut Point Peninsula as inferred from DVDP 1, 2 and 3 drillcores and surface mapping. *Dry Valley Drilling Project, Antarctic Research Series* 33. American Geophysical Union, Washington D.C.

MacCulloch, R.J., 1996. *The Microclimatology of Antarctic Soils*. Unpublished MSc (Hons.) Thesis, University of Waikato, Hamilton, New Zealand.

Oke, T.R., 1987. *Boundary Layer Climates*. University Press, Cambridge.

Thompson, D.C., A.M. Bromley & R.M.F. Craig, 1971. Ground temperatures in an Antarctic dry valley. *New Zealand Journal of Geology and Geophysics* 14:477-483

Weyant, W.S., 1966. The Antarctic climate. In: Tedrow, J.C.F. (Ed.), *Antarctic soils and soil forming processes. Antarctic Research Series* 8. American Geophysical Union, Washington. pp. 47-59.

Ecosystem Processes in Antarctic Ice-free Landscapes, Lyons, Howard-Williams & Hawes (eds)
© *1997 Balkema, Rotterdam, ISBN 90 5410 925 4*

Survival of low temperatures by the Antarctic nematode *Panagrolaimus davidi*

D.A.Wharton
Department of Zoology, University of Otago, Dunedin, New Zealand

ABSTRACT: Research on the ability of the Antarctic nematode *Panagrolaimus davidi* to survive low temperatures is described. Its growth characteristics in relation to temperature suggest that it lies dormant for most of the year and grows rapidly when conditions are favourable. It is freezing tolerant and has been shown to survive intracellular freezing. Survival of intracellular freezing may be related to the rapid freezing of a high proportion of the body water of the animal and to the presence of a substance which inhibits recrystallisation. Ideas for future studies of this phenomenon are outlined.

1   INTRODUCTION

Panagrolaimus davidi is a free-living nematode which lives associated with moss and algal growth in the meltwater from glaciers in the McMurdo Sound region of Antarctica, where it appears to be restricted to coastal sites (Wharton & Brown 1989). *P. davidi* was isolated and established in culture in 1988, feeding on bacteria, and since then has been the subject of a number of investigations, particularly on its low temperature biology. This paper will summarise what has been learned about its ability to live and survive at low temperatures.

2   DISCUSSION

2.1   *Effect of low temperatures on growth rates*

Growth rates at various temperatures have been determined by transferring single eggs to small nutrient agar cultures and monitoring at daily intervals until the death of the adult worm which developed from the egg (Brown 1993). These experiments indicate that the intrinsic rate of natural increase ("r") is maximal at 25°C and that there is no population growth at temperatures below 6.8°C. Other population parameters (e.g. fecundity, total productivity, rate of population increase) also indicate an maximal temperature of 25°C (Brown 1993). The relationship between temperature and growth rate is similar to that expected for a temperate species. *P. davidi* is not adapted to grow at low temperatures. The strategy adopted by this nematode appears to be to lie dormant for most of the year and then to reproduce rapidly when conditions are favourable.

Microenvironmental temperature data for the moss habitat of the nematode at Cape Bird, Ross Island (Block 1985) indicate that *P. davidi* could complete 1.8 generations in a year (Brown 1993).

2.2   *Freezing tolerance in Panagrolaimus davidi*

The life cycle strategy of this nematode focuses attention on its ability to survive environmental stresses. The nematode can survive the complete loss of its body water and enter into a state of anhydrobiosis (Wharton & Barclay 1993). Most research, however, has concentrated on the nematode's ability to tolerate subzero temperatures. Cold tolerant nematodes, in common with other animals, tolerate subzero temperatures by one of two strategies. They may be freeze avoiding; preventing inoculative freezing from the surrounding water and allowing the body contents to supercool (remain liquid) at temperatures below their melting point. Alternatively the nematode may be freezing tolerant; freezing being triggered by inoculative freezing from the surrounding water but the nematode survives the freezing of its body contents (Wharton 1995). Nematodes are aquatic organisms and are likely to be frozen by inoculative freezing when the water surrounding them freezes. They can only adopt a freeze avoiding strategy if they possess a structure, such as a sheath or an eggshell, which prevents inoculative freezing (Wharton 1995). The eggshell of *P. davidi* prevents inoculative freezing and allows the embryo or larva within the egg to supercool in the presence of external ice and thus freeze avoid (Wharton 1994).

Other stages in the life cycle of *P. davidi* (four larval stages and the adult) cannot prevent inoculative freezing but survive the freezing of their body contents (Wharton & Brown 1991). The ability to survive freezing appears to be related to the age and/or nutrient status of the culture. *P. davidi* has recently been grown in a liquid culture medium (Sulston & Hodgkin 1988). This yields nematodes with a high degree of freezing tolerance and 89% survival after exposure to -40°C has been recorded (Wharton & Block, in press). The nematodes will also survive exposure to -80°C (Wharton & Brown 1991).

Some authors have suggested that nematodes in soil may dehydrate rather than freeze due to freeze concentration effects (Forge & MacGuidwin 1992; Pickup 1990). Freeze concentration results from ice crystal growth removing liquid water and concentrating the salts in the non-frozen portion of the solution thus raising its osmolality. We have observed freezing of *P. davidi* at osmolalities up to 1138 mosmol l^{-1}. Water extracted from a variety of moss samples indicate that the nematodes are naturally exposed to an osmolality of about 9 mosmol l^{-1}. This suggests that freeze concentration could raise salt concentrations by a factor of 120 and still result in nematode freezing. Freeze concentration effects do not appear to be sufficient to prevent the nematode from freezing (Wharton & To 1996).

2.3   *Survival of intracellular freezing*

It is usually considered that freezing tolerant animals will only survive the freezing of their extracellular compartments (Lee 1991). This has, however, rarely been tested. Survival of intraceullular freezing has been demonstrated in some insect fat body cells (Salt 1959; Salt 1962; Morason 1994). Nematodes are transparent, enabling ice formation in their bodies to be observed using a cryomicroscope. We have observed freezing and melting in all body compartments; including intracellular compartments (Wharton & Ferns 1995). Our strain of *P. davidi* is parthenogenetic. Nematodes which have been mounted singly on the cryomicroscope and observed to freeze intracellularly can be transferred to an agar culture and their subsequent survival and reproduction recorded. The presence of intracellular ice has also been demonstrated using freeze fracture (Wharton & Ferns 1995) and freeze substitution (unpubl.) techniques. The ability of this nematode to survive intracellular freezing is of great interest. Cryopreservation generally involves preventing the formation of intracellular ice, which is usually fatal to cells. If we can understand how this Antarctic nematode can survive intracellular freezing it may result in new procedures for the cryopreservation of a variety of biological materials.

Measurements using Differential Scanning Calorimetry (DSC) indicate that 82% of the body water of the nematode freezes (Wharton & Block, in press). This is high in comparison to most other freezing tolerant animals and is consistent with the presence of intracellular freezing. The freezing of the nematodes is extremely rapid: 13 s as measured by DSC and 0.2 s measured by cryomicroscopy. Cryomicroscopy gives the more accurate measurement since DSC has an inherent thermal lag. Other freezing tolerant animals (frogs, hatchelling turtles, insects) take hours or days to reach their maximum ice content. Rapid freezing of all body compartments may be important for surviving intracellular freezing since it would minimise the osmotic stresses which would result if different body compartments froze at different times.

A sample of ice held at a constant temperature below its melting point will recrystallise. This involves the growth of slightly larger ice crystals in the sample at the expense of smaller ice crystals, forming fewer larger crystals. This redistribution of ice could be harmful to a frozen animal (Knight & Duman 1986); particularly if the ice is located intracellularly. The supernatant from an homogenised sample of *P. davidi* shows some ability to inhibit recrystallisation. The recrystallisation inhibition is destroyed by heating, which may indicate that a protein is responsible (Ramløv *et al.,* in press).

2.4   Future work

Further work on the biology of this nematode will focus upon determining the location of ice and the disposition of cellular constituents in frozen nematodes, using freeze substitution techniques and transmission electron microscopy. The presence and function of low molecular weight cryoprotectants, such as polyols and sugars, will be determined. The possible involvement of protein ice active compounds, which perhaps manipulate the way in which ice forms in the body to prevent damage, will be investigated. Studies in Antarctica will determine whether nematodes fresh from the field also freeze intracellularly, as do those from our laboratory cultures. The isolation and culture of other species of Antarctic nematodes will be attempted to see if the ability to survive intracellular freezing is widespread amongst these animals. Comparisons will also be made with nematodes that do not survive freezing to further elucidate the mechanisms involved. Long-term studies on nematode ecology are being conducted in the McMurdo Dry Valleys (Freckman & Virginia 1990). Our physiological studies may indicate how these animals are able to survive the extreme environmental stresses they face in Antarctic habitats.

REFERENCES

Block, W., 1985. Ecological and physiological studies of terrestrial arthropods in the Ross Dependency 1984-85. *British Antarctic Survey Bulletin* 68:115-122.
Brown, I.M,. 1993. *The Influence of Low Temperature on the Antarctic Nematode* Panagrolaimus davidi. PhD Thesis, University of Otago, Dunedin, New Zealand.
Forge, T.A. & A.E. MacGuidwin, 1992. Effects of water potential and temperature on survival of the nematode *Meloidogyne hapla* in frozen soil. *Canadian Journal of Zoology* 70:1553-1560.
Freckman, D.W. & R.A. Virginia, 1990. Nematode ecology of the McMurdo Dry Valley ecosystems. *Antarctic Journal of the United States* 25:229-230.
Knight, C.A. & J.G. Duman, 1986. Inhibition of recrystallization of ice by insect thermal hysteresis proteins: a possible cryoprotective role. *Cryobiology* 23:256-262.
Lee, R.E., 1991. Principles of insect low temperature tolerance. In R.E. Lee & D.L. Denlinger (Eds.), *Insects at Low Temperatures*. Chapman & Hall, New York. pp. 17-46.
Lee, R.E., J.J. McGrath, R.T. Morason & R.M. Taddeo, 1993. Survival of intracellular freezing, lipid coalescence and osmotic fragility in fat body cells of the freeze-tolerant gall fly *Eurosta solidaginis*. *Journal of Insect Physiology* 39:445-450.
Pickup, J., 1990. Strategies of cold-hardiness in three species of antarctic dorylaimid nematodes. *Journal of Comparative Physiology B* 160:167-173.

Ramløv, H., D.A. Wharton & P. Wilson, in press. Recrystallisation inhibition in a freezing tolerant Antarctic nematode, *Panagrolaimus davidi,* and an alpine weta, *Hemideina maori* (Orthoptera; Stenopelmatidae). *Cryobiology.*

Salt, R.W., 1959. Survival of frozen fat body cells in an insect. *Nature* 184:1426.

Salt, R.W., 1962. Intracellular freezing in insects. *Nature* 183:1207-1208.

Sulston, J. & J. Hodgkin, 1988. Methods. In Wood, W.B. (Ed.) *The Nematode* Caenorhabditis elegans. Cold Spring Harbour Laboratory, New York. pp. 587-606.

Wharton, D.A., 1994. Freezing avoidance in the eggs of the antarctic nematode *Panagrolaimus davidi. Fundamental and Applied Nematology* 17:239-243.

Wharton, D.A., 1995. Cold tolerance strategies in nematodes. *Biological Revues* 70:161-185.

Wharton, D.A. & S. Barclay, 1993. Anhydrobiosis in the free-living antarctic nematode *Panagrolaimus davidi. Fundamental and Applied Nematology* 16:17-22.

Wharton, D.A. & W. Block, in press. Differential Scanning Calorimetry studies on an Antarctic nematode (*Panagrolaimus davidi)* which survives intracellular freezing. *Cryobiology.*

Wharton, D.A. & I.M. Brown, 1989. A survey of terrestrial nematodes from the McMurdo Sound region, Antarctica. *New Zealand Journal of Zoology* 16:467-470.

Wharton, D.A. & I.M. Brown, 1991. Cold tolerance mechanisms of the antarctic nematode *Panagrolaimus davidi. Journal of Experimental Biology* 155:629-641.

Wharton, D.A. & D.J. Ferns, 1995. Survival of intracellular freezing by the Antarctic nematode *Panagrolaimus davidi. Journal of Experimental Biology* 198:1381-1387.

Wharton, D.A. & N.B. To, 1996. Osmotic stress effects on the freezing tolerance of the Antarctic nematode *Panagrolaimus davidi. Journal of Comparative Physiology B* 166:344-349.

Ecosystem Processes in Antarctic Ice-free Landscapes, Lyons, Howard-Williams & Hawes (eds)
© *1997 Balkema, Rotterdam, ISBN 90 5410 925 4*

Moisture content in soils of the McMurdo Sound and Dry Valley region of Antarctica

I.B.Campbell & G.G.C.Claridge
Land and Soil Consultancy Services, Nelson, New Zealand

M.R.Balks & D.I.Campbell
Department of Earth Sciences, University of Waikato, Hamilton, New Zealand

ABSTRACT: The moisture content in soils, and underlying permafrost, has been investigated from 216 sites in coastal McMurdo Sound and the Dry Valleys. Moisture content in the active layer of well drained soils in the warmer coastal McMurdo regions averages around 5%, but diminishes to values of around 1% or less inland in soils on dry valley floors and valley sides. Values are intermediate in higher elevation upland valleys. In coastal regions, ice-cemented permafrost is typically found 25 to 60 cm below the surface, the depth to which the summer 0°C isotherm penetrates, but many soils, especially further inland, are dry frozen (below 0°C but contain insufficient moisture to be cemented) to considerable depth. Moisture content varies because of site factors such as microtopography, site aspect and energy inputs including albedo and windiness. Soil moisture, a critical factor in all processes in the dry valleys terrestrial ecosystem, thus varies appreciably both regionally, as well as over small distances, due to climatic and micro site characteristics.

1   INTRODUCTION

The aridity which characterises the Antarctic cold desert is a function of the prevailing very low temperatures, low precipitation and low humidity. While there is an abundance of water in Antarctica in the form of ice, liquid water is only available for the terrestrial ecosystem and soil processes from localised thaw near streams and lakes for short periods, or from melting snowfall. Most snowfall that falls on the surface is lost, either by sublimation or through removal by wind.

Soils of the Antarctic cold desert are characterised by a stony surface desert pavement, by underlying bouldery sandy/silty gravel textures and by weakly weathered profiles (Campbell & Claridge 1987; Bockheim 1980). With increasing age, there is some differentiation of properties within soil profiles, especially the extent of oxidation and salinisation, the two primary soil forming processes in cold deserts.

Permafrost is the lower part of the soil profile which remains always below 0°C. In the McMurdo Sound region, permafrost occurs at depths of up to about 60 cm, depending on climatic and site factors. The material between the soil surface and the underlying permafrost is the active layer, which is subject to diurnal and annual freeze and thaw cycles. The seasonally thawed active layer and the underlying permafrost are generally easily distinguishable from each other during summer months. The active layer is typically loose and unconsolidated at the time of maximum thaw, while the underlying permafrost is usually very hard and ice-cemented. The abrupt change is usually accompanied by a marked increase in water content. However, where soil moisture levels are low and there is insufficient moisture to cement the soil, both the active layer and the permafrost are loose and the boundary is not easily recognised.

Up until recently, little has been known of hydrological properties of the active layer of soils, the water content relationships between the active layer and the permafrost, and the extent of local site or regional differences in soil moisture content. However Campbell & Claridge (1969, 1982) recognised regional differences in soil properties as a consequence of climate and soil moisture differences in the Transantarctic Mountains. Some measurements of soil moisture contents have occasionally been made, for example, during biological investigations (Cameron *et al.* 1970), while Cheng *et al.* (1991) measured soil water content in the active layer when studying frost heave phenomena on King George Island.

Water is essential for the operation of all terrestrial ecosystem functions, including chemical and physical alteration of the environment, leaching and nutrient transport and biological activity and interactions. In the cold desert environment, where liquid water is typically only present in small amounts for periods ranging from a few days to four to six weeks each year, an understanding of soil water attributes is seen as critical to increasing our understanding of the functioning and interaction of Antarctic terrestrial ecosystems. For example, Freckman & Virginia (in press) reported results of extensive studies of nematodes in the Dry Valleys. They were however unable to define the critical factors influencing the distribution of nematodes because of scarce information on soil properties, including soil moisture contents. In this paper, we report the results of observations and measurements of soil moisture contents that we have made since 1990 in the McMurdo Dry Valleys region of Antarctica and we show how soil moisture contents differ both locally and regionally.

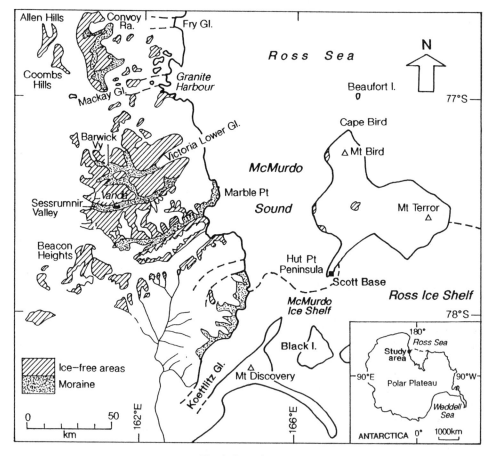

Fig. 1. Location map.

2   METHODS

2.1   Site locations

Measurements of water content of soils and underlying permafrost were made in coastal areas of the McMurdo Sound region, which falls within the Coastal Antarctic Soil Climatic Zone previously defined by Campbell & Claridge (1969, 1982, 1987). Later, investigations were extended into the Central Mountains Soil Climatic Zone where soil temperatures and precipitation regimes are different (Campbell *et al.* 1994).

Sites in the Coastal Antarctic Climatic Zone at Marble Point, Cape Roberts, Cape Evans and Scott Base (Fig. 1) were examined. Central Mountain Zone sites were studied around and in the vicinity of Vanda Station, at Barwick Valley, Sessrumnir Valley, Dias, Beacon Heights, the Coombs Hills and from Greenville Valley in the Convoy Range. At most localities, sites were sampled at a variety of locations on landforms with differing age, topography and intrinsic drainage attributes, but at some localities only a single site was sampled.

2.2   Sampling methods

At each locality, pits were excavated to the surface of the ice-cemented permafrost, usually between 20 and 60 cm, but where ice cement was not encountered the pits were excavated to depths in excess of 1 m where possible. Samples approximately 1 kg in weight were collected from the side of the pit, using a trowel, from all horizons recognised within the profile. The samples were placed in clean plastic bags, immediately sealed, field weighed, then double bagged to prevent moisture loss. Ice-cemented material was sampled to depths commonly around 1.2 m by incremental core drilling with a motor-driven rotary diamond-tipped corer. In most instances, measurements of soil temperatures were made at the time of sampling to assist with the identification of the permafrost-active layer boundary. Temperatures were measured at 10 cm depth intervals using thermistor probes inserted into the soil.

Measurement of soil water content was made at Scott Base where the samples were weighed again, oven dried at 100°C for 24 hours, then re-weighed to measure water loss. Moisture contents were calculated as a percentage of the dry weight of the sample. The determination of water content gravimetrically was considered to be the only practical means of obtaining consistent and comparative results. This was done because field time was limited, the soils were commonly bouldery and difficult to sample, while sampling at depth, or the presence of hard ice-cemented soil prevented volumetric sampling

For all samples, the proportion of material <2 mm in size was determined by sieving and for a small number of samples, bulk density was assessed by repacking excavated soil into a container of standard volume and weighing.

3   RESULTS

A total of 216 sites in the coastal McMurdo and Dry Valley region were examined. Moisture contents for sampling localities with essentially similar site characteristics were summed and averaged to provide an indication of regional or local trends in moisture content (Table 1). The results are discussed on a local and regional basis. Permafrost depths were estimated from observations of thawing depth at sampling time around the period of maximum thaw, as well as from soil temperature data. Results of repeated samplings from a number of sites showed that for differing sites or areas, water contents were usually within a narrow and distinctive range.

Table 1 Summary of site data, active layer and permafrost gravimetric moisture measurements from 216 sites in coastal McMurdo, dry valley and upland and inland valley areas. The moisture data are averaged from groups of profiles.

Locality	Date sampled	Site features sampled	Horizon	No. of sites	Average ice cmt	Average depth to permf	Average Grav H$_2$0 %: AL	ICG/DFG	Water cont range %: AL	ICG/DFG
Marble Point	Dec 1990	undulating moraines	AL, ICG	5	60	-60	4.7	37.0	0.5-9.1	15.0-133.0
Marble Point	Dec 1990	mechanically disturbed surfaces	AL, DFG	7	62	-65	5.8	6.9	2.1-8.8	4.7-20.0
Scott Base	Dec 1990	hill slope, well drained	AL, ICG	5	26	-35	3.4	29.0	1.2-5.3	4.0-62.0
Scott Base	Dec 1990	hill slope, moist sites	AL, ICG	9	10	-25	6.3	36.0	2.2-10.0	7.0-153.0
Scott Base	Dec 1990	mechanically disturbed surfaces	AL, DFG	4	26	-30	7.0	8.5	1.5-15.6	4.0-27.0
Marble Point	Nov 1991	undulating morainaes	AL, ICG	10	54	-60	5.5	35.0	0.5-12.0	10.0-144.0
Marble Point	Nov 1991	seasonally wetted surfaces	AL, ICG	4	36	-45	9.5	23.0	1.0-22.0	7.0-80.0
Marble Point	Nov 1991	mechanically disturbed surfaces	AL, DFG	2	57	-60	3.7	17.7	1.0-10.0	3.0-64.0
Marble Point	Nov 1991	ridge/flat, topo sequence	AL, ICG	3	47	-55	5.7	63.0	1.0-16.0	13.0-414.0
Marble Point	Dec 1991	beach ridge surfaces	AL, ICG	5	60	-65	1.9	26.0	0.3-7.0	6.0-122.0
Marble Point	Dec 1991	wetting front	AL	22	NA	-	5.6	-	0.8-15.0	-
Scott Base	Dec 1991	hill slope, well drained	AL, ICG	4	27	-35	3.6	53.0	0.4-9.3	6.0-129.0
Scott Base	Dec 1991	hill slope, well drained	AL, ICG	4	20	-25	12.0	15.0	8.0-29.0	7.0-38.0
Cape Evans	Dec 1991	beach and moraine surfaces	AL	3	29	-35	4.6	-	0.2-10.5	-
Scott Base	Dec 1992	hill slope, well drained	AL, ICG	5	33	-35	4.4	28.0	0.6-12.7	10.0-105.0
Vanda	Jan 1993	valley floor ridges	AL DFG	20	NP	-30	0.8	-	0.2-3.9	-
Vanda	Jan 1993	valley floor, moist sites	AL	14	-	-25	4.3	-	0.3-18.0	-
Vanda	Jan 1993	valley side moraine	AL DFG	1	NP	-20	0.6	2.4	0.3-1.3	1.9-2.7
Vanda	Jan 1993	wetting front	AL DFG	4	-	-25	3.8	-	0.1-11.0	-
Cape Roberts	Jan 1993	coastal beach surfaces	AL	3	57	-60	4.1	-	0.2-10.8	-
Marble Point	Jan 1993	undulating moraines	AL	1	50	-60	5.4	-	2.8-12.0	-
Scott Base	Jan 1993	hill slope, moist sites	AL, ICG	6	20	-25	9.4	24.2	6.5-11.0	5.3-61.0
Dias	Jan 1994	upland moraine surface	AL, DFG	1	NP	-15	1.8	2.2	0.6-	1.0-2.9
Vanda	Jan 1994	valley side moraine	AL, DFG	3	NP	-25	0.5	1.3	0.1-1.7	1.1-1.7
Vanda	Jan 1994	wetting front	AL, DFG	8	-	-30	3.5	-	0.2-11.7	-
Vanda	Jan 1994	fan/beach surface	AL, DFG	3	NP	-40	1.0	-	0.3-2.2	-
Beacon Heights	Jan 1994	upland valley moraines	AL, ICG	13	40	-10	1.2	13.0	0.3-2.8	1.3-39.0

Table 1 (continued)

Locality	Date sampled	Site features sampled	Horizon	No. of sites	Average depth to		Average Grav H$_2$0 %:		Water cont range %:	
					ice cmt	permf	AL	ICG/DFG	AL	ICG/DFG
Beacon Heights	Jan 1994	upland valley moraines	AL, DFG	1	NP	-10	1.3	1.9	0.6-2.5	1.3-2.7
Barwick Valley	Jan 1995	valley floor moraine	AL	1	50	-50	0.9	6.9	0.2-1.3	6.9 -
Vanda	Jan 1995	fan\beach surface	AL, DFG	12	NP	-40	0.4	-	0.1-0.9	-
Asgard Range	Jan 1995	upland valley moraine	AL	1	40	-40	1.4	5.7	0.9-1.8	5.7 -
Mt. Brook	Jan 1995	upland valley moraine	AL, DFG	6	NP	-10	2.1	3.3	0.4-4.5	2.4-7.7
Mt. Brook	Jan 1995	upland valley moraine	AL, ICG	4	44	-10	1.6	14.8	0.7-2.8	1.4-33.8
Greenville Valley	Jan 1995	upland valley moraine	AL, DFG	2	NP	-10	3.9	5.4	0.9-6.7	2.6-18.0
Greenville Valley	Jan 1995	upland valley moraine	AL, ICG	4	25	-10	2.7	17.7	0.6-5.4	3.2-48.0
Marble Point	Jan 1995	undulating moraine	AL	1	60	-60	4.2	-	3.1-6.3	-
Vanda	Jan 1995	fan/beach ridge surface	AL	13	NP	-40	0.7	-	0.4-2.4	-

AL   active layer
ICG   ice-cemented ground
DFG   dry frozen ground
NP   not present
ICE CMT   ice cement

65

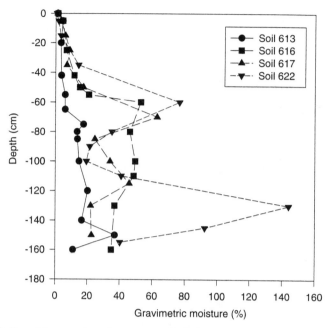

Fig. 2. Soil moisture content for some well drained coastal McMurdo Sound region soils. Ice cemented permafrost is at about 60 cm. The peaks below this represent individual ice lenses.

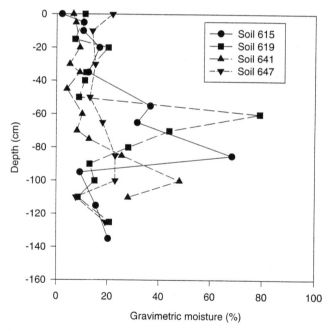

Fig. 3. Moisture content for soils from wet sites. Active layer moisture contents are greater than in well drained soils. The ice-cemented permafrost is around 30 cm and permafrost moisture contents are lower than in well drained sites

3.1   Coastal McMurdo Sound regions

Sites were investigated from Marble Point, Scott Base, Cape Roberts and Cape Evans. At these sites, soils were examined from a variety of surfaces including dry sites, typically morainal surfaces remote from any obvious stream water flows; wet sites which received some liquid water during summer thaw; sites that had been subjected to mechanical disturbances (Campbell *et al.* 1994); from sites with differing topographic characteristics; and from sites which were subject to short term changes in moisture content.

The well drained coastal sites have active layer water contents averaging around 5% with values of <1% at the soil surface increasing up to 12% at the frozen ground surface (Fig 2). Water contents in the ice-cemented material beneath are much higher, averaging about 35 to 40%, rising to >140% where there are ice lenses and the weight of water exceeds that of the soil. At moist sites, water contents are higher in the active layer (Fig. 3) but lower in the ice cement and permafrost beneath, by comparison with dry sites.

The mechanically disturbed sites have similar water contents to the well drained sites in both the active layer and the permafrost, indicating that there has been no significant accumulation of moisture in the permafrost since the ground disturbances occurred (Campbell *et al.* 1994).

Results from the two sequences of sites investigated with differing topographic aspects indicate that there are marked differences in moisture content from site to site, particularly in the permafrost (Fig. 4), which correspond with landscape changes.

Two coastal sites where soil moisture changes with time and over distance were investigated, showed similar results (Fig. 5). The soils beside a thawing snow patch which provided some liquid water had moisture contents of 12 and 15%, while after 14 days the moisture content was around 1% or similar to ambient values. Likewise, a soil which was sampled at 50 cm intervals away from a melting snow patch showed a rapid decline in moisture content from 14% at the edge of the snowbank to <1% 4 m away from the water source.

3.2   Dry Valley floors and sides

The localities investigated in the dry valleys included well drained and moist sites on the valley floor in the vicinity of Vanda Station, moist lake edge wetted zone soils, soils on till on the valley floor and side and on Dias.

Compared with the coastal McMurdo Sound region, the well drained sites around Vanda Station all have low soil moisture contents in the active layer of less than 1% (Fig. 6). The permafrost depth is around 30-40 cm and below this, the soils are usually dry frozen with moisture contents similar to those measured in the active zone. Ice cement is usually present only in those sites that are moist and where there is an obvious water source. At moist sites, moisture contents of the active layer resemble those of moist sites of the coastal McMurdo Sound region.

Two sequences of samples from an experimental site on a beach/fan surface on the north side of Lake Vanda had average moisture contents of 0.4% in January 1995 and 0.7% in December 1995 (Fig. 7). Soils on till deposits (Fig. 8) had similar moisture contents to those on the beach/fan deposits, 0.5% in the active layer. The dry-frozen permafrost, which commenced at 20 cm depth, had moisture contents ranging from 1.1% to 2.7%. The soils on till from the Dias and Barwick Valley had slightly higher moisture contents but the Barwick Valley site was underlain by ice-cemented permafrost in which the moisture content increased to 7%.

At the edge of Lake Vanda, sequences of samples were taken from the lake edge, through the moist soil zone and into the dry soil beyond (Fig. 5). The soils were saturated at the edge of the lake edge and the water content progressively diminished to <1% at a distance of 8 m from the lake edge.

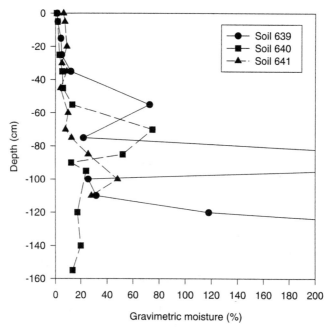

Fig. 4. Moisture contents for soils on a ridge crest (639), side slope and toe slope (641) of a small knoll illustrating the influence of microtopography. The ridge crest soil has the highest permafrost moisture content and the toe slope soil, greater active layer moisture content but lowest permafrost moisture content.

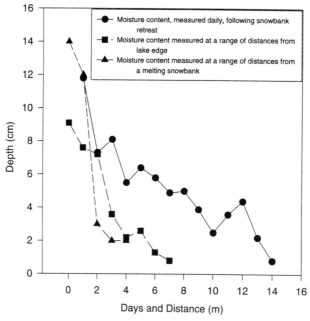

Fig. 5. Moisture content for sequences of soils showing time and distance relationships. Water sources such as streams, lakes or melting snow patches provide water over a distance of only a few meters, while soils that are wetted by thawing and retreating snow patches remain moist only for short periods.

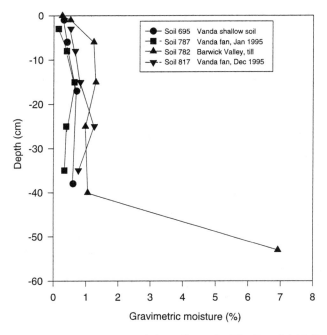

Fig. 6. Moisture contents for a range of dry valley soils. The Barwick Valley soil is underlain by ice-cemented permafrost. The others are dry frozen and have low active layer and permafrost moisture values.

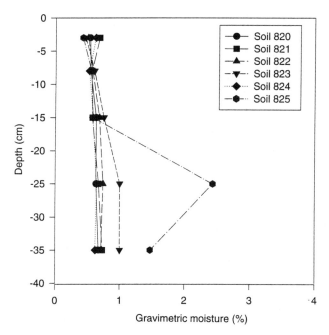

Fig. 7. Moisture content values for soils on a fan surface near Lake Vanda in December 1995. The values are over 75% higher than a similar series of sampling in January 1995.

3.3   Upland soils of the Central Mountain Zone

Soils of upland regions of the Central Mountain Zone were investigated in detail at Beacon Heights in the upper Taylor Glacier region, at the Coombs Hills near Mt. Brook and from Greenville Valley in the Convoy Range (Fig. 1). A single site was also examined in the Asgard Range in Sessrumnir Valley.

   A distinguishing feature of many of these soils was the absence of ice cement. At Beacon Heights, one site out of 13 was found without ice cement; at the Coombs Hills, 6 out 10 sites were without ice cement and at Greenville Valley, 2 out 6 sites were without ice cement. Soil temperature measurements indicate that the 0°C isotherm and permafrost depth is likely to be around 10 cm depth.

   Moisture contents in the active layer averaged 1.3% at Beacon Heights (Fig. 9), a little higher at Mt. Brook (Fig. 10) and around 3 to 4% at Greenville Valley (Fig. 11). They increased sharply when ice-cemented ground was encountered (Figs. 9, 12 & 13) but were still 2 to 3 times lower than in ice-cemented ground of the coastal Dry Valley region. The occurrence of ice-cemented ground in the upland region did not coincide with the permafrost table and most soils were dry frozen between the upper surface of the permafrost and the ice-cemented ground below, where present (Figs. 9 & 12).

   In many soils of the upland region and on till in the Wright Valley there is a distinct increase or a peak in moisture values between about 5 to 15 cm depth. This is commonly associated with the presence of salic soil horizons.

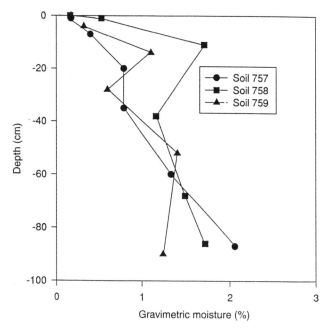

Fig. 8. Moisture content for three soils on valley side till near Lake Vanda. The permafrost is probably about 30 cm depth and the very low moisture content continues in the dry frozen ground below.

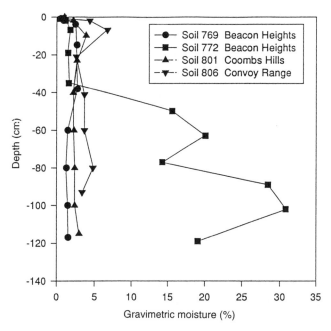

Fig. 9. A comparison of moisture contents for upland valley soils from Beacon Heights, Coombs Hills and Greenville Valley in the Convoy Range. All soils except 772 are dry frozen and without ice cement.

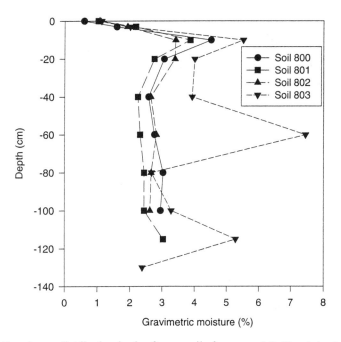

Fig. 10. Soil moisture distribution in dry frozen soils from near Mt. Brook in the Coombs Hills. The peaks at around 10 to 15 cm correspond with a salt horizon in the soils.

71

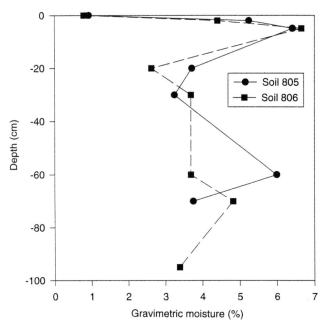

Fig. 11. Soil moisture distribution in dry frozen soils from Greenville Valley in the Convoy Range. Soil moisture contents at this lower altitude, slightly warmer and more moist location are greater than nearby Mt. Brook.

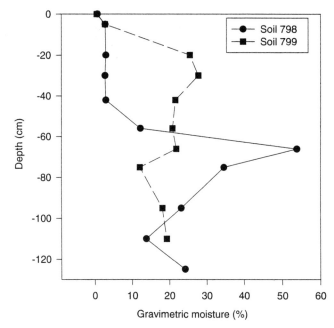

Fig. 12. Moisture content in soils with ice-cemented permafrost at Mt. Brook. Permafrost moisture content values are much lower than in coastal areas and the commencement of the ice-cement does not coincide with the $0^{\circ}C$ isotherm.

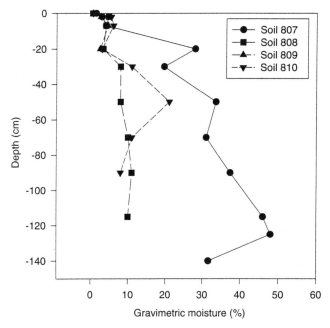

Fig. 13 Moisture content of ice-cemented soils in Greenville Valley, Convoy Range.

4   DISCUSSION

4.1   *Regional soil moisture differences*

The results indicate that there are clear regional differences in the moisture content of soils of the Dry Valley region. In the coastal McMurdo Sound area where snowfall is greatest, the moisture contents of the active layer, averaging around 5%, are up to 5 times more than those of many inland dry valley sites where snow fall is less. For most soils, the moisture content is usually around 0.5% at the soil surface, increasing to around 8% near the top of the ice cement. There is considerable variation from site to site, however.

The very low moisture contents of the soils in the Wright Valley and Vanda area are possibly a result of high evaporation rates due to the persistent up-valley and down-valley winds, while the somewhat greater values at some higher elevations are probably a result of the increased precipitation in the upland areas. In the upland areas, soil moisture contents were lowest at Beacon Heights, a little higher at the Mt. Brook sites and greatest in the Greenville Valley sites. This trend appears to mirror climatic conditions and a precipitation gradient.

In many of the sites examined the soils were found to be dry frozen to depths exceeding 1 m and moisture values are generally low throughout the soil. No clear pattern emerged as to the distribution of the sites which lacked ice-cemented permafrost. It is thought, however, that this may be a result of local factors, particularly site windiness and increased evaporation rates.

Based on work by Weyant (1966) and on soil properties, Campbell & Claridge (1969, 1987) defined five soil climatic regions reflecting the availability of moisture for soil processes. The results reported here provide a basis for quantification of the soil climate of two of these regions, the Coastal Mountain and the Coastal Antarctic Soil Climatic Zones. Soils of the warmer and moister Maritime Antarctic Zone can be expected to have higher soil moisture contents than the soils we have discussed here. At the other extreme, soils of the more arid Inland Mountain and Central Mountain Zones can be expected to have much lower soil moisture contents.

4.2   Micro site soil moisture differences

The greatest moisture differences occur between those sites which receive some summer water flow and those which remain dry. At wet sites, the soil becomes saturated within the water source zone while adjacent soils, up to a distance of several metres, become moistened by capillary flow. The wetted zone adjacent to a water source shows a strong moisture gradient from the saturated soil to low soil moisture values over a distance of a few metres. Soils that are moistened from a transitory source, such as a thawing snow patch, will return to an ambient moisture state within a few days once the moisture source has disappeared. The soils at wet or seasonally moistened sites generally have lower permafrost water contents than the adjacent soils, with downward movement of water apparently being restricted by ice cemented soil.

Smaller differences in moisture content between landscape elements such as ridge crests, side slopes and toe slopes, were also measured. These differences are probably a reflection of differences in site energy relationships including soil temperatures, albedo and evaporation potential.

In the upland region where temperatures are cooler and liquid water flows are rare, the major soil moisture difference observed is between those sites where ice cement is present at some depth within the soil, but unrelated to the permafrost table and those sites which are dry frozen.

4.3   Temporal differences in soil moisture content

For the most part, the results reported here represent a single measurement made at the time of sampling and they provide little indication of changes in the soil moisture content with time, especially on a seasonal basis. However, when 3 sites at Marble Point were sampled again one year later there were some small differences in moisture content. At Marble Point, some sampling was carried out shortly after a snowfall and the moisture values within the upper 20 cm of the soil were greater than the average for the previous sampling, suggesting that some moisture from the snowfall was moving into the soil. At another sampling area near Vanda where 12 sites (two replicates with individual sites within 1 m of each other) were sampled in January 1995 and again in December. The average soil moisture content over 12 sites was 75% greater in December than in January.

4.4   Mechanisms of moisture transfer

We have little data relating to the mechanisms and process of moisture transfer within Antarctic soils. It is evident from field observations that apart from the limited areas of soils wetted by liquid water flows, most soils receive only limited moisture from snow fall, as the snow often ablates. Vapour probably constitutes an important component of the total moisture present in the soil and there are probably significant vapour fluxes with soil temperature changes and following snowfalls. Salt accumulations within the soil also appears to influence the soil moisture content. Salts lower the freezing point of the soil solution and therefore extend the time that liquid water is available. The extent to which moisture is lost from, or accumulates within, the permafrost is also largely unknown. The absence of significant moisture accumulations within the permafrost of soil fill materials over the past forty years at Marble Point suggests that the transfer of moisture is very slow. Likewise, the persistence of stagnant glacial ice from old ice bodies which remain beneath some old and weathered soils for more than a million years suggests that loss or interchange of moisture by diffusion may be negligible.

Throughout most of the exposed ground surfaces of the Dry Valleys and other cold desert soils, potential evaporation exceeds moisture availability. The exceptions are the relatively small areas of soils which are periodically saturated, such as the peripheral zones of transient stream flows or thawed lake margins, etc. Away from these zones, there is no evidence for any hydrological continuity over either small or large areas of exposed bare ground (Claridge *et al.,* this volume). The moisture characteristics of the soils at any particular site are, therefore, likely to be governed by the general characteristics of the climate of the region (amount of snowfall, summer temperatures, windiness, etc.), and also specific site features such as proximity to a patch of snow, surface stoniness and colour, slope and aspect, soil age and salinity. Soil moisture content is also influenced by factors such as the amount of energy available for driving soil heating and evaporation (Campbell *et al.,* this volume; MacCullough 1996). The soils are, therefore, extremely spatially variable in respect of both their physical and soil moisture properties.

It is unlikely then that biological species will be present contiguously over large or significant areas. As reported by Freckman & Virginia (in press), the distributions of invertebrates in the Dry Valleys are extremely patchy. Soil invertebrate distributions are likely to reflect the variation in soils properties from place to place, especially soil moisture and site energy relationships. Antarctic cold desert ecosystems may therefore function more as a series of disparate units than as an interlinked ecological system.

5   CONCLUSIONS

1. There are clear regional differences in the moisture content of soils within coastal McMurdo Sound and the McMurdo Dry Valley region. Coastal soils have the greatest moisture content, but the moisture content of soils diminishes inland. The permafrost water content is greatest in the coastal regions, and the upper surface of the ice-cemented permafrost typically corresponds with the $0°C$ isotherm.

2. Soils of inland regions at higher altitudes have lower moisture contents, both in the soil and the underlying ice-cement, water content values. The depth to the ice-cement generally does not correspond with the $0°C$ isotherm and the soils are commonly dry frozen between the surface of the permafrost and the underlying ice cement.

3. At some sites, the soils are dry frozen, with very low moisture contents, to considerable depth. Ice-cemented permafrost lies well below the depth that can be achieved by excavation with hand tools. The absence of ice cement may be a function of low available water and increased evaporation due to site windiness.

4. Although there is a general and reasonably clear relationship between soil moisture content and climate, the mechanisms of moisture transport under limited moisture conditions are unclear. Indications are that there are small seasonal differences in soil moisture content within the active layer. However, within the ice-cemented permafrost, changes in moisture content may be very slow as soils that were disturbed by construction activity 40 years ago show little sign of reforming an ice-cemented permafrost.

5. As well the regional differences in soil moisture content, there are appreciable differences in soil moisture over short distances which are probably a result of local site energy relationships, including ground surface colour, stoniness, aspect and site albedo.

6. On the soils of older surfaces, the distribution and behaviour of moisture within the soil profile is influenced by the soil salinity, more particularly, the presence of salic horizons within the soil.

7. Soils that are moistened by periodic snow fall, or which are wetted from transitory moisture sources, dry out very quickly and return to ambient soil moisture state within hours or a few days. In areas adjacent to summer water sources, the soils are moistened over a restricted distance by capillary flow.

8. It is expected that ultraxerous soils, which occur further south on the inland portions of the Transantarctic Mountains, would have exceedingly low soil moisture contents and provide the most difficult conditions for life to exist within the soil.

REFERENCES

Balks, M.R., D.I. Campbell, I.B. Campbell & G.G.C. Claridge, 1995. Interim results of 1993/94 soil climate, active layer and permafrost investigations at Scott Base, Vanda and Beacon Heights, Antarctica. *University of Waikato, Department of Earth Sciences Special Report* 1.

Bockheim, J., 1980. Properties and classification of some desert soils in coarse-textured glacial drift in the Arctic and Antarctic. *Geoderma* 24:487-493.

Cameron, R.E., J. King & C.N. David, 1970. Soil microbiology of Wheeler Valley, Antarctica. *Soil Science* 109:110-120.

Campbell, I.B. & G.G.C. Claridge, 1969. A classification of frigic soils - the zonal soils of the Antarctic Continent. *Soil Science* 107:75-85.

Campbell, I.B. & G.G.C. Claridge, 1982. The influence of moisture on the development of soils of the cold deserts of Antarctica. *Geoderma* 28:221-238.

Campbell, I.B. & G.G.C. Claridge, 1987. *Antarctica: Soils, weathering processes and environment*. Elsevier, Amsterdam. 368 p.

Campbell, I.B., G.G.C. Claridge & M.R. Balks, 1994. The effect of human activities on moisture content of soils and underlying permafrost from the McMurdo Sound region, Antarctica. *Antarctic Science* 6:307-314.

Campbell, D.I., R.J.L. MacCullough & I.B. Campbell, this volume. Thermal regimes of some soils in the McMurdo Sound region, Antarctica.

Cheng, Z., C. Zhijiu & X. Hei-Gang, 1991. Characteristics of the active layers on Fildes Peninsula of King George Island, Antarctica. *Antarctic Research* 2:24-37.

Claridge, G.G.C., I.B. Campbell & M.R. Balks, this volume. Ionic migration in soils in the dry valley region.

Freckman, D.W. & R.A. Virginia, in press. Ecological interactions between soil organisms and their environment. In: Priscu, J. (Ed.), *The McMurdo Dry Valleys, Antarctica: A Cold Desert ecosystem*. American Geophysical Union, Washington.

Weyant W.S. 1966. The Antarctic climate. In: Tedrow J.C.F. (Ed.), *Antarctic soils and soil-forming processes. Antarctic Research Series* 8. American Geophysical Union, Washington D.C. pp. 47-59.

Ecosystem Processes in Antarctic Ice-free Landscapes, Lyons, Howard-Williams & Hawes (eds)
© 1997 Balkema, Rotterdam, ISBN 90 5410 925 4

Moisture and habitat structure as regulators for microalgal colonists in diverse Antarctic terrestrial habitats

D.D.Wynn-Williams
British Antarctic Survey, Cambridge, UK

N.C.Russell & H.G.M.Edwards
Chemistry Chemical Technology, University of Bradford, UK

ABSTRACT: The availability of water is the major regulator for primary terrestrial microalgal colonists (cyanobacteria and eukaryotic algae) in Antarctica. A comparative study of lithosolic soils at seven diverse sites ranging from the wet Maritime Antarctic at Signy Island to the arid Continental McMurdo Dry Valley region has shown up to 92% of variation in microalgal colonisation to be potentially attributable to soil moisture. The water-holding capacity of the soils studied was potentially accountable for 70% of the variation in colonisation observed. Correlation between colonisation and particle size and morphology was not demonstrable, probably because the fine lithosols of the Taylor Valley were as dry as the very different, coarse, volcanic soils of the slopes of Mt. Melbourne, Edmonson Point. The production of large amounts of exopolysaccharide (EPS) sheaths (as much as 46% of the cell volume of *Oscillatoria* in culture) is proposed as a water conservation mechanism for microalgae in Antarctic lithosols. However, in extremely arid parts of the McMurdo Dry Valleys region, the soil distance from sources of melt-water is so dry that it is uncolonized and unstable. In such regions, microalgae adopt the avoidance strategy of the cryptoendolithic habitat in translucent sandstone such as found in Beacon Heights near the edge of the Polar Plateau at the head of Taylor Valley. Here, microalgal EPS may have a role in water conservation and bio-weathering. The bio-weathering aspect of habitat modification during colonisation processes has been studied here by a new application of FT-Raman microscopic spectroscopy. This non-destructive laser-based technique has shown the differential production of oxalate in the rock profile. The potential for the involvement of oxalate in the dissolution of the endolithic substratum is discussed.

1 INTRODUCTION

The primary colonists of Antarctic lithosols (mineral soils) are cyanobacteria and eukaryotic algae (referred to jointly as microalgae). Their photosynthetic activity fixes carbon, providing organic nutrients for associated heterotrophic organisms. This central role is augmented by their production of exopolysaccharide mucilages. These cement and stabilise the soil (Wynn-Williams 1993b) and help to regulate the water balance of the cells and the whole community. Microalgae also provide photosynthetically-fixed carbon for the endolithic microbial ecosystems inside translucent rocks in the McMurdo Dry Valleys (Friedmann 1982) and northern Victoria Land. Their fungal symbionts and commensals also produce organic acids which contribute to the bio-weathering of rock.

Available water is the major regulator for these terrestrial microbial colonists (Wynn-Williams 1993b). They are deprived of essential water by low temperatures associated with winter conditions in the Maritime Antarctic, or by the negative water balance (whereby ice sublimes before it can melt) prevailing in continental ecosystems. The latter conditions occur notably in the McMurdo Dry Valleys (MDV) where snow-lie is transitory, but also occur in porous volcanic soils (as in the Mount Melbourne Bay region) and in high latitude periglacial soils such as those of southern Alexander Island, Antarctic Peninsula. Extreme water deprivation results in the avoidance strategy of the endolithic habitat.

For emergence from dormancy and metabolic activity, the duration of the water supply is as critical as the amount available. Eukaryotic algae can use water vapour as well as liquid water, but either phase must be retained within the habitat for a significant period of time for growth to occur. Water retention depends on the circulation of water or vapour in the rock or soil profile. The particle size of the substratum defines pore size and continuity. This can be modified beneficially by the synthesis of water-retaining exopolysaccharide (EPS) sheaths by microalgae. Endolithic EPS production optimises water retention whilst their fungal symbionts synthesise organic acids which mobilise inorganic nutrients (Johnston & Vestal 1993) and disrupt the rock structure by bio-weathering. This stratified habitat generally includes a black-pigmented lichen zone, a white lichen layer with hyaline fungi, and a green microalgal zone, although not all these zones are conspicuous in all communities (Friedmann *et al.* 1988; Nienow & Friedmann 1993). The microalgal zone may contain either algae or cyanobacteria depending on location and water regimes. The rigidity of this endolithic habitat makes chemical analysis of extracts crude and difficult (Johnston & Vestal 1993). However, in this paper we used the laser-based technique of FT-Raman spectroscopy to overcome this problem by non-intrusive *in situ* analysis of the fractured profile (Edwards *et al.* 1995).

This paper presents results of studies on the role of water availability and particle size in the colonisation of lithosols by microalgae. It also discusses the role of mucilages in this process and their action, together with organic acids, in the biodegradation and exfoliation of endolithic habitats to create and inoculate lithosols in Antarctic cold deserts.

2   SITE DESCRIPTIONS

Eight sites in four climatically diverse regions of Antarctica were selected to compare and contrast the role of water and substratum in primary colonisation processes and habitat creation. The Maritime Antarctic research site at Jane Col, Signy Island (60°43'S, 45°35'W, Fig.1) is located on a level saddle 150m asl. (Wynn-Williams 1990a). It is an oligotrophic lithosolic habitat composed of frost-sorted quartz-mica schist soil polygons whose edaphic characteristics were shown to have little inter-polygon variation (Davey & Rothery 1993). Research primary colonisation processes was initiated at Jane Col in 1984/85 (Wynn-Williams 1992) and at two contrasting continental sites on Alexander Island in 1992/93 (Wynn-Williams 1993a). These were Ablation Valley (70°49'S) and an oasis on Two Step Moraine (71°53'S). Ablation Valley is a cold desert region with artesian melt-water streams emerging from its flanks to irrigate frost-sorted polygons and support small patches of bryophytes (Smith 1988). Two Step Moraine, at the intersection of Mars Glacier and King George VI Sound, has relatively stable homogeneous ground containing cyanobacteria and even supporting patches of mosses in moister areas (Wynn-Williams 1993a). In the summer of 1993/94 Cloches were installed on lithosol fines there and at nearby Mars Oasis.

Although Edmonson Point in Wood Bay (74°20'S, 165°08'E), is at a similar latitude to Two Step Moraine, its volcanic ash soil derived from Mt. Melbourne makes it a much drier habitat. Most of the area has a coarse soil with very low proportion of fines (<0.063 mm fraction, usually <2%) (R. Bargagli, pers. comm.). However, an area now referred to as the

"Colonisation site" contained much more silty material and has a consequently greater water retaining potential. Being only *ca.* 1 km inland, the site was potentially exposed to maritime influence. However, sea-ice was present throughout most of the 1995/96 summer.

The Taylor Valley Long Term Ecological Research site (LTER) was a truly cold desert habitat, being well inland away from marine influences, and having a very limited input of precipitation (Wharton 1994). The terrestrial site is being studied for its nematode populations, using open-topped cloches (Powers *et al.* 1994) which were supplemented in 1995/96 by the closed BAS-type system (Wynn-Williams 1992, 1996).

The endolithic sandstone community studied by FT-Raman spectroscopy was obtained from a north-west facing outcrop of Beacon Heights orthoquartzite at 2200 m altitude on East Beacon (77°50'S, 160°52'E) composed of well rounded quartz grains in a white clay mineral and quartz cement (McElroy & Rose 1987).

3   MATERIALS AND METHODS

3.1   *Sampling, microscopy and image analysis (IA) of soil crusts*

In midsummer when microalgal growth was maximal (Davey 1991), five replicate cores 15 mm in diameter taken from central fine silt were examined directly under a cover-glass by green-

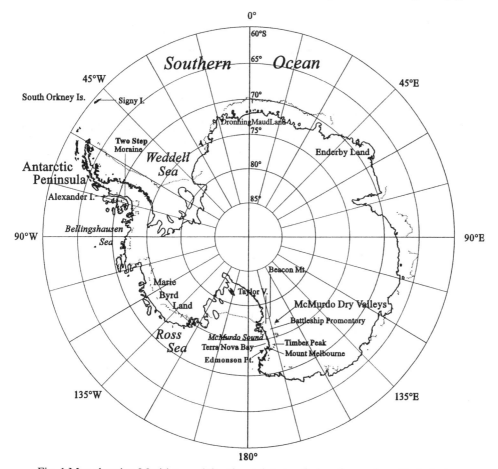

Fig. 1 Map showing Maritime and Continental Antarctic locations referred to in text.

light epifluorescence using a Leitz Laborlux microscope system fitted with a Ploemopak N2.1 filter block (excitation at 515 to 560 nm) and a x10 objective (total x125). Images were digitised and analysed using Seescan Solitaire and Sonata systems (Seescan PLC, Cambridge, UK) to quantify autofluorescing microalgal colonists (Wynn-Williams 1990b, 1996).

The length , breadth and total area of cover of microalgal colonists was derived from their perimeter and area, assuming their morphology to approximate cylinders of various lengths with parallel sides and hemispherical ends (Wynn-Williams 1990b).

3.2 *Quantification of mucilaginous sheaths*

For comparative purposes, pure cultures of dominant fellfield cyanobacteria and algae were grown under identical conditions of light and temperature (15°C) in Chu-10 medium containing nitrate and recommended for the growth of the dominant cyanobacterium *Phormidium autumnale* (Davey *et al.* 1991). Suspensions of larger cells or trichomes were mixed 1:1 or 2:1 with 10% aqueous Nigrosin, and smaller cells were mixed at 1:1 with 1% Nigrosin. The protoplast was quantified by IA under green light epifluorescence and the total dimensions, including the sheath, were quantified by transmitted light. Sheath dimensions were determined by subtraction.

3.3 *Fourier Transform-Raman spectroscopy of endolithic communities*

Endolithic communities were fractured vertically and the laser beam was directed on to successive layers from the surface to the abiotic interior of the rock (Wynn-Williams 1986). This stratified habitat can be broadly analysed in five zones: an abiotic iron-rich crust (0 to 1 mm); a black lichen zone (*ca.* 1 to 3 mm); a white fungal layer (*ca.* 3 to 5 mm); a green microalgal zone (*ca.* 5 to 8 mm) and an inner zone of accumulated iron leachate. In the sample described here, the green zone was not conspicuous. The grain size of the East Beacon orthoquartzite analysed here by Raman spectroscopy, was determined by transmission microscopy (at magnification x40) and image analysis of gently crushed rock. The size classes for feret diameter (integrated mean of 36 diameters) were as follows (Wynn-Williams, unpubl.): 32% up to 200µm; 33% from 200 to 400 µm; 14% from 400 to 600 µm; 21% >600 µm.

In Raman spectroscopy, the sample is exposed to a laser beam of monochromatic radiation. Elastic collisions of photons with molecules result in light scattered intensely at the incident frequency (Rayleigh scattering). However, during concurrent inelastic collisions a small fraction of the back-scattered radiation shows shifted frequencies which correspond to vibrational transitions in molecular components of the sample. These frequencies can be lower or higher than that of the incident light (Stokes and anti-Stokes lines respectively). Spectral lines which are shifted to energies lower than the laser source are produced by ground-state molecules. Lines at higher frequencies are due to molecules in higher excited vibrational states. The frequencies (λ) of light of different wavelengths are so large that the parameter is expressed as the equivalent wave number cm^{-1}. This is the number of waves per centimetre path in a vacuum (*ca.* 15 000 cm^{-1} for red light) and numerically is $1/\lambda$. To present the Raman spectrum, the Rayleigh scattering line at the excitation wavelength of the laser is standardised to a wave number = 0. The shift of frequencies above or below the excitation wavelength indicate the chemical bonds characteristic of the organic and inorganic molecules comprising the sample.

Spectra were recorded using a Fourier-Transform Bruker IF66 instrument with FRA Raman module attachment with 350 mW Nd/YAG laser excitation at 1064 µm and a liquid nitrogen-cooled germanium detector (Edwards *et al.* 1995). This was coupled via a TV camera to a Raman microscope with a x100 objective to give a resolution of *ca.* 10 μm at the sample. About 10 000 scans at 4 cm^{-1} resolution were needed to obtain good spectra with wavenumbers accurate to ±1 cm^{-1}. Twenty replicate spectra were taken down a 10 mm profile).

3.4   Water content of soils

Where initial moisture content permitted, large mineral debris was removed with a 1.73 mm aperture Endecott sieve. Samples were dried at 105°C to constant weight before re-hydrating to saturation. After decanting immediate free-draining water, they were re-weighed to determine the percentage water-holding capacity (%WHC).

3.5   Particle sizing of soils

Emphasis was placed on the soil crust fines that provide the habitat for the surface-dwelling cyanobacteria. The method selected therefore excluded the larger particles which constituted the coarse support for the crust. The upper 1 to 2 cm of the soil was shaken vigorously in *ca.* 10x its volume of water. Aliquots of the suspension immediately above the sediment visible after 10 s settlement were pipetted on to a microscope slide and covered with a cover slip to produce a thin film of particles. Fields were selected at random within the optimum density of permitting enumeration of separate particles by transmitted light (at x400 magnification) and image analysis. Particle size classes based on Feret diameter (mean of 36 diameters) were quantified by transmitted light microscopy at x400 magnification combined with image analysis (Seescan PLC). Circularity ($4\pi A/P^2$, where A = area and P = perimeter of a particle) was determined concurrently as a measure of microalgal deviation from spherical morphology (circularity = 1) with structural implications for the binding of soil particles.

4   RESULTS AND DISCUSSION

4.1   Moisture and microalgal colonisation processes

Water is essential to life, and avoidance of desiccation and the cellular damage it induces is a major feature of microbial survival in Antarctica. Avoidance strategies may be ecological as in habitat selection or habitat modification, or physiological with the aid of compatible solutes. A latitudinal gradient of diverse terrestrial sites from the wet Maritime Antarctic (Signy Island) to dry Continental Antarctica (Alexander Island and Terra Nova Bay) was established within the remit of the SCAR programme BIOTAS (Biological Investigations of Terrestrial Antarctic Systems) (Smith & Wynn-Williams 1992) . Quartz mica-schist soils at Signy Island typically have a high content of available water derived from *ca.* 315 mm of precipitation (mainly as rain) during the summer growing season (Wynn-Williams 1993c). This water does not readily drain out of the silty profile which has a *ca.* 50% WHC (Table 1). Soil polygon fines in this habitat are therefore rarely water-limited and support a typical microalgal cover of *ca.* 8% under the analytical conditions used here. This cover comprises mainly cyanobacterial filaments whose total trichome length is *ca.* 22 000 μm mm^{-2}. Under these moist maritime conditions, temperature becomes rate-limiting *per se* and through its effect on extending the growing season. This can result in a population increase to *ca.* 75% cover in three years when the surface temperature is elevated *ca.* 3% above ambient in a field cloche to simulate climate warming (Wynn-Williams 1996).

Comparative studies of primary colonisation were made at diverse sites described in Table 1. The soil texture of frost-sorted polygons in Ablation Valley was similar to that at

Table 1. Colonisation of soil by microalgae relative to soil moisture

Location	Site	Sample type	Lat.°S	Colonisation (% area[a] ±cv%[b])	Total length (μm mm^{-2}) ±cv%	Water content (% dry wt.)	Water holding capacity (%)	Particle size classes as mean Feret diameter (μm): Percent. frequency (n = 5):				Mean particle circularity
								0-1	1-2	2-3	>3	
Signy Island	Jane Col	polygon fines	60°42'	8.30±29	21,704±38.5	22.9	48.5	43	29	11	17	0.49
Ablation Valley	colonisation	polygon fines	70°49'	7.86±6.7	6280±25.6	21.7	45.5	59	5	7	29	0.50
Two Step Moraine	preliminary	mineral fines	71°54'	10.70	6768	10.1	38.5	45	31	10	14	0.53
Taylor V LTER	sand wedge	mineral fines	77°38'	0.0005±223	Negl.[c]	0.3	17.0	76	20	3	1	0.58
Edmondson Point	primary	Dec. moist	74°20'	ND[d]	ND	18.3	38.5	32	31	12	26	0.47
	primary	Jan. dry	74°20'	0.04±30.6	234±28.3	0.2	37.6	28	19	11	42	0.48
	colonisation	Dec.	74°20'	ND	ND	0.7	29.3	42	28	10	20	0.46
		Jan. Area 1	74°20'	0.08±52.7	33.6±48.7	1.1	26.6	45	23	11	21	0.48
		Jan. Area 2	74°20'	3.76±83.7	14,380±75.9	4.1	39.5	44	22	10	25	0.48
		Jan. Area 3	74°20'	4.85±46.7	19,842±38.3	9.0	44.8	47	22	9	21	0.51
	secondary	Pond edge (1)	74°20'	100	ND	135.8	146.6	40	25	10	25	0.46
		Pond edge (2)	74°20'	100	ND	97.8	183.1	40	23	7	31	0.47
		Pond edge (3)	74°20'	100	ND	69.6	127.8	35	26	12	27	0.47

Particle sizing done after 10 second settlement from a soil suspension. Wet smears enumerated microscopically at x400 magnification using image analysis.
[a] Percentage area of colonisation; [b] coefficient of variation % = (standard deviation/mean)x100 for five replicate samples; [c] Negl. - negligible; [d] ND - not determined; [e] single bulked sample

Signy Island with abundant fine silt (Table 1). Despite its general characteristics of a cold desert, the moisture content of 21% in polygon fines in Ablation Valley was similar to that at Signy Island because of artesian melt-water and downslope drainage. Despite its much more southerly latitude, this results in a percentage cover similar to the maritime site. Latitude *per se* is not a climatic factor. The proportion of cyanobacterial filaments in this soil was substantial but lower than that at Signy Island. However, there was significant spatial variation in population composition both at Signy (Davey & Clarke 1991; Davey & Rothery 1993) and Ablation Valley, suggesting a stochastic element in the colonisation process.

At Two Step Moraine soil fines were not frost-sorted into polygons. As a continental site, their midsummer water content was only 10.1% with a substantially lower WHC than at Signy Island, despite their similar content of finer particles. Nevertheless, their *ca.* 11% cover suggests that this moisture content was adequate for significant colonisation by cyanobacteria. Moreover, small moss shoots were frequently observed in this soil, culminating in a large moss patch in a nearby hollow (Wynn-Williams 1993a).

The homogeneous but coarse volcanic soil of the "Primary Site" at Edmonson Point, Wood Bay, was a marked contrast to Two Step Moraine despite its similar latitude (Table 1). In December at the edge of a receding snow pack, the water content was similar to that of Signy Island. However, despite its substantial WHC it desiccated rapidly to a typical summer moisture content of 0.23% in January. This was more like that of cold desert soils in the McMurdo Dry Valleys (Claridge & Campbell, this volume). The coarse texture of the soil, with consequently large pore spaces, allowed free circulation of air in the upper profile with consequent desiccation and minimal upwards percolation of moisture from the soil profile. This effect was accentuated by diurnal heating and cooling which causes the soils air to expand and contract, thus flushing out the moisture. This habitat was therefore severely water-limited for much of the growing season, with a consequently minimal population of cyanobacterial colonists. This negligible microflora was of little value as a baseline population for assessing the effects of climate change. Cloches on the site were therefore moved to a nearby site (Colonisation Site) characterised by areas with a much higher silt content. A dry, raised area of this site, devoid of snow in December, was nearly as dry as the Primary Site. However, in the hollow of a shallow drainage gully, the moister representatives of a downslope cloche series (from 1 to 9% moisture content) had increasingly abundant cyanobacterial populations (Table 1). The first area (area 1) was, at 1% moisture content, too dry for manipulative colonisation as its microalgal cover was <0.1%. However, the moister areas had a percentage cover of 3.8 to 4.9% which is adequate for determining responses to influential factors.

A "Secondary Site" at Edmonson Point was established near the end of the summer in 1994 at the periphery of a pond (S. Onofri, pers. comm.). Unfortunately, melt-water caused the water level of the pond to rise substantially during the following spring, and the cloches installed were engulfed. However, this plentiful supply of water resulted in a complete cover of microalgae dominated by *Nostoc* spp. and *Phormidium* spp. The water content of the soil at the time of sampling varied according to location, and one sample was near saturation (Table 1).

Excluding the saturated Edmonson Point Secondary Site which had a continuous cover of cyanobacterial felt and the preliminary bulk sample from Two Step Moraine, regression analysis (y=a+bx) of percentage colonisation area (%Col) on percentage moisture content of soil (dry weight basis) suggested that 92% of the variation in microalgal colonisation at these seven climatically and edaphically diverse sites (mainly cyanobacterial) might be accounted for by soil moisture (Table 2). Although this finding does not imply causation, the predicted negligible colonisation in completely dry soil was confirmed by the lowest cover observed (0.0005% at the Dry Valley LTER). Nearly 70% of this relationship could be attributable to the water-holding capacity of the soil which was correlated with the moisture observed at the time of sampling. Although filamentous forms (mainly cyanobacteria) dominate soil pioneering communities, their direct dependence on soil moisture was not demonstrable. There was no direct correlation between microalgal colonisation and particle size or morphology, probably

Table 2. Regression analysis of microalgal colonisation with lithosol moisture and particle descriptors for seven diverse Antarctic sites.

y parameter	x parameter	Correlation coefficient, r	r^2 (%)	Regression intercept, a	Regression slope, b
Colonisation area (%)	% moisture	0.961***	92.42	-0.835	262
Colonisation area (%)	% WHC[a]	0.829*	68.7	27.9	2.59
Colonisation area (%)	total length	0.749*	56.1	1992	1961
Colonisation area (%)	circularity	-0.233	NS	NS	NS
Colonisation area (%)	particles <3 μm	-0.050	NS	NS	NS
Total microalgal length	% moisture	0.596	NS	NS	NS
% moisture	% WHC	0.746*	55.6	29.8	0.86
% moisture	circulatory	-0.183	NS	NS	NS
% WHC	circulatory	-0.607	NS	NS	NS
% WHC	particles <3 μm	-0.523	NS	NS	NS
% WHC	particles <3 μm	0.523	NS	NS	NS

n = 7 replicates
[a] water-holding capacity
* P <0.05
*** P <0.001
NS - not significant

because the dry valley soils and volcanic soils on the slopes of Mt. Melbourne at Edmonson Point were equally dry but for very different reasons. Although of low porosity because of its fine texture, the dry valley soil was desiccated by katabatic winds whereas coarse the volcanic soil was so porous that moisture evaporated very quickly whatever the wind conditions. Neither moisture content at the time of sampling nor the WHC were directly attributable to the particle size at the surface of the lithosol. The extent of precipitation, timing and duration of snow lie, and the interaction between the permafrost and overlying lithosol are likely to be the most influential factors for soil moisture content at the surface.

4.2   The role of microalgal mucilages in colonisation of soil and rocks

As the microalgal crust develops, its typically filamentous mesh and mucilaginous sheaths will progressively conserve scarce moisture by restricting its diffusion (Wynn-Williams 1993b). This will seal the crust and fill near-surface pores in the profile so that the ecosystem thus becomes self-stabilising. However, when water availability becomes severely growth-limiting, as in cold desert areas distant from melt-water sources, lithosols become too desiccated and unstable to support significant populations of microalgal colonists. Desiccation-avoidance strategies are then necessary in alternative lithic habitats. One of us has found the crustose lichen *Acarospora gwynnii* in cavernously weathered sandstone boulders at Battleship Promontory (76°55'S). The same organism was truly chasmolithic in Balham Valley (77°25'S) with rhizines penetrating up to 15 mm into the translucent living rock (Wynn-Williams 1986). These organisms exhibit a range of avoidance strategies culminating in the cryptoendolithic habitat within the fabric of the rock under extremely arid conditions (Friedmann 1982).

The volume of water-conserving exopolysaccharide (EPS) produced by cyanobacteria isolated from Antarctic fellfield habitats is substantial (Table 3). The EPS sheath surrounding filaments of *Phormidium autumnale*, dominant in moist soils, comprises 6.9% of the cell volume in culture. The sheaths of fine filaments of *Pseudanabaena catenata* are up to 37.3% of trichome volume whilst those of the larger *Oscillatoria* sp. constitute a substantial 46.8% of the

Table 3. Dimensions of Antarctic fellfield microalgae and their mucilaginous in culture

Isolate	Total breadth[a] (mean±cv%[b]) (µm)	Maximum length (L) observed (µm)	Sheath volume (as % of cell volume)
Cyanobacteria			
Pseudanabaena catenata	2.97±5.39	269	37.3
Oscillatoria sp.	3.84±2.47	2438	46.8
Phormidium autumnale	5.92±4.90	1094	6.86
Eukaryotic algae			
Sphaerocystis sp.	3.49±9.17	5.63	undetected
Pinnularia sp	13.54±3.01	42.6	undetected
Planktosphaerella sp.	29.91±9.50	29.9	8.6
Zygnema sp	31.86±3.92	>5000	undetected

[a] n = 10 replicates
[b] coefficient of variation % = (standard deviation/mean)x100

cell volume. EPS produced by free-living and symbiotic endolithic cyanobacteria would conserve moisture around the cells and restrict moisture loss by occluding pore spaces. During rehydration, EPS would also exert strong hydrostatic forces on the rock constraining them. This would weaken the endolithic rock structure at the microalgal level, resulting in fracturing and exfoliation of the lithic community. Despite these moisture-conservation strategies, there are locations where, despite apparently suitable translucent sandstone rocks, the cold dry katabatic winds from the Polar Plateau are excessively harsh for endolithic growth, resulting in death of pre-existent communities, evident as fossils (Friedmann & Weed 1987).

4.3   Raman spectroscopy of endolithic communities

During optimisation of their habitat under climatic stress, endolithic microbes also produce organic acids with the potential to modify the habitat structurally and nutritionally. Johnston & Vestal (1993) had shown the production of oxalate in the endolithic habitat but were unable to show the distribution of organic acids in the undisturbed, living communities. Now *in situ* analysis by FT-RS of the distribution of organic and inorganic components of an endolithic community sampled from Beacon sandstone at East Beacon Mountain, upper Taylor Valley in southern Victoria Land (Wynn-Williams 1993b) is shown in Fig. 2. The outer iron oxide rich crustal zone (zone 1) shows only the strong quartz vibrational bands of sandstone (465 cm^{-1} and 264 cm^{-1}) and no evidence of microbiota. However, the black lichen layer (zone 2) has a strong signal for calcium oxalate dihydrate (1472 cm^{-1}) together with a broad feature from 1350 to 1300 cm^{-1} indicative of chlorophyll in the phycobiont. This broad feature is missing in the white hyaline fungal lichen layer (zone 3) which, instead, includes strong vibrational bands at 1462 and 1489 cm^{-1}, characteristic of calcium oxalate monohydrate. These oxalate bands show the region where organic acids of fungal origin are combining with calcium from the substratum as part of the bio-weathering process. Another aspect of this process is shown in the accumulation zone (zone 4) which has only traces of calcium oxalate monohydrate (the twin peaks in Fig. 2) but accumulates leached iron oxide to give it a reddish colour at the surface. The inner zone (zone 5) is beyond this leaching process and has a Raman spectrum similar to that of the surface crust.

The differentiation between the dihydrate and the monohydrate reflects microclimatic moisture gradients in the profile. Formation of the dihydrate occurs at the lower temperatures

(<5°C) which prevail near the wind-chilled surface (Nienow *et al.* 1988a), distal from the thermal buffering of the rock mass. It is also associated with more humid conditions (nearer the surface source of water from transient lying snow. The monohydrate is formed in drier, more acid conditions which accounts for the leaching of iron into the profile. Organic acids produced in this way may mobilise nutritive ions such as potassium and chelate toxic ions such as Boron which has been shown to be inhibitory to microbial growth in Dry Valley soils (Cameron *et al.* 1968).

The sample illustrated in Fig. 2 did not have a conspicuous green algal layer, but this, when present, is characterised not only by the broad chlorophyll feature but also by a vibrational band at 1525 cm^{-1} indicative of carotenoid accessory pigments (Edwards *et al.* 1995). Accessory pigments help to optimise the absorption of very low light levels within the rock, to which the algae and cyanobacteria are so well adapted (Nienow *et al.* 1988b). The diverse free-living algae or cyanobacteria and phycobionts of these communities (Friedmann *et al.* 1988) characteristically produce mucilaginous sheaths which can exert hydrostatic forces during the rehydration process. This accentuates the biochemical weathering carried out by endolithic fungi and mycobionts. Calcium oxalate monohydrate accumulates in the algal zone, where present (Edwards *et al.* 1995). Hence, in combination with the physical weathering by heating and cooling, bio-weathering by acid dissolution and EPS expansion potentially results in exfoliation of the sandstone at the accumulation zone level and disintegration of its structure. This releases the microbiota and their organic products into the local environment, with the potential to create soil communities in sheltered niches or propagules for long range dispersal. Endolithic microbiota, including associated bacteria (Siebert *et al.* 1996) and yeasts (Vishniac 1993, 1996), may be a primary source of soil inocula in the upper reaches of the Dry Valleys. If these organisms then enter an anhydrobiotic state, they constitute a propagule bank primed for re-hydration during seasonal precipitation which may be enhanced by changed weather patterns

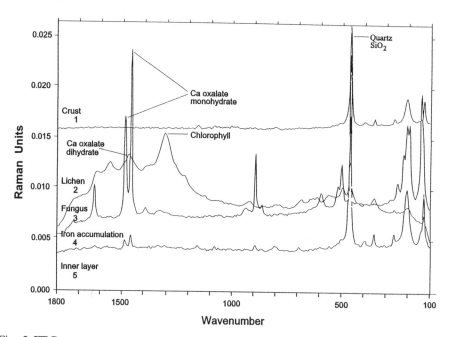

Fig. 2 FT-Raman spectrum of the profile of an endolithic microbial community in Beacon sandstone from East Beacon Mountain, upper Taylor Valley, Southern Victoria Land, showing the differential production and accumulation of organic materials in different layers (see text for details).

during climate warming (King 1994). This FT-Raman technique permits precise biochemical analysis without disrupting community structure. This allows the monitoring of biochemical responses of the same community to manipulated growth conditions such as increased humidity.

5 CONCLUSIONS

The findings presented here support the proposal that moisture is a much more influential parameter for colonisation by Antarctic terrestrial microalgae than the composition and particle size of the lithosol substratum. The common production of abundant exopolysaccharide sheaths by dominant microalgae in pioneer communities on Antarctic lithosols may a protective asset during desiccation stress. In extreme arid conditions in Antarctic Dry Valleys near the Polar Plateau where colonisation by microalgae and microfungi is restricted to the endolithic habitat, Raman spectroscopy demonstrated biosynthesised oxalate in the rock profile. This has the potential to modify the structure of the substratum by bio-weathering, thereby contributing to the colonisation of the region by promoting exfoliation and consequent release of microbiota into favourable local and distant habitats.

6 ACKNOWLEDGEMENTS

David Wynn-Williams is grateful to the following for the opportunity to carry out fieldwork: the British Antarctic Survey at Signy Island and at Rothera station over several seasons; the New Zealand Antarctic Programme in 1982/83 during initial collections of endolithic material from East Beacon Mountain; the US Antarctic Program (especially Dr. Polly Penhale, NSF and Prof. Diana Freckman, Colorado State University) for a repeat visit to this site in 1995/96; and the Italian Antarctic research programme ENEA (especially Ing. Mario Zucchelli and Prof. Roberto Bargagli, Universita di Siena) at Terra Nova Station and Edmonson Point in 1995/96. Nicola Russell acknowledges the support of Prof. Howell Edwards (University of Bradford) during Raman spectroscopy of endolithic material.

REFERENCES

Cameron, R.E., C.N. David & J. King, 1968. Soil toxicity in Antarctic dry valleys. *Antarctic Journal of the United States* 3:164-166.

Davey, M.C., 1991. The seasonal periodicity of algae in Antarctic fellfield soils. *Holarctic Ecology* 14:112-120.

Davey, M.C. & K.J. Clarke, 1991. The spatial distribution of microalgae on Antarctic fellfield soils. *Antarctic Science* 3:257-263.

Davey, M.C. & P. Rothery, 1993. Primary colonisation by microalgae in relation to spatial variation in edaphic factors on Antarctic fellfield soils. *Journal of Ecology* 81:335-343.

Davey, M.C., H.P.B. Davidson, K.J. Richard & D.D. Wynn-Williams, 1991. Attachment and growth of Antarctic soil cyanobacteria and algae on natural and artificial substrata. *Soil Biology and Biochemisty* 23:185-191.

Edwards, H.G.M., N.C. Russell, M.R.D. Seaward & D.D. Wynn-Williams, 1995. FT-Raman spectroscopic studies of environmental biodeterioration. In: Armstrong, R.S. (Ed.), *Proceedings of the First Australian Conference on Vibrational Spectroscopy.* University of Sydney Press, Sydney. pp. 41-42.

Friedmann, E.I., 1982. Endolithic microorganisms in the Antarctic cold desert. *Science. New York.* 215:1045-1053.

Friedmann, E.I. & R. Weed, 1987. Microbial trace fossil formation, biogenous and abiontic weathering in the Antarctic cold desert. *Science* 236: 703-705.

Friedmann, E.I., M.S. Hua & R. Ocampo-Friedmann, 1988. Cryptoendolithic lichen and cyanobacterial communities of the Ross Desert, Antarctica. *Polarforschung* 58:252-260.

Johnston, C.G. & J.R. Vestal, 1993. Biogeochemistry of oxalate in the Antarctic cryptoendolithic lichen-dominated community. *Microbial Ecology* 25:305-319.

King, J., 1994. Recent climatic variability in the vicinity of the Antarctic Peninsula. *Intenational Journal of Climatology* 14:357-369.

McElroy, C.T. & G. Rose, 1987. Geology of the Beacon Heights area, Southern Victoria Land, Antarctica. *Miscellaneous Map Series, Map 15*. New Zealand Geological Society, Wellington:

Nienow, J.A. & E.I. Friedmann, 1993. Terrestrial lithophytic (rock) communities. In: Friedmann, E.I. (Ed.), *Antarctic Microbiology*. Wiley-Liss, New York. pp. 343-412.

Nienow, J.A., C.P. McKay & E.I. Friedmann, 1988a. Cryptoendolithic microbial environment in the Ross Desert of Antarctica: light in photosynthetically active region. *Microbial Ecology* 16:271-289.

Nienow, J.A., C.P. McKay & E.I. Friedmann, 1988b. The cryptoendolithic microbial environment in the Ross Desert of Antarctica: Mathematical models of the thermal regime. *Microbial Ecology* 16:253-270.

Powers, L.E., D.W. Freckman, M. Ho & R.A. Virginia, 1994. McMurdo LTER: Soil and nematode distribution along an elevational gradient in Taylor Valley, Antarctica. *Antarctic Journal of the United States* 29:228-229.

Siebert, J., P. Hirsche, B. Hoffmann, C.G. Gliesche, J. Peissl, & M. Jendrach, 1996. Cryptoendolithic microorganisms from Antarctic sandstone of Linnaeus Terrace (Asgard Range): diversity, properties and interactions. *Biodiversity and Conservation* 5: 1337-1364.

Smith, R.I. Lewis, 1988. Bryophyte oases in ablation valleys on Alexander Island, Antarctica. *The Bryologist* 91:45-50.

Smith, R.I. Lewis & D.D. Wynn-Williams, 1992. Introduction to the BIOTAS programme. In: Wynn-Williams, D.D. (Ed.), *BIOTAS manual of methods for Antarctic terrestrial and freshwater research*. Scientific Committee on Antarctic Research, Cambridge. pp. 1-10.

Vishniac, H.S., 1993. The microbiology of Antarctic soils. In: E.I. Friedmann (Ed.), *Antarctic Microbiology*. Wiley-Liss, New York. pp. 297-341.

Vishniac, H.S., 1996. Biodiversity of yeasts and filamentous microfungi in terrestrial Antarctic ecosystems. *Biodiversity and Conservation* 5: 1365-1374..

Wharton, R.A., 1994. McMurdo Dry Valleys LTER: An overview of 1993-1994 research activities. *Antarctic Journal of the United States* 29: 224-226.

Wynn-Williams, D.D., 1986. Microbial colonisation of Antarctic fellfield soils. In: Megusar, F. & M. Cantar (Eds.), *Perspectives in Microbial Ecology, Proceedings of the Fourth Symposium on Microbial Ecology, Ljubljana*. Slovene Society for Microbiology, Ljubljana. pp. 191-200.

Wynn-Williams, D.D., 1990a. Microbial colonisation processes in Antarctic fellfield soils - An experimental overview. *Proceedings of the NIPR Symposium on Polar Biology* 3: 164-178.

Wynn-Williams, D.D., 1990b. The application of image analysis to natural terrestrial ecosystems. *Binary* 2:15-20.

Wynn-Williams, D.D., 1992. Plastic cloches for manipulating natural terrestrial environments. In: Wynn-Williams, D.D. (Ed.), *BIOTAS manual of methods for Antarctic terrestrial and freshwater research*. Scientific Committee on Antarctic Research, Cambridge. pp. 1-3.

Wynn-Williams, D.D., 1993a. Is there life on Mars (Glacier)? *Natural Environment Research Council News* 26:24-25.

Wynn-Williams, D.D.,1993b. Microbial processes and initial stabilisation of fellfield. In: Miles, J. & W.H. Walton (Eds.), *Primary Succession on Land. Special Publication No. 12 of the British Ecological Society*. Blackwell Scientific Publications, Oxford. pp. 17-32.

Wynn-Williams, D.D.,1993c. Soil crust microbes as indicators of environmental change in Antarctica. In: Guerrero, R. & C. Pedros-Alio (Eds.), *Proceedings of the Sixth International Symposium on Microbial Ecology, Barcelona, September 1992*. Spanish Society for Microbiology, Barcelona. pp. 105-108.

Wynn-Williams, D.D., 1994. Potential effects of ultraviolet radiation on Antarctic primary terrestrial colonisers: Cyanobacteria, algae and cryptogams. In: Weiler, C.S. & P.A. Penhale (Eds.), *Ultraviolet Radiation in Antarctica: measurements and biological effects. Antarctic Research Series* 62. American Geophysical Union, Washington DC. pp. 243-257.

Wynn-Williams, D.D., 1996. Response of pioneer soil microalgal colonists to environmental change in Antarctica. *Microbial Ecology* 31:177-188.

Ecosystem Processes in Antarctic Ice-free Landscapes, Lyons, Howard-Williams & Hawes (eds)
© 1997 Balkema, Rotterdam, ISBN 90 5410 925 4

Micro-scale distribution of photoautotrophic micro-organisms in relation to light, temperature and moisture in Antarctic lithosols

J.Cynan Ellis-Evans
British Antarctic Survey, Cambridge, UK

ABSTRACT: The distribution of photoautotrophic organisms in inorganic soils (lithosols) at the Larsemann Hills, Princess Elizabeth Land, Continental Antarctica, was studied with respect to gradients of light, moisture and temperature. During the course of the study (4 January to 2 February), incoming radiation generally approximated the theoretical curve for the latitude. A strong diurnal temperature cycle was evident at the soil surface. An inverse gradient was observed for water content and this was particularly evident in the top centimetre due to the effects of solar radiation and the daily katabatic winds. The most pronounced gradient was for light, and PAR penetrated further (4 mm) into moist lithosols than into dry. Wavelengths below 600 nm were strongly absorbed resulting in an orange-red light regime. Peak biomass in moist lithosols was found at 2.5 to 3.0 mm depth where PAR was in the range 1.0 to 0.01% of incident. In dry lithosols, the peak was lower in the profile and the light levels correspondingly lower. Community diversity was limited to cyanobacteria, a few species of diatoms and green unicells. The use of cloches demonstrated that desiccation stress prevented development on the lithosol surface, rather than temperature or light conditions.

1 INTRODUCTION

The streams, lakes and pools of Antarctica frequently appear oases of life within the surrounding landscape of rock, ice and raw mineral lithosols. Antarctic soils are primitive, but nevertheless exhibit considerable structural variation (Claridge & Campbell 1985) which provides a range of potential habitats for soil organisms (Walton 1984). Given the apparent barren nature of many of these soils, terrestrial communities have received comparatively little attention from Antarctic researchers over the years, with most interest being focussed on moss peat and ornithogenic soils, or the more unusual terrestrial situations, such as cryptoendolithic or hypolithic environments (reviewed by Wynn-Williams 1990a; Vishniac 1993; Nienow & Friedmann 1993).

Several papers (Broady 1979; Davey 1989, 1991; Davey & Clarke 1991; Wynn-Williams 1990b) have described the distribution and composition of photo-autotrophic micro-organisms in various soil and rock environments in the milder, moister, Maritime Antarctic region (the western side of the Antarctic Peninsula and off-lying island groups), and indicated that whilst broad-scale distribution can be attributed to factors such as light, nutrients or water availability, vertical distribution in bare fellfield soils appears to be due primarily to water availability.

In the more extreme environment of the Continental Antarctic region, which constitutes the greater part of Antarctica, the most detailed studies of algal and cyanobacterial distribution in soils have been those of Broady (notably Broady 1986, 1989) which were reviewed by

Nienow & Friedmann (1993). The continental region is characterized by consistently lower temperatures and humidity than the maritime region and these studies have indicated that water availability again exerts a major influence on the broad-scale distribution of algae and cyanobacteria with much lower species diversity and biomass being associated with drier soil sites.

With the exception of the very specialized cryptoendolithic and hypolithic habitats (see Nienow & Friedmann 1993), there has been no attempt to closely examine the vertical zonation of photosynthetic micro-organisms in Continental Antarctic terrestrial habitats and only one study, by Davey & Clarke (1991), for a Maritime location. Most samples appear to have been surface scrapes or bulked samples which do not provide the fine detail necessary to investigate sub-surface organisms present in relatively low numbers. The present study examined the physical environment of intact lithosols in the Larsemann Hills, Ingrid Christensen Coast, Antarctica in relation to the vertical distribution of microbial communities.

2   SITE CHARACTERISTICS

Lithosols are widespread in the Larsemann Hills and are one of the few alternatives to weathered rock as substrata for colonization in this barren region. The lithosols are broadly composed of various gneisses, which, in contrast to rocks of the nearest other substantial ice-free areas (Rauer Island group and the Vestfold Hills), are light coloured, though obviously

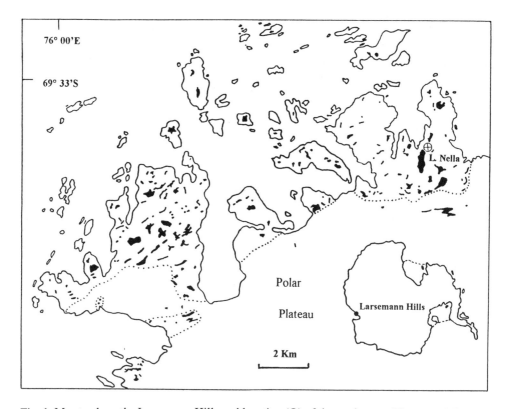

Fig. 1. Map to show the Larsemann Hills and location (⊗) of the study area. The many lakes are shown in black fill. Dotted line shows the extent of ice free ground emerging from the Plateau ice cap.

iron-pigmented to varying degrees. Many of the particles examined were polished from abrasion and had acquired surface chemical coatings. Whilst the lithosols at the study site had a substantial sand component, gravel (approximately 3 to 7%) and small stones (<2%) also contributed to the particle size composition.

The sampling sites were located (Fig. 1) near Lake Nella where a retreating snow bank provided a progressively drying location. The area has a gently undulating topography running east-west, across the main slope and two seepage areas became discernible as the snow cover retreated. Both areas drained from several snow patches, only one of which (the largest) persisted throughout the study period. Between the two seepage areas the terrain had a convex profile and, as a result, exhibited a drying gradient. The whole site was covered with snow at the start of the field programme (mid-December) but the snow quickly receded to the eastern side of the main slope before the lithosol study started in early January.

3   MATERIAL AND METHODS

Cores of lithosol were sampled using a corer of either 5 or 10 cm in diameter and 10 to 12 cm length. These were either probed (see below) or extruded at 0.5 to 1 cm intervals into separate containers for subsequent analysis of chlorophyll a and water content.

Microscopical examination of algae (by autofluoresence or phase contrast) was undertaken at magnifications of up to 1600x using a Swift field portable inverted microscope or a Zeiss epifluoresence compound microscope. All identifications were based on fresh material as it was not practicable to carry out more detailed cultural work during the study period.

Extruded sections for chlorophyll a were frozen and held in the dark to maximise extraction efficiency (as suggested by Hansson 1988), extracted into 95% methanol and absorbance measured in a Schimadzu Model UV-2100 dual beam spectrophotometer, using a 4-cm cuvette. Chlorophyll a content cm^{-2} of lithosol was then calculated as described in Davey & Clarke (1991). Where necessary, samples from several cores were pooled to ensure adequate amounts of pigment for analysis.

Distribution of cells was examined in whole cores using the fracture method of Davey & Clarke (1991). Cores, collected from various locations, were frozen to -20°C, scored, fractured and the edge trimmed flat whilst the core remained frozen. The samples were then placed on a slide and the fracture surface examined using epifluorescence microscopy, making counts in bands 0.25 mm thick.

Incident light was averaged over 10 minute intervals using a Skye Instruments PAR sensor connected to a Campbell Scientific Instruments CR10 data logger. Surface temperature fluctuations at the study sites were recorded using catheter-mounted bead thermistors connected to a Grant Instruments Squirrel data logger (resolution 0.1°C).

Water content (expressed as % dry weight) of core sections was measured in replicate 10 cm diameter cores (five per site) by slicing into 1 cm sections and drying these at 105°C for 24 h. Temporal variation in water content of a moist lithosol in a drying situation was monitored by taking duplicate cores each day for 5 days and measuring water content in 1 cm slices. On Days 2 and 3 the sampling rate was increased to every four hours, but only the top 1 cm was processed.

Temperature profiles were recorded *in situ* using a thermocouple mounted in a fine hypodermic needle for greater rigidity and positioned with a manual micromanipulator. Downwelling light profiles within lithosol cores were measured using an optical fibre micro-probe (after Jorgensen & Des Marais 1986). The probe was driven upwards, through each core, by a motorised micromanipulator (Märzhauser), and profiles measured at 0.1 to 0.2 mm vertical intervals. The silica-step index optical fibre used had a core of 105 μm diameter, an acceptance angle in water of 20%, collected >96 % of light within this acceptance angle and had a transmission value per metre, in the spectral range 400 to 700 nm, of >98%. Light channelled

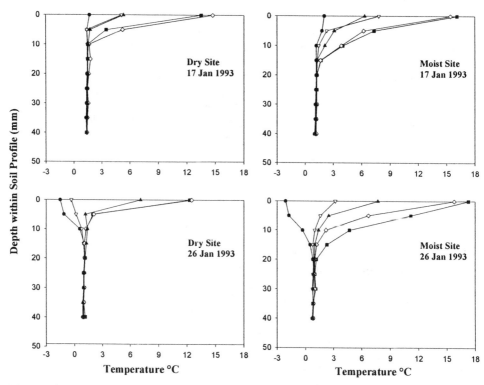

Fig. 2. Temperature profiles over 24 hour periods for dry and moist lithosol sites on two sampling occasions. Sampling times are represented by ● (08:00), ∇ (12:00); ■ (16:00), ◊ (20:00), and ▲ (24:00).

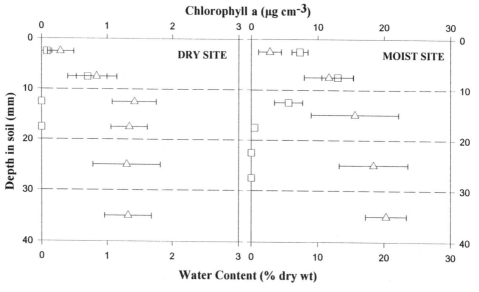

Fig. 3. Vertical profiles of percentage water content (Δ) and chlorophyll-*a* per unit area (❑) in: a - a dry lithosol, and b - a moists lithosol.

92

through the micro-probe (tapered 60 μm tip) was directed into an S1000 fibre optic CCD spectrometer (World Precision Instruments) which had a useable range of 350 to 750 nm and was used here over the visible light range (400 to 700 nm). White light was provided by a collimated fibre-optic cool light source. Spectral curves were integrated over the range 400 to 700 nm to provide estimates of down-welling photosynthetic photon flux density (PPFD). Time constraints prevented measurement of spherically integrated PPFD.

To establish if the removal of the water gradient from a wet soil would favour surface growth of organisms two small (10 cm x 10 cm) perspex cloches were placed on a drying site for four weeks and compared to a control site with no cloche. Time did not permit a more rigourous statistical design. One cloche had four sides and a top which effectively sealed the system. A second cloche had two sides and a top to allow free circulation of air through the apparatus. Temperature was measured hourly at the soil surface of both cloches and the control. Because of the destructive effect of sampling for water content and for cell counts these measurements were only made at time zero and after four weeks. Water content was assessed as described above. Algal/cyanobacterial coverage was assessed at the core surface and at 2 mm depth, by image analysis using a Seescan Instruments Solitaire system and the methodology described by Wynn-Williams (1989).

4   RESULTS

The largely cloud-free conditions experienced in the Larsemann Hills throughout the study period (18 December to 1 February) resulted in virtually model diurnal curves for incident radiation. It was unfortunately not feasible to measure light incident to sites under snow cover. Clearly such sites received reduced radiation inputs and spectral characteristics of this radiation would have been modified.

Surface temperatures at the main study site showed a strong diurnal pattern with maximum daily temperatures frequently exceeding 15°C during the course of the day, particularly at the moist site, and generally falling to less than 2°C each night. The pattern broadly followed the pattern of solar radiation input. The moist sites absorbed more radiation and were usually several degrees warmer than virtually adjacent dry sites where much of the radiation input was reflected by the light-coloured soil surface. There was an increasing frequency of sub-zero temperatures (minimum -3°C) at night, later in the month, as the brief summer period came to an end.

Soil temperature depth profiles are illustrated in Fig. 2. In the example shown, a daily surface temperature range of up to 19°C (26 Jan) was measured for a moist soil whereas the range was 13°C for a dry site only 1.5 m distant. At 5 mm depth the temperature ranges were 13°C and 3°C respectively, and at 10 mm the ranges were 5°C and 1°C so significant temperature variation predominantly occurred in the top centimetre. The stability of sub-surface temperatures was in large part attributable to the presence of permafrost at approximately 25 to 30 cm depth at the study site.

Water content at the dry site (Fig. 3) never exceeded 2% and water was undetectable in the top 5 mm with the available techniques. At the moist site (Fig. 3) a strong gradient was evident with less than 3% water content in the top 3 mm and approximately 30% at 15 to 20 mm depth. Matric forces can probably be ignored in such large grain size inorganic soils so the observed gradients would seem to reflect the interaction of water table and evaporative effects. Salt excrescences were observed in certain dry areas and mineral salt loadings would have further influenced water availability, but the salt burden was distributed very unevenly, particularly if considered on a scale relevant to microbial habitats.

Profiles of chlorophyll *a* in both dry and moist soils (Fig. 3) revealed similar sub-surface peaks, but an order of magnitude difference in values, as was seen in the moisture data. The low resolution possible (5 mm) prevented accurate location of the pigment peak, but it appeared to be within the top 10 mm. No chlorophyll was detected below 20 mm depth.

Temporal changes in water content within a moist lithosol profile undergoing drying were studied over a 5 day period (15 to 19 January). The results (Fig. 4) indicate that whilst water content decreased over the entire 4 mm profile, the changes represented only 5 to 15 % (w/w) in the region below 10 mm depth, but almost the entire 40% content of the top 10 mm. More detailed examination of changes occurring in this top 10 mm are shown in the inset histogram of Fig 4. Greatest water loss was associated with the daily radiation maxima (0800 to 1600 h) and not with the period when the dry katabatic winds were active (2200 to 1000 h). Indeed there is a suggestion of slight water gain during this latter period, though replication was not adequate to confirm this possibility.

For purposes of comparison, light penetration measurements with optical fibre microprobes (Fig. 5) were undertaken in not only dry and moist lithosols, but also in a diatom film and a cyanobacterial mat both overlying water-saturated lithosols. Reflection of incident light occurred at the surface of the dry and moist lithosols (28%±4% and 19%±3% respectively) with greatest reflectivity being associated with the dry site. In contrast virtually no reflected light (1 to 3%) was observed at the wet sites with their organic surface layers. Light penetration (taking a lower limit of 0.01% of incident) occurred to greater depth in the dry and moist lithosols than at the wet sites due to high absorption by the microbial communities in the latter case. Light penetrated further in the moist lithosol than at the dry site as wetting of

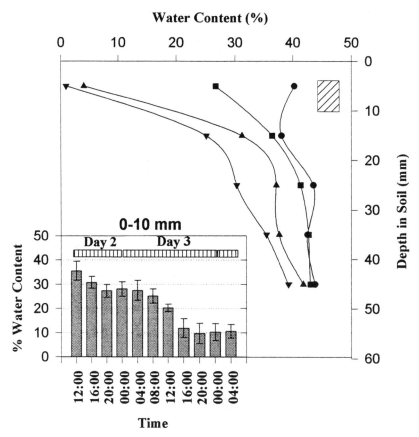

Fig. 4. Vertical profiles of percentage water content on day 2 (●), 3 (■), 4 (▲) and 5 (▼) of a drying moist lithosol. The hatched rectangle (top right) shows the approximate vertical distribution of sub-surface biomass. The inset box (bottom left) shows detailed monitoring of water content at 0 to 10 mm on days 2 and 3 with error bars (±1 se)

particles changes their refractive index relative to the surrounding environment and thus increases transparency. The results presented here represent the profiles, at each site, that yielded greatest light penetration. Variation of up to 1.5 mm (more typically 0.5 mm) in the 0.01% light depth were observed at all four sites during the series of measurements but the relationship of light penetration depth between sites, as shown in Fig. 5, remained unchanged.

Spectral analyses of two locations, the moist lithosol and the cyanobacterial mat, representing extremes in terms of light penetration, are presented in Fig. 6 and show markedly different curves. Light in the region 400 to 600 nm is absorbed by the lithosols more readily than light in the region 600 to 700 nm and this effect becomes more pronounced with depth. By 3 mm depth the light environment is heavily biased towards orange/red light and overall light levels are in the region of 0.1 to 0.01% of incident. The profiles for the cyanobacterial mat are more complex. Clear evidence of absorption of wavelengths associated with chlorophyll *a* (430, 670 nm) and and to a lesser extent, carotenoids (450 to 520 nm) was seen at 0.5 mm. By 1 mm depth this had been further developed with a broad absorption maximum between 430 and 500 nm. The absorption of wavelengths in the range 550 to 625 nm at greater depth indicate the presence of phycobilins (probably phycocyanin) in the sub-surface mat. Only red light (700 nm) and, to a lesser extent, violet light (400 nm) penetrate to significant depth in the cyanobacterial mats. The photic zone within these mats is much shallower than in the soils due to the high concentrations of efficient light-trapping pigments within the mat.

At both study sites photoautotrophs were rarely observed within the top 1 mm and a chlorophyll *a* peak was observed within the section 0 to 10 mm (Fig. 3). The relatively crude sectioning of these samples restricted more precise identification of chlorophyll peaks so to more accurately map the distribution of micro-organisms in soil profiles, epifluorescence microscopy of frozen cores was undertaken. The large particle size and low water content of

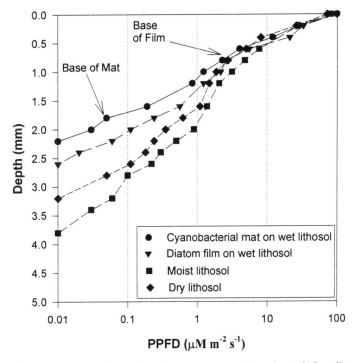

Fig. 5. PPFD penetration profiles for dry and moist lithosols and for diatom films and cyanobacterial mats overlying wet lithosols.

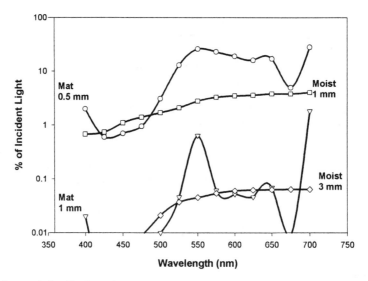

Fig. 6. Spectral distribution of PAR at 1 mm and 3 mm in a moist lithosol and at 0.5 mm and 1 mm in a cyanobacterial mat. Data are expressed as a percentage of incident light at each wavelength.

some samples caused far more problems than saturated soil fines previously examined by this method. The results (Fig. 7) showed a broad distribution of organisms (0.5 to 9.0 mm) at the moist site with the peak occurring at 2.5 to 3.0 mm depth corresponding to a light regime of 1.0 to 0.01% of incident light (using data from Fig. 5). In contrast, the dry site showed a much tighter distribution (1.5 to 6.0 mm), bearing out the chlorophyll data, with a peak at 3.0 to 3.5 mm, corresponding to a light regime of 0.1 to 0.01% of incident light. Over 60% of the dry site

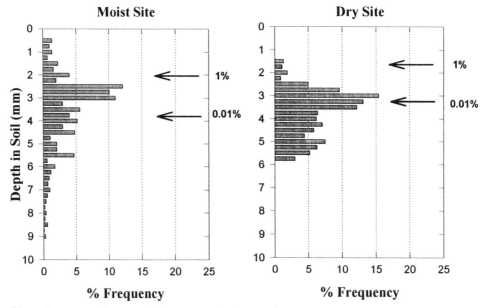

Fig. 7. Percentage frequency histograms to illustrate the distribution of micro-organisms in dry and moist lithosols. Arrows indicate 1% and 0.01% light levels in the lithosols.

community lay below this zone compared to 32% in the moist soil. No detailed examination of heterotrophs was made during this study but incidental observation of bacterial cells during autotroph counts suggested a similar distribution pattern to that observed for algae and cyanobacteria.

Community diversity in the lithosols was very restricted in all thirty sites examined. The bulk of the biomass in the moist lithosols comprised cyanobacteria, notably filamentous oscillatoriaceans with narrow trichomes (possibly two distinct species), *Gleocapsa* spp. and *Nostoc* sp., together with at least one species of unicellar green alga and several diatom genera, including *Pinnularia*, *Achnanthes* and *Navicula*. The dry sites comprised two species of filamentous Oscillatoriaceae, a *Nostoc* species (rarely), one *Gleocapsa* species and diatoms from the genera *Pinnularia* and *Navicula*. Microscopical examination of community structure revealed that, in the moist sites, filaments formed the framework of the community and that, in such circumstances, the unicells were closely associated with these matrices. At dry sites, unicells became numerically more significant and loosely associated filaments were scattered randomly. Many of the cells and filaments occurring at both moist and dry sites showed obvious mucilage production which was presumeably a pre-requisite for the desiccating environment.

Over the course of the cloche experiment (Table 1), the enclosed cloche caused a marked increase in water content of the top 10 mm of lithosol relative to the control (+9% compared to -0.2%) and filamentous cyanobacteria were observed both at the 2 mm stratum and (albeit only in small numbers) at the soil surface. There was no significant change in community composition, but an increased area of filaments was measured by the end of the experiment. The cloche without sides and the control both showed essentially unchanged vertical profiles for soil organisms over the same period. Mean surface temperatures in the enclosed cloche were approximately 2.2°C higher than in the open cloche which closely followed the temperature fluctuations of the control site. However minimum temperatures were very similar in all three situations, indicating that the frequency of freeze-thaw cycles would also have been comparable.

Table 1. Effect of Cloches on Lithosol Populations

	Enclosed cloche	Open cloche	Control site
Temperature (°C)			
Maximum	25.2	21.6	19.9
Mean	9.2	7.4	6.9
Mininum	-0.5	-0.3	-0.3
Water Content (% dw)			
Start	1.9	2.0	1.7
Finish	11.0	2.3	1.5
Change	+9.1	+0.3	-0.2
Filaments - Surface			
Start	29±5	104±39	58±27
Finish	255±74	88±28	61±12
% Change	+779%	-15%	+5%
Filaments - (2 mm)			
Start	259±81	127±39	105±52
Finish	564 ± 126	185 ± 69	92±20
% Change	+117.7%	+46%	-12%
Other Cells - (2 mm)			
Start	826±218	562±133	630±174
Finish	980±244	634±190	594±138
% Change	+19%	+13%	-6%

Filament length expressed as μm^2 per mm^{-2} lithosol surface (±se, n = 10)
Cell numbers expressed as cells per mm^{-2} (±se, n = 10)

5   DISCUSSION

This study has established that a discrete sub-surface community exists in certain lithosol environments, of low water content, in the Larsemann Hills. This coastal Continental Antarctic location has a climate that is somewhat milder than inland areas of the continent, such as the Dry Valleys of southern Victoria Land, the location of many previous terrestrial studies (Vincent 1988; Wynn-Williams 1990a). The lithosol environment of the present study is nevertheless relatively harsh and the growing season extends to only 2 to 3 months maximum, before the environment freezes once again. The discrete distribution of the photoautotrophs reflects the environmental constraints and in this sense broadly parallels the distribution of micro-organisms seen in endolithic and sub- (or hypo-) lithic niches. In all three instances, the surface is an extreme environment with high radiation levels (including high UV; Wynn-Williams 1994), fluctuating temperatures, freeze-thaw cycles, and high desiccation stress. Where water is readily available such as in seepage areas ("soaks") or streams, surface growth, of cyanobacteria in particular, can be luxuriant, but where water is limited, stress-avoidance strategies such as the endolithic habit are frequently encountered in microbiological studies (Nienow & Friedmann 1993). Retreating sub-surface can reduce desiccation stress, but subjects the micro-organisms to other more extreme gradients, notably of light, and this is demonstrated both here and in previous work on the endolithic environment by Nienow et al. (1988a).

These stress-avoidance strategies inevitably require compromises and again this is evident in the present study with respect to moisture, temperature and light and have been documented for other environments (Nienow & Friedmann 1993). Comparisons with other sub-surface environments are instructive, but have to be made with care. An example is that of the cryptoendolithic environment which has gradients of light and moisture primarily driven from a single direction, the atmosphere, whereas in lithosols the gradients of light and moisture operate in opposing directions

In dry lithosols, the water content was always <2%, but photoautotrophs were nevertheless observed in these conditions, only disappearing when water levels fell below 1%. Studies of moist lithosols demonstrated how rapidly soil moisture conditions changed with the large average particle size facilitating rapid gaseous exchange with the dry atmosphere. The high daily radiation input had a significant impact as evidenced by the strong diurnal temperature fluctuations in the top 5 mm of the profiles, and further assisted moisture loss. The endolithic environment is buffered to a greater depth as the desert varnish of fine mineral material present on the rock surface restricts moisture and thus heat loss through evaporation (Kappen et al. 1981).

The temperature data from the Larsemann Hills study suggested that, in the summer period, temperatures within the lithosol profile below 15 mm depth were constantly low, but always above zero, at least down to 50 mm depth, the maximum depth of measurements. Water conditions in the 15 to 50 mm region were significantly higher than in the 0 to 15 mm region, and yet virtually no photoautotrophs were found in the lower region as it lay well outside the photic zone. Light penetration by PAR showed significant attenuation with greatest penetration occurring in moist/wet conditions when particles became more transparent. Comparison with cryptoendolithic environments (Nienow et al. 1988a) suggested greater attenuation in the lithosols because of the greater light transmission qualities and more ordered structure of quartz grains in the former compared to gneiss particles within the lithosol. Even greater light transmission is possible within quartz stones, with 0.6 to 2.7% of incident radiation penetrating 13 to 80 mm (Broady 1981). As a result sublithic algae have been recorded as much as 61 mm sub-surface, whereas the lithosol communities never exceeded 10 mm and even then, those organisms found below 4 mm were probably placed at depth by physical processes and physiologically inactive. The base of the cryptoendolithic community rarely exceeded 5 mm in the data presented by Nienow et al. (1988a).

Peak biomass distribution in the moist lithosols was in the region 2.5 to 3.0 mm which broadly corresponds to light levels of 1.0 to 0.1% of incident light, whereas the dry lithosol had peak biomass distribution in the region 3.0 to 4.0 mm (equivalent to light levels of 0.1 to 0.005% of incident light). Sublithic algae have been reported from soil depths equivalent to 0.01% of incident light in hot desert environments (Berner & Evenari 1978) and Nienow *et al.* (1988a) reported that between 1% and 0.005% of incident light constituted the light environment for the *Hemichloris* zone (2.6 to 4.0 mm depth) at the base of the cryptoendolithic community within well colonized Beacon sandstone at Linneaus Terrace, in the Dry Valleys. The different photo-autotrophic communities would therefore appear to exist within broadly the same low light conditions. Nienow *et al.* (1988a) established that the iron oxide crusts on the surface of sandstone colonized by the endolithics were responsible for reducing the total photon flux to about 20% in the space of a few microns. Reflectivity of dry colonized sandstone rocks was roughly 20 to 40%, with wetting decreasing these values by about 65%. Average values of 28% (dry) and 19% (wet) for Larsemann Hills lithosols were of the same order. Overall the light penetration of the gneissic lithosols was equivalent to the light gradients calculated by Nienow *et al.* (1988a) for well colonized sandstone, but whereas much of the photon flux was absorbed by the pigmented fungi in the upper layers of the endolithic community, the comparatively poorly colonized lithosols of the Larsemann Hills probably contributed rather less to the process than the gneissic particles. Comparison of light penetration through cyanobacterial mats or diatom films overlying a wet lithosol, with the profiles recorded for the moist and dry lithosols, indicated that whereas the diatoms appeared to favour a higher light environment than that of the dry/moist sub-surface lithosol communities, the light environment of the latter niches was, in light penetration terms, comparable to the lower regions of the cyanobacterial mat. The low light environment of the dry/moist lithosols and endolithic algae is clearly not exceptional when put alongside, for example, the light regime seen by benthic cyanobacterial mats in ice-covered lakes (Priddle 1980).

The spectral distribution within the lithosols was broadly in agreement with that of the cryptoendolithic *Hemichloris* zone measured by Nienow *et al.* (1988a) with red/orange light penetrating significantly deeper than blue/green light. The ratio of red (700 nm) to blue (400 nm) altered perceptibly with depth, being 1.62 at the surface, 5.95 at 1 mm and 9.14 at 3 mm, significantly higher than the 4.8 value reported by Nienow *et al.* for the spectral distribution at 3 mm, just above the *Hemichloris* zone. The spectral balance was therefore shifted further to the red end in the dry/moist lithosols than in the endolithic environment. This implies that the photosynthetic emphasis in the lithosol communities will be on photosystem II activity, since no obvious absorption peak was seen for chlorophyll at 670 nm and virtually all wavelengths below 470 nm were at <0.01% of incident by 3 mm depth where peak biomass was found. Clearly an action spectrum would be required to confirm this as given the low pigment content of the lithosols, the ability to detect significant spectral changes attributable to absorption by specific pigments, has to be questioned. Attempts to measure photosynthetic oxygen production for action spectra, using oxygen microelectrodes (after Jorgensen *et al.* 1987) did not yield statistically acceptable data. By contrast, the complementary profiling of a wet lithosol cyanobacterial mat showed the now well accepted spectral profiles first reported by Jorgensen & Des Marais (1986). Substantial absorption of wavelengths 670 nm and 430 nm, seen as troughs in Fig. 8 indicate high concentrations of chlorophyll *a* and the pronounced shoulder on the trough from 430 to 500 nm at 0.5 mm would seem to indicate high concentrations of carotenoids, a feature normally only developing strongly with depth in the profiles of Jorgensen & Des Marais, whereas carotenoids are often in high concentrations in surface layers of Antarctic soil mats. These profiles are in strong contrast to those of the lithosols and the differences become even more marked at 1 mm depth. Comparisons would suggest that the spectral distribution of the lithosols reflects their physical and chemical coating characteristics rather than biological influences.

Combining the profiling data with that from the preliminary cloche experiments would suggest that desiccation avoidance imposes a sub-surface existence on the lithosol communities and that whilst the depth of maximum biomass development is not within the most stable moisture or temperature regime, the zone would appear to have acceptable limits in terms of moisture and temperature and still allows access to light levels comparable to those experienced in endolithic and sublithic environments. In the dry lithosol, the zone has lower light levels than in the moist lithosol as it is situated that much lower in the profile, which may indicate that the zone lies close to the moisture content limit. Being moved further into the profile by moisture limits results in a tighter biomass distribution band. In the Maritime Antarctic the lithosols are darker schists and a higher relative humidity in the atmosphere reduces the region of desiccation stress, so the peak of biomass is virtually at the surface and then declines rapidly with depth (Davey & Clarke 1991). A considerable number of organisms located below any reasonable photic zone in both maritime soils and moist lithosols of the Larsemann Hills are difficult to explain as many are motile, but may reflect physical disturbance, washdown, senescent cells or a heterotrophic metabolism.

Given the well documented tolerance, by soil micro-organisms, of a wide temperature range, being mainly psychro-tolerant, it would seem that minimizing exposure to freeze-thaw cycles would be the main requirement of a specific niche. In the ideal cryptoendolithic environment, temperatures would have to be held just below zero to achieve this objective (Nienow et al. 1988b). Temperature profiling from the present study suggest that temperatures in the lithosols will be subject to freeze-thaw cycling towards either end of the growing season, but detailed temperature monitoring (data not shown) over the entire season (mid-December to early February) indicated only twelve freeze/thaw events during that time in the top 50 mm of lithosol (moist or dry) and most of these were not severe. Davey et al. (1992) concluded that in such circumstances freeze-thaw cycling is not important to organism survival during the summer. It would therefore appear that water and light availability govern the micro-scale distribution of lithosol communities at the Larsemann Hills and that moisture is the major controlling factor.

REFERENCES

Berner, T. & M. Evenari, 1978. The influence of temperature and light penetration on the abundance of the hypolithic algae in the Negev Desert of israel. *Oecologia* 33:255-260.

Broady, P.A., 1979. The terrestrial algae of Signy Island, South Orkney Islands. *British Antarctic Survey Science Report* 98:1-117.

Broady, P.A., 1981. The ecology of sublithic terrestrial algae at the Vestfold Hills, Antarctica. *British Phycological Journal* 16:231-240.

Broady, P.A., 1986. Ecology and taxonomy of the terrestrial algae of the Vestfold Hills. In: Pickarad, J. (Ed.), *The Vestfold Hills: An Antarctic Oasis.* Academic Press, Sydney. pp. 165-202.

Broady, P.A., 1989. Broadscale patterns in the distribution of aquatic and terrestrial vegetation at three ice-free regions on Ross Island, Antarctica. *Hydrobiologia* 172:97-113.

Claridge, G.G. & I.B. Campbell, 1985. Physical geography - soils. In: Bonner, W.N. & D.W.H. Walton (Eds.), *Key Environments - Antarctica.* Pergamon Press, New York. pp. 62-70.

Davey, M.C., 1989. The effects of freezing and dessication on photosynthesis and survival of terrestrial Antarctic algae and cyanobacteria. *Polar Biology* 10:29-36.

Davey, M.C., 1991. The seasonal periodicity of algae on Antarctic fellfield soils. *Holarctic Ecology* 14:112-120

Davey, M.C. & K.J. Clarke, 1991. The spatial distribution of microalgae on Antarctic fellfield soils. *Antarctic Science* 3:257-263

Davey, M.C., J. Pickup & W. Block, 1992. Temperature variation and its biological significence in fellfield habitats on a maritime Antarctic island. *Antarctic Science* 4:383-338.

Hansson, L-A., 1988. Chlorophyll-a determination of periphyton on sediments: identification of problems and recommendation of method. *Freshwater Biology* 20:347-352.

Jørgensen, J.J. & D.J. Des Marais, 1986. A simple fibre-optic microprobe for high resolution light mesurements: application in marine sediments. *Limnology and Oceanography* 31:1376-1383.

Jørgensen, J.J., Y. Cohen & D.J. Des Marais, 1987. Photosynthetic action spectra and adaptation to spectral light distribution in a benthic cyanobacterial mat. *Applied Environmental Microbiology* 53:879-886.

Nienow, J.A. & E.I. Friedmann, 1993. Terrestrial lithophytic (rock) communities. In: Friedmann, E.I. (Ed.), *Antarctic Microbiology,* Wiley-Liss, New York. pp. 343-412.

Kappen, L., E.I. Friedmann & J. Garty, 1981. Ecophysiology of lichens in the Dry Valleys of Southern Victoria Land, Antarctica. I. Microclimate of the cryptoendolithic lichen habit. *Flora* 171:216-235.

Nienow, J.A., C.P. McKay & E.I. Friedmann, 1988a. The cryptoendolithic microbial environment in the Ross Desert of Antarctica: light in the photosynthetically active zone. *Microbial Ecology* 16:271-289.

Nienow, J.A., C.P. McKay & E.I. Friedmann, 1988b. The cryptoendolithic microbial environment in the Ross Desert, Antarctica: mathematical models of the thermal regime. *Microbial Ecology* 12:253-270.

Priddle, J., 1980. The production ecology of benthic plants in some Antarctic lakes. II laboratory physiology studies. *Journal of Ecology* 68:141-153.

Vincent, W. F., 1988. *Microbial Ecosystems of Antarctica.* Cambridge University Press, Cambridge. 304 p.

Vishniac, H., 1993. The microbiology of Antarctic soils. In: Friedmann, E.I. (Ed.), *Antarctic Microbiology.* Wiley-Liss, New York.. pp. 297-342

Walton, D.W.H., 1984. The terrestrial environment. In: Lawes, R.M. (Ed.), *Antarctic Ecology.* Academic Press, London. pp. 1-60.

Wynn-Williams, D.D., 1989. TV image analysis of microbial communities in Antarctic fellfields. *Polarforschung* 58:239-250.

Wynn-Williams, D.D., 1990a. Ecological aspects of Antarctic microbiology. *Advances in Microbial Ecology* 11:71-146.

Wynn-Williams, D.D., 1990b. Microbial colonization processes in Antarctic fellfield soils - an experimental overview. *Proceedings of the NIPR Symposium on Polar Biology* 3:164-178.

Wynn-Williams, D.D., 1994. Potential effects of ultra-violet radiation on Antarctic primary terrestrial colonizers: cyanobacteria. algae and cryptogams. In: Weiler, C.S. & P.A. Penhale (Eds.), *Ultravoilet Radiation in Antartica: measurements and biological effects. Antarctica Research Series* 62. American Geophysical Union, Washington, D.C. pp. 243-257.

Ecosystem Processes in Antarctic Ice-free Landscapes, Lyons, Howard-Williams & Hawes (eds)
© *1997 Balkema, Rotterdam, ISBN 90 5410 925 4*

Lichens and the Antarctic environment: Effects of temperature and water availability on photosynthesis

B. Schroeter – *Botanisches Institut der Universität Kiel, Germany*

L. Kappen – *Institut für Polarökologie und Botanisches Institut der Universität Kiel, Germany*

T.G.A. Green – *Biological Sciences, University of Waikato, Hamilton, New Zealand*

R.D. Seppelt – *Australian Antarctic Division, Kingston, Tas., Australia*

ABSTRACT: In this paper we demonstrate to what extent lichens are able to profit from snow as a water source for rehydration under the extreme environmental conditions in Continental Antarctica. Net photosynthesis of the foliose macrolichen *Umbilicaria aprina* occurred at subzero temperatures in the field even when snow was the only water source. Laboratory experiments revealed that in dry thalli of *U. aprina*, metabolic activity was initiated during rehydration from snow at subzero temperatures. The degree of rehydration of dry thalli of *U. aprina* from snow at subzero temperatures was strictly dependent on the temperature, with the final water content decreasing with decreasing temperature. In the crustose epilithic lichen *Buellia frigida* overheating of the rock surface and the black thallus by strong insolation leads to snow melt and therefore to an activation of photosynthetic activity by liquid water at a small-scale. This phenomenon occurred even if air temperature was down to -10°C. It was demonstrated that activation of different thalli depended on the micro-topography of the rock surface. The possible influence of global climatic changes on water availability and the consequences for lichens in polar desert ecosystems are discussed.

1 INTRODUCTION

The vegetation of Antarctic polar desert ecosystems is dominated by lichens and bryophytes; vascular plants are not present. Typical cryptogamic deserts can be found in the ice-free areas around the Antarctic continent. Comparatively rich cryptogamic floras were described for Birthday Ridge, northern Victoria Land (Kappen 1985), the Granite Harbour region, southern Victoria Land (Schroeter *et al.* 1992, 1994; Seppelt *et al.* 1995, 1996), the Windmill Island region (Melick *et al.* 1994) and for the area of Syowa Station (Inoue 1991) amongst others. A low biodiversity and a structural simplicity are key features of these ecosystems, which are determined by climatic constraints and by their geographical isolation. The two most important physical factors influencing survival and growth of organisms in Antarctic polar desert ecosystems are water availability and temperature (Kennedy 1993; Block 1994; Wynn-Williams, this volume).

Lichens are well adapted to cold temperatures. Lichens which have the green alga *Trebouxia* as the photobiont are known to be extremely cold-resistant in the dry as well as in the hydrated state (Kappen & Lange 1972; Nash *et al.* 1987) and several lichen species are known to be able to attain significant net photosynthetic rates at subzero temperatures (see Kappen 1993; Schroeter *et al.* 1994). Lichens are able to sustain unfavourable periods in the dry and inactive state due to their poikilohydric nature and they are able to withstand long periods of drought. If water is available either as vapour, liquid, or solid, lichens are able to rehydrate and to gain metabolic activity.

The aim of this paper is to demonstrate to what extent the photosynthesis of two widespread continental Antarctic lichen species is adapted to a subzero temperature regime and, if so, to what extent they benefit from the specific environmental water relations at subzero temperatures.

2   MATERIAL AND METHODS

2.1   *Habitat and locality*

The Granite Harbour region (77°01'S, 162°32'E), southern Victoria Land, continental Antarctica, is known for its exceptional richness in lichens and mosses (Schroeter *et al.* 1992, 1994; Seppelt *et al.* 1992, 1995, 1996). Its outstanding vegetational features were already mentioned in Taylor's account on Scott's last expedition (Taylor 1913, 1916) but not further investigated. The flora of Granite Harbour includes now about 31 lichen species, seven mosses and one hepatic (Seppelt *et al.,* unpubl.).

All field measurements were carried out at Cape Geology, Granite Harbour, during two expeditions in January/February 1992 and in November/December 1994. In January/February 1992 the lichens were activated mainly by meltwater, only rarely snowfall provided sufficient water for thallus rehydration and metabolic activity (see Schroeter *et al.* 1992). In November/December 1994 the seasonal aspect was different: the boulder beach was covered by snow of up to 1 m in depth, and only a few lichen-covered rocks were free of snow when the expedition arrived at Granite Harbour in mid-November. Because no meltwater streams occurred until mid-December, the only sources for lichen hydration were occasional snowfall events and the slow melt of superficial snow deposits on the boulders.

2.2   *The species*

Two lichen species were selected which represent the most prominent species in the Granite Harbour area:

Buellia frigida Darb. is an epilithic crustose lichen which is probably the most widespread and abundant lichen in continental Antarctica (see Seppelt *et al.* 1995). It is the dominant lichen species in the Granite Harbour area where it grows on boulder beaches and glacial moraines. In this region its major water supply is from superficial snow deposits (Schroeter *et al.* 1992).

Umbilicaria aprina Nyl. is the most prominent macrolichen in the Granite Harbour area. Its toliose, umbilicate thallus grows monophyllus on rock faces frequently with a northerly aspect and on boulders in meltwater runoffs. At Granite Harbour it is morphologically extremely variable and reaches diameter up to 20 cm (Schroeter *et al.* 1994; Seppelt *et al.* 1995).

2.3   *Gas exchange measurements*

Field measurements of the CO_2 gas exchange were carried out using a minicuvette system (CMS400, Walz GmbH, Germany) in the open flow as described by Schroeter *et al.* (1994). The system used comprised a differential IRGA (Binos 1004P, Fisher-Rosemount, Germany) with a cooling trap to remove water from the gas stream. Temperatures in the cuvette were tracked to ambient by a feedback control system with ventilated air temperature sensors inside and outside of the cuvette. If necessary, temperatures in the cuvette could be fixed by an electronic controller with an accuracy better than 0.1°C. Thallus temperatures were monitored by a very fine thermocouple that was inserted into the lichen thallus. A homogemeous illumination was provided by a specially designed light source with glass fibre optics (FL-400, Walz GmbH, Germany). Depending on the experimental setup the lichens were artificially hydrated with melt water or were allowed to rehydrate from snow at subzero temperatures. Water content of the thalli was determined gravimetrically by an electronic balance (BA110, Sartorius, Germany).

2.4 Chlorophyll a fluorescence measurements

Measurements of the chlorophyll a fluorescence of the photobionts of the lichens were carried out using a PAM-2000 fluorometer (Walz GmbH, Germany) as described in Schroeter et al. (1992). The glass fibre optics were pointed at the lichen thallus from a distance of 2 to 4 mm at an angle of 60°, taking care not to shade the thallus. At ambient light conditions usually ten measurements of the fluorescence parameters Ft and Fm' were taken randomly over the thallus regardless if colour changes indicated partial hydration or dehydration of the thallus. From these measurements the mean of the parameter ▲F/Fm' (Ft-Fm'/Fm') was calculated. Air temperature and thallus temperature were taken immediately before a series of fluorescence measurements using a very fine thermocouple (FL-2030, Walz GmbH, Germany). Photosynthetic photon flux density (PPFD, μmol m^{-2} s^{-1}) incident at the surface of the lichen thallus was also measured immediately before the fluorescence measurements using a PPFD glass-fibre microprobe (FL-2030, Walz GmbH, Germany). The PPFD values were used to calculate the apparent electron transport rate through PSII (▲F/Fm' x PPFD: see Genty et al. 1989; Bilger et al. 1995). Neither the PPFD incident at the algal level inside of the lichen thallus nor the absorption coefficient of the lichen cortex layer could be measured. However, the values calculated for the electron transport rate by using the surface PPFD values give a reasonable measure of the metabolic activity of the photobionts of the lichen if not used for a comparison of activity rates of different thalli.

2.5 Microclimatic measurements

A northwest facing granite boulder colonized by Buellia frigida was the site for automatic microclimatic measurements of thallus temperatures, air temperatures and PPFD (see Fig. 1). Temperatures were measured using very fine thermocouples which were inserted in or attached to the lichen thalli. Five different thalli were measured simultaneously. Air temperature near the granite boulder was also measured with a very fine thermocouple sheltered from direct sunlight by an aluminium tube (see Fig. 1). PPFD was measured by means of a quantum flux sensor (Li-190SB, LI-COR, USA). The sensor was orientated in the same direction as the surface of the lichen thalli. Data were collected every five min by a datalogger (Squirrel, Grant, UK). Hydration of the thalli was checked visually in hourly intervals over the measuring period. Water flows from the melting snow cover and a subsequent colour change of the thalli from grey to black indicated hydration of the thalli (Kappen et al., unpub.).

3   RESULTS

3.1   Small-scale patterns of microclimate and metabolic activity in Buellia frigida

Figs. 1 and 2 show the granite boulder at Cape Geology, Granite Harbour, where microclimatic measurements were carried out in the second half of November 1994. The surface of the granite boulder was partly covered by snow then. A patchy distribution of thalli of B. frigida of various size classes covered the slightly inclined upper surface of the boulder (Figs. 1 & 2). In the afternoon when the picture (Fig. 1) was taken sun radiation had heated the open rock surface to

Fig. 1. Microclimatic measurement site at Granite Harbour, Continental Antarctica in November 1994. Direct sunlight heated the rock surface and caused snow melt. Meltwater streams can be seen at the edge of the snow pack.

around +10°C as measured by a thermocouple probe. Small melt water streams ran from the snow pack over the open rock surface surface at this time (Fig. 1). At this time the small meltwater streams met only a total of about 30% of the lichen thalli of *B. frigida* depending on the small-scale topography of the rock surface.

The microclimatic situation for selected thalli (see Fig. 2) at this boulder during a period of six days from 29 November to 4 December is shown in Fig. 3. The diel courses of PPFD indicate clear skies during this period. The immediate decrease of the incident light intensity at *ca.* 9:00 pm was due to the shadow of the mountain ridge south of the measurement site; the rock surface was exposed to direct sunlight between *ca.* 8:00 am and *ca.* 9:00 pm ("daytime") while only diffuse sunlight reached the rock surface between *ca.* 9.00 pm and *ca.* 8:00 am ("night-time"). PPFD gave values up to 1800 µmol m^{-2} s^{-1} as on 29 November and 4 December. Diffusive incident irradiances gave only *ca.* 100 µmol m^{2}s^{-1} during "night-time". 3 and 4 December are examples of cloudy days where the PPFD values during the "daytime" were reduced and due to diffusive light the PPFD levels during the "night-time" were comparatively high (200 µmol m^2 s^{-1} in minimum) and the albedo pattern disappeared.

The diel courses of the air temperature were strongly correlated with the PPFD pattern. Air temperature ranged between +2.2°C (1 December) and -15.1°C (2 December). Within 20 h air temperatures could vary with an amplitude of 17.3°C, the mean air temperature of the complete measurement period was -6.2%.

The diel amplitudes of thalli temperatures of *Buellia frigida* were much more pronounced than those of the air temperature, reaching 22.8°C in thallus 1 and 26.8°C in thallus 4 (Fig. 3). The mean thallus temperatures (complete measuring period) indicate the importance of the small-scale topography: different thalli experience different temperature regimes even at only a few centimetres distance. The mean thallus temperature of the measuring period ranged from -0.9°C for thallus 1 and 3, which were both strongly influenced by the retreating snow

106

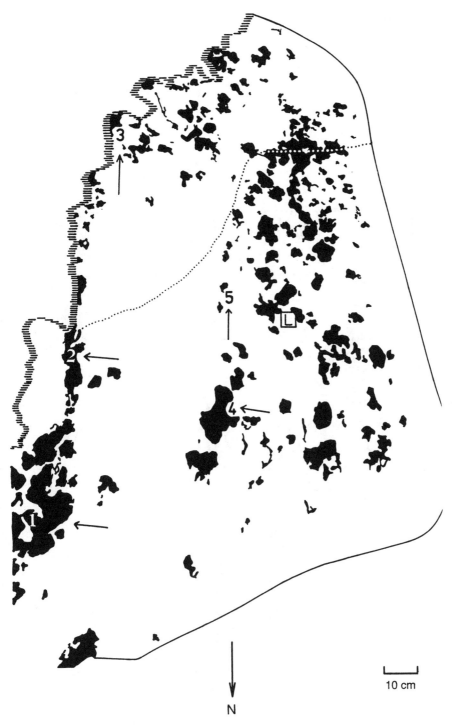

Fig. 2. Schematic graph depicted of the rock surface (see Fig. 1). Arrows point to the spots with thalli of *Buellia frigida* where thallus temperatures were monitored. "L" indicates the position of the PPFD sensor. The dotted line indicates the edge of the snow deposit on 28 November when recording of the microclimatic parameters started. All black areas are thalli of *B. frigida*.

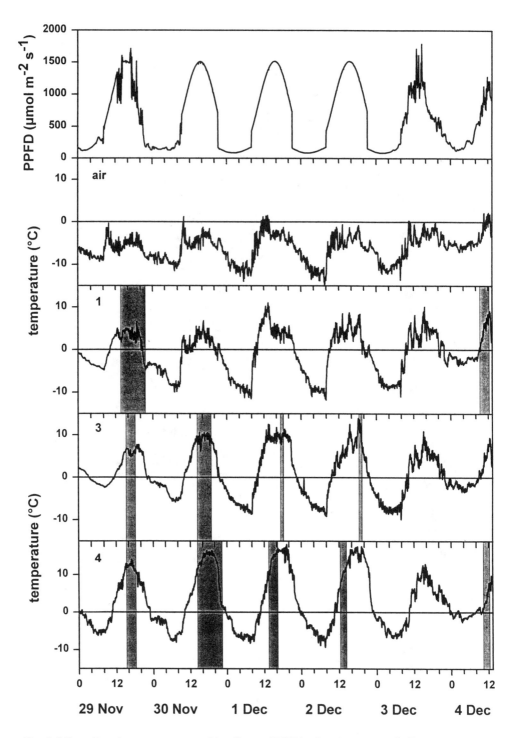

Fig. 3. Microclimatic measurements of irradiance (PPFD), air temperature, thallus temperatures on the three *Buellia frigida* sites, as shown in Figs. 1 and 2, from 28 November to 4 December 1994 at Granite Harbour, continental Antarctica.

cover, to +2.8°C in thallus 4. All thalli were considerably warmer than ambient air temperature, with an extreme difference of >20°C in thallus 4.

The influence of the small-scale topography on the microclimatic differentiation between different thalli in *Buellia frigida* is even more obvious if the pattern of duration and distribution of periods of hydration are compared. The daily hydration period ranged from less than 1 h on 3 December for thallus 3, to more than 9 h for thallus 1 on 29 November, and for thallus 4 on 30 November. Except for 3 December, when all thalli stayed dry and inactive, different thalli were hydrated by meltwater streams for different time periods of the day.

The activation of photosynthesis by meltwater hydration in *Buellia frigida* is demonstrated in Fig. 4 by means of measurements of the chlorophyll *a* fluorescence of the chloroplasts of the lichen algae. The data presented here were measured manually at 2 to 4 h intervals at a single thallus of *B. frigida* growing on a boulder *ca.* 50 m apart from the previously described site. This boulder was *ca.* 2.5 m high. The surface towards north was strongly inclined and patchy colonized by *B. frigida*. The inclined surface and the top of the boulder were covered by a 5 to 20 cm thick snow layer. When the rock surface was heated by the incident radiation the snow deposit melted. Small metwater streams occurred which changed their way on the inclined rock surface with the changing border of the retreating snow cover. The chlorophyll *a* fluorescence parameter ▲F/F$_m$' * PPFD which could be regarded as the apparent electron flow rate through PSII (ETR; see Schreiber *et al.* 1995), indicated whether this thallus of *B. frigida* was photosynthetically active. From 16 to 20 November the thallus was hydrated by meltwater formation or even only by water vapour evaporation along the edge of the snow deposit that covered part of the boulder. Photosynthetic activity was apparent for up to 8 h after noon. In the time course the snow pack slowly retreated by about 2 to 5 cm per day. When the snow border further retreated no melt water activation of the thallus could take place and the lichen thallus stayed dry and inactive for the next five days, except for 22 November when heavy snowfall deposited moisture directly on the lichen thallus.

3.2 Photosynthetic activity at subzero temperatures and rehydration from snow in Umbilicaria aprina

A typical situation for *Umbilicaria aprina* is shown in Fig. 5 when the thalli are frozen after a melt period. This happened, e.g. in February, at the end of the summer period. The fully hydrated thalli were frozen quickly during one night period and stayed frozen for several days.

Fig. 6 shows two diurnal courses of CO_2 exchange of *U. aprina* in the frozen state at ambient subzero temperatures. For this experiment a dry thallus of *U. aprina* was hydrated artificially at about 0°C with melt water and subsequently frozen at -10°C in the measuring cuvette of the Walz minicuvette system. The air temperature in the measuring cuvette tracked ambient and the thallus of *U. aprina* stayed frozen for the next 38 h. Under these conditions it is most likely that only extracellular freezing occurs. However, we were not able to analyse if extracellular freezing of water affects both symbiants in the same way or if it results in different degrees of dehydration in the algal and mycobiont cells. The vertical arrows in the CO_2 exchange plot (Fig. 6) indicate that the cuvette was openend and fresh snow was put on the lichen thallus (at subzero temperatures) in order to prevent thallus dehydration and/or to allow rehydration of the thallus. During the 48 h period shown in Fig. 6, *U. aprina* always showed positive net photosynthetic rates regardless of whether temperatures were below or above 0°C. Low but substantial CO_2 exchange rates were monitored at all temperatures down to -10°C (see 20 November *ca.* 1:00 am to 5:00 am) and -17°C (see 19 November *ca.* 6:00 am). As the system tracked ambient temperatures and PPFD the coldest temperatures were reached when PPFD was less than 100 µmol m^{-2} s^{-1}. The highest CO_2 exchange rates were monitored in the afternoon when a high PPFD up to *ca.* 1700 µmol photon m^{-2} s^{-1} had caused also a substantial increase of the thallus temperatures. Photoinhibitory effects were never apparent (authors, in prep.), instead a decrease of the net photosynthetic rates could be related to water loss of the

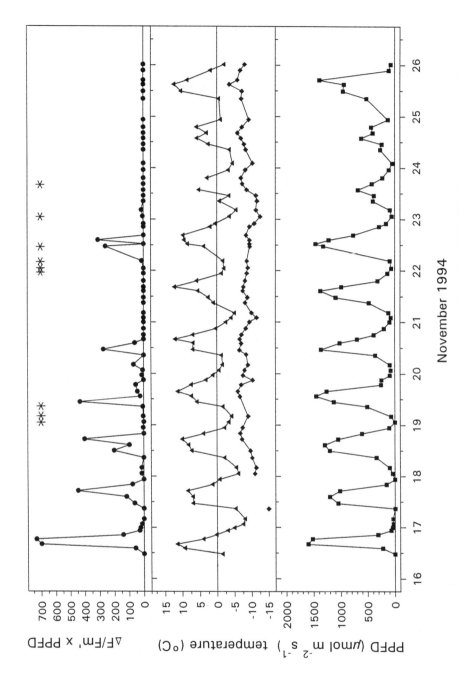

Fig. 4. Diurnal courses of photosynthetic activity (▲F/Fm' x PPFD, ●), air temperature (♦) and thallus temperature (▲) and PPFD (■) in *Buellia frigida* at Granite Harbour, Continental Antarctica. Asterisks indicate snowfall events.

Fig. 5. Frozen thallus of *Umbilicaria aprina* (February 1992) at Granite Harbour, Continental Antarctica.

thallus (see 20 November 14:00 am) because rehydration of the thallus allowed an immediate return to the previous higher levels of net photosynthetic rates. This performance of the lichen was obvious on both days when thallus temperatures were either below or above 0°C.

The rehydration of air-dry lichen thalli at -4°C and 200 μmol photon m^{-2} s^{-1}, and the subsequent activation of photosynthesis is shown in Fig. 7 for a dry thallus of *Umbilicaria aprina* (Schroeter & Scheidegger 1995). At the beginning of this laboratory experiment the thallus of *U. aprina* was air-dry and metabolically inactive. After the thallus had adjusted to -4°C it was covered with snow and held at the same temperature in the gas exchange cuvette. At intervalls the thallus was removed from the cuvette and weighed. The thallus was then again installed in the gas exchange cuvette where it was again covered with snow. The reactivation process of photosynthesis takes place in 4 phases (Fig.7). During the first 90 min (Phase I) the thallus increased weight but neither respiratory nor photosynthetic activity was detectable in the CO_2 exchange measurements. When the thallus further increased weight during the next 150 min (Phase II) metabolic activity of the thallus commenced with respiratory CO_2 evolution. At the beginning of Phase III a further increase in weight led to a net CO_2 uptake rate. This was a

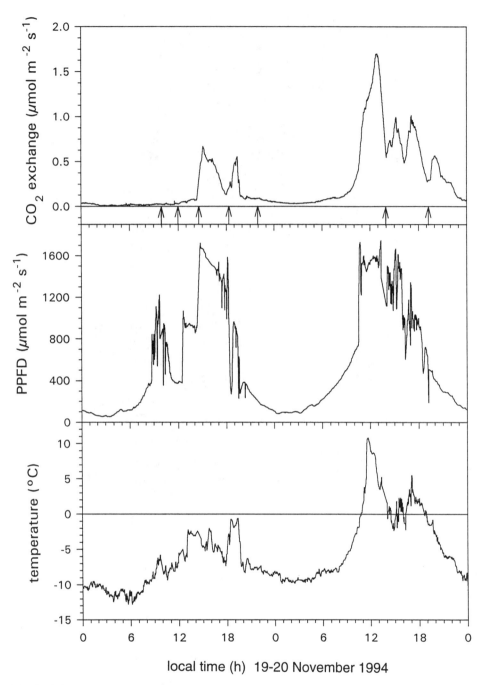

Fig. 6. Field measurements of CO_2 gasexchange, PPFD and thallus temperature in *Umbilicaria aprina* on 19 and 20 November 1994. Temperature and PPFD were tracked to ambient. Arrows in the top graph indicate rehydration of the frozen lichen thallus from snow.

clear indication that the algal symbionts of the lichen had gained enough water to start photosynthetic activity. In Phase III, a linear increase of net photosynthesis over the next 6.5 h was paralleled by a slow increase in thallus weight. A hydration status close to saturation was reached at the end of Phase III. In the next 10 h (Phase IV) only a small increase in the photosynthetic rate was observed and eventually, after a period of 20 h, maximum values were reached for both, photosynthetic rate and water content.

When air-dry thalli (n = 5) of *U. aprina* were covered with snow at subzero temperatures a substantial increase in weight could be measured within a temperature range from -4.5°C to -21°C (Fig. 8; data from Schroeter & Scheidegger 1995). The increase of weight indicates clearly that water uptake by the dry lichens from snow at subzero temperatures is possible. The rate of water uptake was directly correlated with the temperature: At -4.5°C 18% of the maximum water content was gained, at +8°C 300% dry weight could be reached. This percentage decreased with decreasing temperature and, at the lowest temperature tested (-21.5°C) the thalli of *U. aprina* only reached 7% of the maximum water content.

4   DISCUSSION

Climatic constraints in Antarctica are largely responsible for the unique characteristics and structure of its terrestrial ecosystems. The vegetation of Antarctic polar desert vegetation consists only of cryptogams such as lichens, bryophytes, algae and cyanobacteria while the two native flowering plants, *Deschampsia antarctica* and *Colobanthus quitensis*, are restricted to the west coast of the Antarctic Peninsula to 68°43'S (Komárková *et al.* 1990). The two most important environmental factors controlling survival and primary productivity of organisms in Antarctic polar desert ecosystems are water availability and temperature (Kappen 1993; Kennedy 1993; Schroeter *et al.* 1994; Schroeter & Scheidegger 1995).

It has been repeatedly shown from laboratory and field experiments that lichens are well adapted to low temperatures (see Kappen *et al.* 1995). Lichens are not only able to withstand very low subzero temperatures (Kappen & Lange 1972; Larson 1978, 1989) but also are able to achieve substantial rates of CO_2 uptake at temperatures down to -24°C (see Kappen

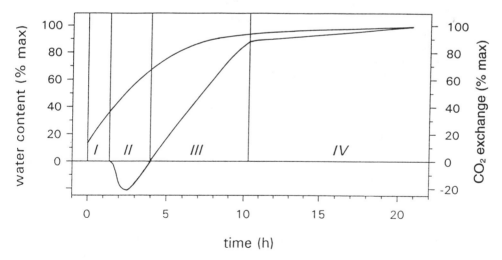

Fig. 7: Water uptake and CO_2 gasexchange in *Umbilicaria aprina* at -4°C and 200 μmol m^{-2} s^{-1}. Water content (WC) and CO_2 exchange (CO_2) as the percentage of the maximum values. The numbers indicate different phases of metabolic activation. Data from Schroeter & Scheidegger (1995).

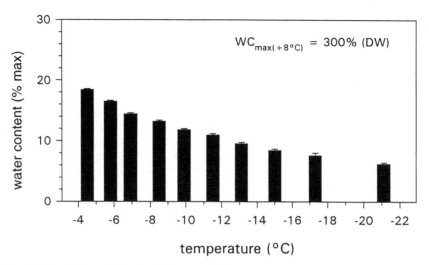

Fig. 8. Rehydration of air-dry thalli of *Umbilicaria aprina* from snow at subzero temperatures. Water uptake was calculated as percentage of the maximum water content at +8°C. n = 5, vertical lines indicate the standard error. Data from Schroeter & Scheidegger (1995).

1993; Schroeter *et al.* 1994). At low PPFD (< 200 µmol m^{-2} s^{-1}), optimum temperatures for net photosynthesis are below +5°C in Maritime Antarctic species and below +3°C in Continental Antarctic species (see Kappen *et al.* 1995). It is, therefore, probably not surprising that several Antarctic lichen species are photosynthetically active at subzero temperatures if the mesoclimatic conditions in Continental Antarctica are considered. However, as liquid water is rarely available in Continental Antarctica, photosynthetic activity at subzero temperatures requires not only an adaptation at the enzyme level but also the capacity to rehydrate from frozen water. Recently, LTSEM micrographs were used to demonstrate the structural changes during rehydration from snow in lichens at subzero temperatures (Schroeter & Scheidegger 1995). Water uptake from snow at subzero temperatures and photosynthetic activation in lichens has been shown as a common phenomenon not only in Antarctic lichen species (Kappen 1989; Kappen & Breuer 1991; Schroeter & Scheidegger 1995), but also in green algal lichens from subalpine regions in Switzerland (Schroeter & Scheidegger, unpubl.), as well as from temperate rainforests in New Zealand (Green & Schroeter, unpubl.). From these experiments it is most probable that rehydration of dry lichen thalli at subzero temperatures, i.e. uptake of water from snow and ice, involves a gaseous phase of water, and that water potential differences between the lichen thallus/air and the air/snow boundaries are the driving force for this process.

As a poikilohydric system, lichens are already known to rehydrate and start photosynthetic activity from water vapour at water potentials of -35 MPa (Nash *et al.* 1990). This is especially true for lichens with green algal photobionts (Lange & Kilian 1985; Scheidegger *et al.* 1995), while cyanobacterial photobionts can only be activated by liquid water (Lange & Kilian 1985; Büdel & Lange 1991; Schroeter 1994). Kappen (1993) and Schroeter *et al.* (1994) suggested that photosynthetic behaviour at subzero temperatures is that of a lichen in equilibrium with water vapour. If one assumes that the major part of the lichen photosynthetic activity in the annual cycle takes place at subzero temperatures and only when frozen water is available, it is not surprising that cyanobacterial lichen species are absent on the Kar Plateau (Seppelt *et al.* 1995, 1996) and Granite Harbour (Seppelt *et al.*, unpubl.) and in other Continental Antarctic sites (Inoue 1991; Kappen 1985; Smith 1986; Melick *et al.* 1994; Seppelt & Ashton 1978).

A different strategy can be observed, as is shown for the crustose, green algal lichen

Buellia frigda, in the beginning of summer in November/December 1994. The strong sun radiation led to substantial heating of the dark black thalli of *B. frigida* which easily reached temperatures of more than 20°C over ambient air temperatures. While the air temperature stayed below zero during the measuring period in November 1994, the thallus temperatures of *B. frigida* often increased above the freezing point. Therefore, the snow pack on the rock surfaces covered by *B. frigida* slowly melted and this free water caused substantial rates of photosynthetic activity, as indicated by chlorophyll *a* fluorescence of the algal symbionts. It is important to realise that the thalli are not always metabolically active at these positive temperatures but remain hydrated for only short periods of time (1 to 6 h; Fig. 3). Not all thalli that covered the rock surface were moistened at one melting event. It is therefore necessary to know not only the thallus temperature, but also the hydration status of the lichens if temperature regimes favourable for primary production or temperature stress effects are to be satisfactorily discussed. However, the distribution of the thalli of *Buellia frigida* at the boulder is controlled by a spatially and temporally uneven moistening pattern. On an almost horizontally exposed boulder surface investigated in our study, a snow pack only a few centimeters deep slowly receded under the prevailing subzero temperature conditions. Consequently, liquid water could repeatedly reach the same lichen thallus. However, the moistening pattern is not homogenous. Various factors such as irradiance, wind speed and the microtopography of the surface in relation to the border of the snow pack decide how often and where such a melt stream could meet a lichen thallus. This results in an apparently stochastic growth pattern of lichen thalli. However, photosynthetic activity and rehydration from snow at subzero temperature may play also an important role for photosynthetic productivity of *B. frigida* as it was shown for *Umbilicaria aprina.*

Our measurements demonstrate that water availability, whether frozen or liquid, controls the existence of the poikilohydric lichens in Antarctic polar desert ecosystems, as was shown in long-term studies for Maritime Antarctic species (Schroeter *et al.* 1991, 1995, 1996). As global climatic change scenarios predict substantial changes of the water balance in Antarctica (Kennedy 1995) this might lead also to dramatic changes in Antarctic polar desert ecosystems. It remains open to what extent cryptograms and especially lichens could benefit from a proposed increase in precipitation related to a temperature increase in polar regions (Schblesinger & Mitchell 1987; Trenberth *et al.* 1987). However, it is most likely that an increase in temperature and water availability will promote different species which are at present limited by the limited water resources and the low-energy environment.

5   ACKNOWLEDGEMENTS

Antarctica New Zealand and VXE-6 are thanked for logistic support, B.S. and L.K. acknowledge the support of the Deutsche Forschungsgemeinschaft (SCHR 473/1-1), TGAG the support of Waikato University and the Foundation of Research, Science and Technology and R.D.S. the support of Australian Antarctic Division.

6   REFERENCES

Bilger, W., U. Schreiber, & M. Bock, 1995. Determination of the quantum efficiency of photosystem II and of non-photochemical quenching of chlorophyll fluorescence in the field. *Oecologia* 102:425-432.

Block, W., 1994. Terrestrial ecosystems: Antarctica. *Polar Biology* 14:293-300.

Büdel, B. & O.L. Lange, 1991. Water status of green and blue-green phycobionts in lichen thalli after hydration by water vapour uptake: do they become turgid? *Botanica Acta* 104:361-366.

Genty, B., J.M. Briantais, & N.R. Baker, 1989. The relationship between the quantum yield of photosynthetic electron transport and quenching of chlorophyll fluorescence. *Biochimica et Biophysica Acta* 990:87-92.

Inoue, M., 1991. Ecological notes on the differences in flora and habitat of lichens between the Syowa Station area in continental Antarctica and King George Island in maritime Antarctica. *Proceedings of the NIPR Symposium on Polar Biology* 4:273-278.

Kappen, L., 1985. Vegetation and ecology of ice-free areas of northern Victoria Land, Antarctica. I. The lichen vegetation of Birthday Ridge and an inland mountain. *Polar Biology* 4:213-225.

Kappen, L., 1989. Field measurements of carbon dioxide exchange of the Antarctic lichen *Usnea sphacelata* in the frozen state. *Antarctic Science* 1:31-34.

Kappen, L., 1993. Plant activity under snow and ice, with particular reference to lichens. *Arctic* 46:297-302

Kappen, L.& M. Breuer, 1991. Ecological and physiological investigations in continental Antarctic cryptogams. II. Moisture relations and photosynthesis of lichens near Casey Station, Wilkes Land. *Antarctic Science* 3:273-278.

Kappen, L.& O.L. Lange, 1972. Die Kälteresistenz einiger Makrolichenen. *Flora* 161:1-29.

Kappen, L., M. Sommerkorn, & B. Schroeter, 1995. Carbon acquisition and water relations of lichens in polar regions - potentials and limitations. *Lichenologist* 27:531-545.

Kennedy, A. D., 1993. water as a limiting factor in the Antarctic terrestrial environment: a biogeographical synthesis. *Arctic Alpine Research* 25:308-315.

Kennedy, A. D., 1995. Antarctic terrestrial ecosystem response to global environmental change. *Annual Revue of Ecology and Systematics* 26:683-704.

Komárková, V., S. Poncet, & J. Poncet, 1990. Additional and revisited localities of vascular plants *Deschampsia antarctica* Desv. and *Colobanthus quitensis* (Kunth) Bartl. in the Antarctic peninsula area. *Arctic Alpine Research* 22:108-113.

Lange, O.L. & E. Kilian, 1985. Reaktivierung der Photosynthese trockener Flechten durch Wasserdampfaufnahme aus dem Luftraum: Artspezifisch unterschiedliches Verhalten. *Flora* 176:7-23.

Larson, D.W., 1978. Patterns of lichen photosynthesis and respiration following prolonged frozen storage. *Canadian Journal of Botany* 56:2119-2123.

Larson, D.W., 1989. The impact of ten years at -20°C on gas exchange in five lichen species. *Oecologia* 78:87-92.

Melick, D.R., M.J. Hovenden, & R.D. Seppelt, 1994. Phytogeography of bryophyte and lichen vegetation in the Windmill Islands, Wilkes Land, continental Antarctica. *Vegetatio* 111:71-87.

Nash III, T.H., L. Kappen, R. Lösch, D.W. Larson, & U. Matthes-Sears, U. 1987. Cold resistance of lichens with *Trebouxia*-photobionts from the North American west coast. *Flora* 179:241-251.

Nash III, T.H., A. Reiner, B. Demmig-Adams, E. Kilian, W.M. Kaiser, & O.L. Lange 1990. The effect of atmospheric desiccation and osmotic water stress on photosynthesis and dark respiration of lichens. *New Phytologist* 116:269-276.

Scheidegger, C., B. Schroeter, & B. Frey, B. 1995. Structural and functional processes during water vapour uptake and desiccation in selected lichens with green algal photobionts. *Planta* 197:399-409.

Schlesinger, M.E. & P.F.B. Mitchell, 1987. Climate model simulations of the equilibrium climate response of increased carbon. *Research in Geophysics* 25:760-798.

Schreiber, U., W. Bilger, & C. Neubauer, 1994. Chlorophyll fluorescence as a nonintrusive indicator for rapid assesssment of in vivo photosynthesis. In: E-D. Schulze & M.M. Caldwell (Eds.), *Ecophysiology of photosynthesis. Ecological studies* 100. Springer Verlag Berlin, Heidelberg, New York. pp. 49-70.

Schroeter, B., 1994. In situ photosynthetic differentiation of the green algal and the cyanobacterial photobiont in the crustose lichen *Placopsis contortuplicata*. *Oecologia* 98:212-220.

Schroeter, B. & C. Scheidegger, 1995. Water relations in lichens at subzero temperatures: structural changes and carbon dioxide exchange in the lichen *Umbilicaria aprina* from continental Antarctica. *New Phytologist* 131:273-285.

Schroeter, B., L. Kappen, & C. Moldaenke, 1991. Continuous in situ recording of the photosynthetic activity of Antarctic lichens - established methods and a new approach. *Lichenologist* 23:253-265.

Schroeter, B., T.G.A. Green, R.D. Seppelt, & L. Kappen, 1992. Monitoring photosynthetic activity of crustose lichens using a PAM-2000 fluorescence system. *Oecologia* 92:457-462.

Schroeter, B., T.G.A. Green, L. Kappen, & R.D. Seppelt, 1994. Carbon dioxide exchange at subzero temperatures. Field measurements on *Umbilicaria aprina* in Antarctica. *Cryptogamic Botany* 4:233-241.

Schroeter, B., M. Olech, L. Kappen, & W. Heitland, 1995. Ecophysiological investiagtions of *Usnea antarctica* in the maritime Antarctic. I. Annual microclimatic conditions and potential primary production. *Antarctic Science* 7:251-260.

Schroeter, B., L. Kappen, & F. Schulz, in press. Long-term measurements of microclimatic conditions in the fruticose lichen *Usnea aurantiaco-atra* in the maritime Antarctic. *Actas del Circo Simposio Espanol de Estudios Antárticos (Barcelona)*.

Seppelt, R.D. & D.H. Ashton, 1978. Studies of the ecology of the vegetation at Mawson Station, Antarctica. *Australian Journal of Ecology* 3:373-388.

Seppelt, R.D., T.G.A. Green, A-M. Schwarz & A. Frost, 1992. Extreme southernmost locations for moss sporophytes in Antarctica. *Antarctic Science* 4:37-39.

Seppelt, R.D., T.G.A. Green & B. Schroeter, 1995. Lichens and mosses from the Kar Plateau, southern Victoria Land, Antarctica. *New Zealand Journal of Botany* 33:203-220.

Seppelt, R.D., T.G.A. Green & B. Schroeter, 1996. Additions and corrections to the lichen flora of the Kar Plateau, southern Victoria Land, Antarctica. *New Zealand Journal of Botany* 34:329-331.

Smith, R.I. Lewis, 1986. Plant ecological studies in the fellfield ecosystem near Casey Station, Australian Antarctic Territory, 1985-1986. *Bulletin British Antarctic Survey* 72:81-91.

Taylor, G., 1913. The western journeys. In: Huxley, L. (Ed.), *Scott's last expedition, Vol. II.* Smith, Elder & Co., London. pp. 182-291.

Taylor, G., 1916. *With Scott: the silver lining.* Dodd, Mead & Co., New York.

Trenberth, K.E., J.R. Christy, & J.G. Olson, 1987. Global atmospheric mass, surface pressure and water vapour variations. *Journal of Geophysical Research* 92:14815-14826.

Ecosystem Processes in Antarctic Ice-free Landscapes, Lyons, Howard-Williams & Hawes (eds)
© *1997 Balkema, Rotterdam, ISBN 90 5410 925 4*

Oases as centres of high plant diversity and dispersal in Antarctica

R.I.Lewis Smith
British Antarctic Survey, Natural Environment Research Council, Cambridge, UK

ABSTRACT: Examples of Continental Antarctic oases, one in East Antarctica and the other in West Antarctica, are compared with regard to their floristic composition. Each possesses an unusually favourable suite of environmental conditions which is considered to be largely responsible for the high plant species diversity and high proportion of fertile bryophytes. A small-scale oasis (a seal carcass, proabably several centuries old) in the northern Antarctic Peninsula region provided data on moss propagule dispersal as determined from the culture of soil sampled along transects radiating from the carcass. The distribution of spores and vegetative propagules of the four principal species appears to be closely correlated to the nature of the diaspores and the hydrological and aeolian characteristics of the site. These data serve to illustrate the potential role of larger-scale species-rich oases as important dispersal centres, and may serve as useful models in helping to elucidate aspects of island biogeography, diversity-stability and patch dynamics theory.

1   INTRODUCTION

The development of terrestrial ecosystems in continental Antarctica is constrained by several critical environmental factors acting on the immigration of propagules and their subsequent establishment and survival. Of these, oceanic isolation from the other Southern Hemisphere continents, distinct dispersal corridors dependent on prevailing wind and storm patterns, climatic severity (especially infrequent moisture availability and short growing or breeding season), and fragmentation of favourable habitats are probably the most important criteria. Despite this, there are few ice-free areas throughout the continent up to *ca.* 2500 m altitude and to within 4 to 5° latitude of the South Pole where microorganisms, cryptogamic plants (notably lichens) and micro-invertebrates do not occur, allbeit in very sparse numbers and usually with very few species represented in the most severe environments. However, there are many areas, mostly small in extent, at low altitude and close to the coast, where a combination of local climatic, topographic and edaphic conditions provide optimal "habitat islands" for colonisation and community development within an otherwise hostile matrix of landscape categories. In the context of the present account, some of these habitat islands support a diverse biota which has resulted in a series of relatively complex small-scale communities within the local ecosystem. These sites are referred to here as "oases", although not in the same context as is generally used in the polar regions (e.g. Korotkevich 1972; Pickard 1986; Freedman *et al.* 1994; Bormann & Fritzsche 1995).

Antarctic oases and their characteristic features have been briefly reviewed by Walton (1984) and Pickard (1986), and the climatic and environmental conditions giving rise to these features have been discussed by Shumskiy (1957), Solopov (1969), Markov *et al.* (1970)

Korotkevich (1972) and Aleksandrov & Simonov (1981). In fact, all the examples given by these and other authors consider oases to be extensive ice-free landscapes, often with lakes, streams and small glaciers. In his review of these large-scale features, Pickard (1986) states "…..but there are small enclaves in these icy wastes where temperatures rise above freezing, where water flows over the rocks and soil, and where plants thrive. These are oases in a vast icy desert." The present author described the vegetation of very small fertile oases (in the traditional sense; see Gary *et al.* 1974; Allaby & Allaby 1990) of only several tens of square metres in Ablation Valley on Alexander Island, regarding an Antarctic oasis as "an area of restricted size where surface water supports vegetation in an otherwise arid and barren region" (Smith 1988). This concept has been developed further for the purpose of this account, and is defined as:

"*An area of restricted size where there is surface water, at least for part of the summer, and where local topography provides a favourable microclimate, allowing a concentration of microbial and plant species to develop a relatively complex vegetation in an otherwise arid and barren desert landscape*".

The hypothesis being tested here is that many habitat islands, where a suite of optimal environmental conditions prevails, serve as the focus for colonisation by a wide range of immigrant propagules, resulting in the development of relatively complex (by Continental Antarctic standards) communities with an unusually large number of species exhibiting high reproductive success. These species-rich oases in turn play a major role as a source of viable propagules which are dispersed to new sites both near and far within the biome.

This study reports on some preliminary findings based on ecological and biodiversity investigations at two meso-scale (i.e. <1 ha) oases, on Alexander Island and in central Victoria Land. Propagule dispersal data, used to illustrate the importance of oases as potential sources of biological material in the colonisation process of new habitats, were obtained from cultured soil samples taken from a micro-scale (i.e. <10 m²) oasis on James Ross Island.

2   RESEARCH SITES

2.1   *Meso-scale oases*

Ablation Valley (70°49'S, 68°25'W) lies about midway along the east coast of Alexander Island, Palmer Land, south-west Antarctic Peninsula. The predominant rock types are conglomerates, sandstones and shales, with lavas and breccias on the valley floor. This extensive ice-free valley (see Clapperton & Sugden 1983) abuts the permanently frozen Ablation Lake, which is formed by the damming of the George VI Sound ice shelf. Although "coastal", the site is about 100 km from the open sea to the north in summer. Very little of the valley is visibly vegetated, but in several places melt water wells up and forms seepage areas and occasional runnels, and here there occur small closed stands of moss-, lichen- and cyanobacteria-dominated vegetation (Smith 1988). The site was visited in January 1995, and the oasis considered here is approximately 625 m² in area and is about 3 to 5 m above the lake shore.

Harrow Peaks (74°06'S, 164°51'E) comprise a group of precipitous mica-rich granite and granodiorite cliffs rising from Wood Bay, central Victoria Land. The study site included a coarse sand- and gravel-covered plateau and an area of block scree. The oasis considered here is at *ca.* 100 to 130 m altitude and occupied an area of *ca.* 0.5 ha. Small patches of late snow provide a source of water intermittently through the summer, allowing a comparatively large number of bryophytes to have colonised the soil and rock crevices. There is biotic influence from a small colony of snow petrels (*Pagodroma nivea*), Wilson's petrels (*Oceanites oceanicus*) and skuas (*Catharacta macormicki*) within and around the site. Many lichens cover the rock and also the terricolous moss cushions. The site was visited in December 1995. It is one of a disjunct series of small species-rich oases in central Victoria Land.

San José Pass (63°55'S, 57°54'W) is on the extensively ice-free northern side of James Ross Island off the north-eastern coast of the Antarctic Peninsula. It lies between St. Martha Cove and Brandy Bay, *ca.* 4 km from the sea at about 200 m altitude. The broad gently sloping ice-free valley running from the pass northwards is abundantly striated by sorted stone stripes with intervening fines. These are very unstable because of their aridity and mobility in strong winds. Visible vegetation, other than occasional lithophytic crustose lichens, is extremely rare and sparse. However, the area (indeed much of the ice-free area of the island) is littered with carcasses and skeletons of seals (mainly crabeaters, *Lobodon carcinophagus*) which had died between about 1 to 2 and several hundred years ago. There appears to have been a major overland route taken from St. Martha Cove, via San José Pass, to Brandy Bay. In 1955, mass mortality of crabeater seals was reported in this region, both on land and on the sea ice (Laws & Taylor 1957). Many of these carcasses have become micro-scale oases, stabilising the soil around them, while their leached nutrients serve as a focus for the establishment of various mosses and lichens. This site (*ca.* 6 m²) was visited in January 1989.

For the purpose of this account, the Antarctic Peninsula includes only the more continental sector between 66°S and 73°S (where it arbitrarily becomes part of Continental Antarctica). This equates it to the true continental Antarctic, all of which lies south of 66°S. Thus, many plant species known from the Antarctic Peninsula north of this latitude are excluded from the floristic tally for the entire region (see Smith, 1996). The southern Peninsula and Continent are referred to here as biomes.

3   METHODS

Information on plant species diversity at each of the study sites was obtained by the author

Fig. 1. Crabeater seal carcass colonised by mosses and lichens. San José Pass, James Ross Island.

during the course of ecological surveys. Additional regional and continental biodiversity and biogeographical data were compiled from personal collections, recent published literature and information held in the British Antarctic Survey's Antarctic Plant Database. The plant groups included in this study are mosses, liverworts and macrolichens (fruticose and foliose taxa). While each site had a large microlichen (crustose) flora, their taxonomic status is still poorly known, making their identity in the field extremely difficult and liable to be very inaccurate.

To determine the composition of the soil propagule bank in the vicinity of a moss and lichen community associated with an ancient crabeater seal carcass on James Ross Island (Fig. 1) soil samples were cultured on a thermogradient-bar incubator (see Smith 1987; Smith & Coupar 1987). Five sub-samples (5 x 5 cm) of fine, mineral surface (0 to 5 mm) soil were taken at positions 1, 2, 5, 10 and 20 m from the carcass along eight radiating transects (north, north-east, east, *etc.*). The samples were kept frozen (-20°C) until the laboratory experiment commenced about five months later. Each set of five sub-samples was bulked and the single sample (total 40) placed in a 9 cm diameter by 2 cm deep Petri dish and maintained at 10°C for ten weeks. Overhead lighting was provided by quartz-halogen lamps giving about 125 mmol m^{-2} s^{-1} photosynthetically active radiation. The dishes were moistened with Bold's nutrient (see Borowitzka 1988) whenever they showed signs of drying. The number of shoots of each moss species that appeared was recorded at weekly intervals, but only data from the final count are reported here.

4   RESULTS

The two sites (Ablation Valley and Harrow Peaks) selected to represent species-rich oases in two widely separated but climatically similar regions (Alexander Island and Victoria Land) both possess floras comprising a high proportion of the regional total. Indeed, they both contain a significant number of the species reported for their respective biomes (Table 1). The James Ross Island seal carcass community growing on the bones and skin, and soil immediately around them, comprised mainly the mosses *Bryum argenteum*, *B.* c.f. *pseudotriquetrum*, *Ceratodon purpureus*, *Encalypta patagonica* and *Tortula princeps*, with occasional *Pottia heimii*. These are all mildly calcicolous species, noted for being opportunistic primary

Table 1. Plant species diversity at three regional scales.

Region	Mosses	Liverworts	Macrolichens
Antarctic Peninsula (66-73°S, including Alexander Island)	44	4	35
Alexander Island (69-73°S)	30 (68%)[1]	1	15 (43%)[1]
Ablation Valley (71°S)	22(50%),73%)[2]	1	12 (34%, 80%)[2]
Continental Antarctica (including Victoria Land)	21	1	12
Victoria Land (72-78°S)	17 (81%)[3]	1	12 (100%)[3]
Harrow Peaks (74°S)	12 (57%: 71%)[4]	1	9 (75%: 75%)[4]

[1]   Number as percentage of Peninsula flora.
[2]   First number as percentage of Peninsula flora; second number as percentage of Alexander Island flora.
[3]   Number as percentage of Continental Antarctic flora.
[4]   First number as percentage of Continental Antarctic flora; second number as percentage of Harrow Peaks flora.

colonists. Several nitrophilous lichens were also prominent, notably *Caloplaca* spp., *Candelariella* sp., *Physcia caesia, Physconia muscigena* and *Xanthoria elegans,* and some unidentified crustose taxa. The alga *Prasiola crispa* and cyanobacterium *Nostoc commune* were also occasional associates. Unicellular green algae occurred on the dry skin and beneath stones.

In the soil culture experiment shoots of four mosses, *B. argenteum, C. purpureus, E. patagonica* and *T. princeps*, were the most abundant. A few other species (*B.* c.f. *pseudotriquetrum, P heimii* and *Bartramia* and/or *Distichium* sp.) occurred very sporadically but have been omitted from the assessment, as were algal growths. The distribution pattern of each species in relation to the assumed source (the carcass) is shown in Fig. 2a-d as "compass stars". Those for *Bryum* and *Ceratodon* indicate a strong "tail" in the north to south-east sector, with much higher numbers of shoots of the latter species, especially close to the source (Fig. 2a & 2b). Shoot numbers were few or absent in the opposite sector. A similar pattern, but with a different orientation was recorded for *Encalypta* and *Tortula*, with the latter species producing a significantly longer tail to the north-west to north-east (Fig. 2c and 2d).

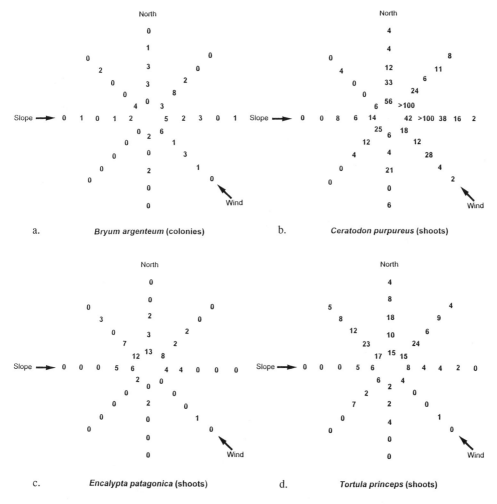

Fig 2. Numbers of moss shoots cultured from soil sampled at 1, 2, 5, 10 and 20 m from the seal carcass at San José Pass, James Ross Island. Positions on the radial transects are on a log scale. a. *Bryum argenteum*; b. *Ceratodon purpureus*; c. *Encalypta patagonica*; d. *Tortula princeps*.

5   DISCUSSION

The apparent discrepancy between the distribution of the two pairs of moss species at the seal carcass micro-site may be readily explained when their respective modes of reproduction are related to local environmental conditions.

Firstly, the prevailing wind at this site is from the south-east, being funneled through San José Pass and downhill over the seal carcass. Secondly, the carcass lay on a gentle slope inclined to the east. Winter snow accumulates on the leeward side of the pass, i.e. uphill from the seal, and when it melts in spring seepage over the then frozen ground is in an easterly direction. Propagules of low mass (e.g. spores) will be more likely to be dispersed by wind towards the north while those of a denser mass (e.g. vegetative structures) may be expected to be dispersed mainly by melt water towards the east.

Both *E. procera* and *T. princeps* produced sporophytes in the source community, *Encalypta* in abundance and *Tortula* less frequently. However, the spore characteristics of these species are different (Table 2). *Encalypta* produces a relatively small number of large, and presumably heavy, spores in contrast to *Tortula* which has a large number of small, presumably light, spores. The pattern of spore deposition may be at least partly explained by the relative numbers of spores produced and their mass. Thus, the *Encalypta* tail is short as few spores are dispersed far from the parent colony, while the *Tortula* tail extends much farther and yielded higher shoot counts. Although spores of both species are almost certainly deposited well beyond the points where they were farthest detected, their numbers must become progressively less with increasing distance from the source.

B. argenteum and *C. purpureus* are two of the commonest and most widely distributed mosses in Antarctica and, indeed, worldwide. However, neither reproduces sexually in Antarctica, but both have specialised vegetative structures capable of developing new plants. *Bryum* produces one to several deciduous buds or bulbils on each stem (see Savicz-Ljubitzkaja & Smirnova 1964) which are readily dislodged from the parent plants and are typically dispersed over short distances by wind or water. This moss commonly occurs in seepage areas, especially along the margin of melt streams. During periods of flooding huge numbers of *Bryum* buds are washed downstream and deposited along the flood-line where the plantlets can establish an ephemeral community within a week or two, often to be washed away during floods early in the following spring (Smith, unpubl.). Local dissemination may also be by wind, as has been shown by Rudolph (1970) who trapped wind-blown *B. argenteum* buds on sticky slides near Hallett Station, north of Harrow Peaks. However, because of their relatively high density, few of these diaspores are likely to be blown far by wind, although they may be transported considerable distances by flowing water. Stems of *Ceratodon* produce large numbers of multicellular gemmae on their rhizoids (Imura & Kanda 1986). Although considerably smaller than *Bryum* buds, these too are readily detached and most probably dispersed mainly by water. Soil propagule bank culture studies at Signy Island, South Orkney Islands, almost invariably yield *C. purpureus* shoots, even though the moss is not always

Table 2. Spore characteristics of *Encalypta patagonica* and *Tortula princeps*.

Spore characters	Encalypta	Tortula
No. Spores/capsule (mean±sd)	10,700±2700	58,800±9000
No. Spores/capsule (range)	6900-14,000	48,000-72,000
Spore diameter (µm)	45.7	15.9
Spore volume (µm^3 x 10^6)	533	123

Data from Convey & Smith (1993); those for *Tortula* are for *T. saxicola*.

present in the immediate area (Smith 1987, 1993; Smith & Coupar 1987). It is particularly abundant in soil below receding icefields. The reason for the widespread distribution of *B. argenteum* and *C. purpureus* may be due, in part, to wind dispersal or to transport on skua feet or feathers. When moist, *Bryum* buds attach very easily to rough surfaces.

This preliminary investigation has demonstrated that certain habitats which exhibit a suite of unusually favourable living conditions in an otherwise inhospitable environment may be colonised by a remarkably high proportion of the regional flora. Some oases also have a similarly diverse invertebrate fauna. For example, the Ablation Valley oasis possessed all the micro-arthropod species so far known from Alexander Island (five mites and two springtails) and about one third of those known in the southern Antarctic Peninsula sector (Table 3). However, the same fauna of the Harrow Peaks oasis is only about one third that of Victoria Land, and one fifth (Collembola) and one thirteenth (Acari) that of Continental Antarctica as a whole. In the latter case, regional endemism is far greater (Pugh 1993).

Despite the high latitude of these sites, both have unusually high numbers of fertile moss species (50% at Ablation Valley, 25% at Harrow Peaks). The liverwort *Cephaloziella exiliflora* also produces sporophytes in the Ablation Valley oasis. This is contrary to previously published statements regarding the decline of sexual reproduction in bryophytes with increasing climatic severity (Longton 1988). However, it is not known whether this is a response to growing in extremely stressful environments whereby spore production may be of greater benefit than asexual reproduction in disseminating a species and establishing new populations, or to local enclaves of atypically favourable microclimate which aids the development of sporophytes. The great majority of Antarctic mosses are monoecious (organs of both sexes borne on the same plant/shoot) and so, if mature organs are produced, there is a strong chance that sexual reproduction will occur. What is not yet known is whether sexual reproduction increases in mosses growing in highly stressed habitats. There is an analogous situation in some Antarctic micro-orthropods which respond to such environmental stress by producing either larger numbers of smaller eggs or smaller numbers of larger eggs, both strategies enabling greater chance of survival (P. Convey, pers. comm.).

The results obtained from the micro-scale oasis (seal carcass) site on James Ross Island provide a fortuitous insight into how a much larger oasis may function as a source of propagules, and how local environmental conditions may play an important role in their dispersal. The occurrence of the four most abundant moss species which developed on the culture dishes illustrates several important aspects of sexual and asexual propagules and

Table 3. Microarthropod species diversity at three regional scales.

Region	Mites* (Acari)	Springtails (Collembola)
Antarctic Peninsula (66-73°S, including Alexander Island)	19	6
Alexander Island (69-73°S)	5 (?7)	2
Ablation Valley (71°S)	5	2
Continental Antarctica (including Victoria Land)	40	11
Victoria Land (72-78°S)	10	7
Harrow Peaks (74°S)	3	2

* Free-living mites only.
Data from Pugh (1993), Greenslade (in press), Convey (pers. comm.).

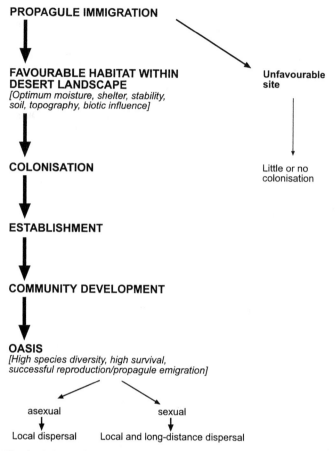

Fig. 3. Schematic model of the development of an Antarctic oasis.

characteristics of their dissemination. However, as yet little is known of the optimal conditions for the production of sporophytes in Antarctica, or what microclimatic or physiological cues may be required to induce spore germination (Smith 1993). Also, although various forms of asexual reproductive structures are known for several bryophytes, no detailed survey has been made and the development and colonisation potential of only a few has been tested. A conceptual model of the development of an Antarctic oasis is proposed in Fig. 3. To develop this further requires a more in-depth approach as addressed by the concepts and theories of island biogeography, species diversity-ecosystem stability relationships, and patch colonisation dynamics (see Tilman 1996; Gustafson & Gardner 1996, and references therein).

The dispersal of bryophytes on a global scale and the hazards they must endure if they are to be functional colonists have been reviewed by Van Zanten & Pócs (1981). However, to understand the mechanism of plant propagule dispersal at the regional scale in Antarctica requires a detailed study of the production and nature of diaspores, as well as of local and regional wind patterns and hydrology. Such research requires focused aerobiological sampling and monitoring, as advocated by Rudolph & Benninghoff (1977) and developing investigations such as those initiated by Kappen & Straka (1988), Smith (1991), Wynn-Williams (1991) and Marshall (1996, 1997). Also, because of the frequent inference to the possible role played by birds in transporting propagules into and within the Antarctic, it would be particularly valuable to conduct a detailed study of certain species, notably skuas, to try to detect attached viable diaspores.

6   ACKNOWLEDGEMENTS

I am grateful to Dr. H.J. Peat for providing data from the B.A.S. Antarctic Plant Database. I am also indebted to many people who made my visits to remote field sites possible: the officers and crew of R.R.S. *John Biscoe* (James Ross Island), Twin Otter pilots operating from the B.A.S. Rothera Station (Alexander Island), and the New Zealand Squirrel helicopter pilots operating from the Italian Stazione Baia Terra Nova (Victoria Land).

REFERENCES

Aleksandrov, M.V. & I.M. Simonov 1981. Intralandscape zoning of low-lying oases of the Eastern Antarctic. (From observations on Molodezhnyi Oasis). In: Govorukha, L.S. & Y.A. Kruchinin, (Eds.), *Problems of Physiographic Zoning of Polar Lands*. Amerind Publishing Company Pvt. Ltd., New Delhi. pp. 223-242. [Translated from original Russian publication, 1971].

Allaby, A. & M. Allaby, 1990. *The Concise Oxford Dictionary of Earth Sciences*. Oxford University Press, Oxford.

Borowitzka, M.A., 1988. Algal growth media and sources of algal cultures. In: Borowitzka, M.A. & L.J. Borowitzka (Eds.), *Micro-algal biotechnology*. Cambridge University Press, Cambridge. pp. 456-465.

Bormann, P. & D. Fritzsche 1995. *The Schirmacher Oasis, Queen Maud Land, East Antarctica, and its surroundings*. Perthes, Gotha. 448 p.

Clapperton, CM. & D.E. Sugden, 1983. Geomorphology of the Ablation Point massif, Alexander Island, Antarctica. *Boreas* 12:125-135.

Convey, P. & R.I. Lewis Smith, 1993. Investment in sexual reproduction by Antarctic mosses. *Oikos* 68:293-302.

Freedman, B., J. Svoboda & G.H.R. Hendry, 1994. Alexandra Fiord - An ecological oasis in the polar desert. In Svoboda, J. & B. Freedman (Eds.), *Ecology of a Polar Oasis*. Captus University Publications, Toronto. pp. 1-9.

Gary, M., R. McAfie & C.L. Wolf, 1974. *Glossary of Geology*. American Geological Institute, Washington, D.C.

Greenslade, P., in press. Collembola from the Scotia Arc and Antarctic Peninsula including descriptions of two new species and notes on biogeography. *Polskie Pismo Entomologiczne*.

Gustafson, E.J. & R.H. Gardner, 1996. The effect of landscape heterogeneity on the probability of patch colonization. *Ecology* 77:94-107.

Imura, S. & H. Kanda, 1986. The gemmae of the mosses collected from the Syowa station area, Antarctica. *Mem. of National Institute of Polar Research, Special Issue No.* 44:241-246.

Kappen, L. & H. Straka, 1988. Pollen and spores transport into the Antarctic. *Polar Biology* 8:173-180.

Korotkevich, Y.S., 1972. *Polar Deserts*. Hydrometeorological Publishing House, Leningrad. [In Russian].

Laws, R.M. & R.J.F. Taylor, 1957. A mass dying of crabeater seals, *Lobodon carcinophagus* (Gray). *Proceedings of the Zoological Society of London* 129:315-324.

Longton, R.E., 1988. *Biology of Polar Bryophytes and Lichens*. Cambridge University Press, Cambridge.

Markov, K.K., V.I. Bardin, V.L. Lebedev, A.I. Orlov & I.A. Suetova, 1970. Periglacial features. In: *The Geography of Antaractica*. Israel Program for Scientific Translations, Jerusalem. pp. 260-318.

Marshall, W.A., 1996. Aerial dispersal of lichen soredia in the maritime Antarctic. *New Phytologist* 134:523-530.

Marshall, W.A., 1997. Laboratory evaluation of a new aerobiological sampler for use in the Antarctic. *Journal of Aerosol Science* 28: 371-380.

Pickard, J., 1986. Antarctic oases, Davis station and the Vestfold Hills. In: Pickard, J. (Ed.), *Antarctic Oasis: terrestrial environments and history of the Vestfold Hills*. Academic Press, Sydney. pp. 1-19.

Pugh, P.J.A., 1993. A synoptic catalogue of the Acari from Antarctica, the sub-Antarctic islands and the Southern Ocean. *Journal of Natural History* 27:323-421.

Rudolph, E.D., 1970. Local dissemination of plant propagules in Antarctica. In: Holdgate, M.W. (Ed.), *Antarctic Ecology Vol.* 2. Academic Press, London. pp. 812-817.

Rudolph, E.D. & W.S. Benninghoff, 1977. Competative and adaptive responses of invading versus indigenous biotas in Antarctica - a plan for organized monitoring. In: Llano, G.A. (Ed.),

Adaptations within Antarctic Ecosystems. Gulf Publishing Company, Houston. pp. 1211-1225.

Savicz-Ljubitzkaja, L.I. & Z.N. Smirnova, 1964. Notula de Bryo argenteo Hedw. ex Antarctida [Note on *Bryum argenteum* from Antarctica]. *Novosti Sistematiki Nizshikh Rastenii* [for 1964]:292-301.

Shumskiy, P.A., 1957. Glaciological and geomorphological reconnaisance in the Antarctic in 1956. *Journal of Glaciology* 3: 56-61.

Smith, R.I. Lewis, 1984. Terrestrial plant biology of the sub-Antarctic and Antarctic. In: Laws, R.M. (Ed.), *Antarctic Ecology Vol.* 1. Academic Press, London. pp. 61-162.

Smith, R.I. Lewis, 1987. The bryophyte propagule bank of Antarctic fellfield soils. *Symposia Biologica Hungarica* 35: 233-245.

Smith, R.I. Lewis, 1988. Bryophyte oases in ablation valleys on Alexander Island, Antarctica. *The Bryologist* 91:45-50.

Smith, R.I. Lewis, 1991. Exotic sporomorpha as indicators of potential immigrant colonists in Antarctica. *Grana* 30:313-324.

Smith, R.I. Lewis, 1993. The role of bryophyte propagule banks in primary succession: case-study of an Antarctic fellfield soil. In: Miles, J. & D.W.H. Walton (Eds.), *Primary Succession on Land.* Blackwell Scientific Publications, Oxford. pp. 55-78.

Smith, R.I. Lewis, 1996. Terrestrial and freshwater biotic components of thewestern Antarctic Peninsula. In: Ross, R., E. Hofmann & L. Quetin (Eds.), *Foundations for Ecological Research West of the Antarctic Peninsula. Antarctic Research Series* 70. American Geophysical Union, Washington D.C. pp. 15-59.

Smith, R.I. Lewis & A.M. Coupar, 1987. The colonization potential of bryophyte propagules in Antartctic fellfield sites. *Comité National français des Recherches Antarctiques* 58:189-204.

Solopov, A.V., 1969. Oases in Antarctica. *National Science Foundation Technical Translation TT 68-50490.* Israel Program for Scientific Translations, Jerusalem. pp. 1-146. [Translated from original Russian publication, 1967].

Tilman, D., 1996. Biodiversity: population versus ecosystem stability. *Ecology* 77:350-363.

Walton, D.W.H., 1984. The terrestrial environment. In: Laws, R.M. (Ed.), *Antarctic Ecology Vol.* 1: Academic Press, London. pp 1-60.

Wynn-Williams, D.D., 1991. Aerobiology and colonization in Antarctica - the BIOTAS Programme. *Grana* 30:380-393.

Van Zanten, B.O. & T. Pócs, 1981. Distribution and dispersal of bryophytes. *Advances in Bryology* 1:479-562.

128

Ecosystem Processes in Antarctic Ice-free Landscapes, Lyons, Howard-Williams & Hawes (eds)
© *1997 Balkema, Rotterdam, ISBN 90 5410 925 4*

RAPD profiling of genetic variation in Antarctic mosses

M. L. Skotnicki, P. M. Selkirk & T. M. Dale
Department of Biological Sciences, School of Science and Technology, University of Waikato, Hamilton, New Zealand

ABSTRACT: Vegetation in icefree areas of Continental Antarctica comprises mosses, liverworts, lichens and algae. Populations of mosses are limited in extent and discontinuous around the coastal regions of Antarctica, and until now no genetic studies of these populations have been possible. The RAPD (Random Amplified Polymorphic DNA) technique allows investigations of genetic diversity in species and populations where no other genetic information is available. We have used RAPDs to study the extent of genetic variation in several species of Antarctic moss (mainly from the Ross Sea region), including *Bryum argenteum, Bryum pseudotriquetrum. Campylopus pyriformis, Ceratodon purpureus, Pottia heimii,* and *Sarconeurum glaciale.* Using single moss shoots, we have shown that RAPD results are reproducible, with identical results consistently obtained for shoots joined at the base, and that genetic variation does occur within populations of Antarctic moss species. Genetic variation may occur within single clumps of a moss species, and the extent of genetic variation differs between Antarctic, New Zealand and Australian isolates of the same moss species. RAPDs can also be used to differentiate between species, assisting taxonomic distinctions.

1 INTRODUCTION

In Continental Antarctica, vegetation comprising mosses, liverworts and algae is confined to ice-free coastal areas. Some fifteen species of mosses have been reported from non-Peninsular Continental Antarctica (Smith 1984), in nine genera (Seppelt & Selkirk 1984). Mosses are of interest for many reasons, including; the basic biology of these haploid organisms, colonisation history of Antarctic moss populations, their physiological survival strategies and mutation rates in the extreme climate, and isolated location of Antarctica. Genetic studies of moss populations (from anywhere in the world) have been limited by the lack of simple techniques applicable to large numbers of isolates.

Recently, the RAPD (Random Amplified Polymorphic DNA) profiling technique has proved to be very useful for analysis of genetic variation in a wide variety of plant species, especially where no other genetic information is available (Williams *et al.* 1990). The method is simple, quick, and relatively inexpensive, and only minute amounts of plant material are required (Rafalski & Tingey 1993). Other methods for studying population genetics are usually more expensive and time-consuming, and require either more DNA sequence information (e.g. for microsatellites), more starting material (e.g. isozymes) or both (restriction fragment length polymorphisms, DNA sequencing). Synthetic DNA molecules of 10 nucleotides are used to prime DNA synthesis of segments of the sample genome by the polymerase chain reaction; DNA fragments are produced of specific and characteristic sizes which can then be scored as present or absent for each sample. These DNA polymorphisms can be analysed to produce dendrograms showing phylogenetic relationships between different isolates (Armstrong *et al.* 1995). RAPDs have already been used successfully to examine the extent of genetic variation in two Antarctic moss species, *Sarconeurum glaciale* and *Bryum argenteum* (Adam *et al.,* in press; Selkirk *et al.,* in press).

We now report the use of RAPDs to further analyse the extent of genetic variation in Antarctic mosses, and as a taxonomic tool to distinguish moss genera and species in specimens collected from the Ross Sea region of Continental Antarctica. This technique is especially useful as an extra aid in moss identification because in the harsh conditions of Antarctica many mosses display phenotypic plasticity depending on the climatic and habitat conditions (Seppelt & Selkirk 1984; Longton 1988); this has often made precise identification difficult.

2 METHODS

2.1 *Collection of moss isolates*

Specimens were collected into paper bags, dried, and stored frozen or at room temperature until used for morphological examination and RAPD analysis.

For the investigation of reproducibility of RAPDs, variation within moss clumps, variation between clumps and between populations, specimens of *Bryum argenteum* and *Pottia heimii* were collected from various locations in Southern Victoria Land, Antarctica (from near Lake Fryxell in the Eastern Taylor Valley and from several drainage channels in the Garwood Valley). Samples of *Bryum argenteum* were also collected from towns in the North Island of New Zealand and in New South Wales, Australia.

For the experiments with RAPDs and taxonomic identification, isolates of *Bryum argenteum, Bryum pseudotriquetrum, Ceratodon purpureus* and *Pottia heimii* were collected at Granite Harbour, Ross Sea region. Isolates of *Sarconeurum glaciale* were collected from Granite Harbour, the Eastern Taylor Valley and the Vestfold Hills; those of *Campylopus pyriformis* were from Mt Melbourne. Five or six isolates of each species were used in this study, although many more isolates of four species (*B. argenteum, C. pyriformis, P. heimii, a*and *S. glaciale*) have been tested with the RAPDs technique and found to give results similar to those reported here.

2.2 *Isolation of DNA*

DNA was isolated from single shoots of each moss clump collected; this amount of DNA was sufficient for many RAPD reactions. Previously, we had observed that within-clump variation can occur in several moss species (Selkirk *et al.,* in press; Skotnicki, unpubl. obs.), and therefore single shoots were used in all subsequent RAPD analyses.

A single shoot, 2 to 3 mm in length, was ground to a fine suspension in 20 ml DNA extraction buffer (200 mM Tris-HCl pH7.5, 250 mM NaCl, 25 mM EDTA, 0.5% w/v SDS; Edwards *et al.* 1991) using an electric drill with a glass bit shaped to fit a 1.5 ml Eppendorf tube. 70 ml extraction buffer was added, and the mixture was incubated at room temperature for 30 m. The suspension was extracted with 150 ml 1:1 phenol:chloroform, the supernatant was recovered after 2 m centrifugation in an Eppendorf centrifuge at 15000 rpm, and DNA was precipitated by incubation with 150 ml ethanol at -20°C for 10m. The DNA was collected by centrifugation at 15000 rpm for 5 m in an Eppendorf centrifuge, and the pellet was resuspended in 250 ml sterile distilled water.

Moss DNA prepared by this method, using a single 3 mm shoot, contained approximately 5 to 10 ng DNA/ml, and was clean enough to use directly for PCR reactions.

2.3   RAPD DNA amplification

Moss DNA was amplified in a polymerase chain reaction (PCR) (Williams *et al.* 1990), using DNA primers OP-A1, OP-A13, OP-A17, OP-B10, OP-C4, OP-C7, OP-P1, OP-P6, OP-P7 or OP-P16 (Operon Technologies). Each PCR reaction contained 1ml DNA, 0.3 mM primer, 0.25 mM each dATP, dCTP, dGTP and dTTP, PCR reaction buffer (Boehringer Mannheim), 2 mM $MgCl_2$, and 0.5 ml Taq DNA polymerase (0.5 Units; Boehringer Mannheim), in a final volume of 10 ml.

PCR reactions were sealed in capillary tips, and incubated in a Corbett Capillary Thermal Cycler, with a program of one cycle of 94°C 3 m, 40°C 2 m, 72°C 3 m, 43 cycles of 94°C 10 s, 40°C 10 s, 72°C 50 s, and one cycle of 72°C 3 m.

The reactions were then electophoresed through a 1.5% agarose gel (Seakem LE) at 3 V cm^{-1} for 3 h. Gels were stained in a dilute solution of ethidium bromide, and photographed under UV light.

RAPD bands were scored as present or absent on gels, with PCR reactions being done in duplicate or triplicate to confirm reproducibility of banding patterns.

2.4   Analysis of RAPD bands

Using the RAPDistance computer program (Armstrong *et al.* 1995), the patterns of DNA fragments obtained using several different primers with each moss sample were compared pairwise to produce a distance matrix. From the distance matrix the computer programs NJTREE (Jin 1988) and TDRAW (Ferguson 1990) were used to calculate a binary dendrogram and an unrooted tree, based on the neighbour-joining method of Saitou & Nei (1987) and Studier & Keppler (1988).

A Permutation Tail Probability (PTP) test (Faith & Cranston 1991) was used to determine whether the total branch length of the tree was significantly shorter than the average of 20 randomly generated trees.

3   RESULTS

3.1   Reproducibility of RAPDs

To ensure that results obtained with moss shoots were reproducible between experiments and reactions, DNA was isolated from two or more shoots which were joined at the base. Four sets of three joined shoots were tested from Antarctic isolates of *Bryum argenteum*, with primers OP-A13, OP-A17, OP-C4, OP-C7 and OP-P7. All RAPD reactions done with these samples gave identical results consistent within the groups of shoots, although some variation was observed between groups. Similar data were obtained with these primers for joined pairs of shoots from several isolates of *Bryum argenteum* from New Zealand and Australia, and for *Campylopus pyriformis* (using primers OP-A17, OP-C4, OP-P1, and OP-P16).

3.2   Variation within moss clumps

Within-clump variation was tested on DNA from three or more shoots taken from different sides of a single clump. In the majority of cases, shoots from the same clump gave identical RAPD banding patterns under the conditions used, and occasional variant shoots could be readily distinguished.

Similar results were obtained for Antarctic isolates of *Bryum argenteum, Pottia heimii*,

Campylopus pyriformis, and *Ceratodon purpureus*. For each species, at least three shoots from each of six clumps were tested, with at least three primers for each species.

3.3 Variation between clumps and among populations

The extent of genetic variation between clumps of *Bryum argenteum* was tested using RAPDs, with primers OP-A13, OP-A17, OP-B10, OP-C4, OP-C7 and OP-P16.

In isolates collected in the Garwood Valley, it was apparent that significant variation occurred between different clumps. No two isolates of the 35 tested gave identical RAPD banding patterns with all five primers tested (Fig. 1a), although isolates from the same drainage channel were generally more closely related to each other than to isolates from different drainage channels. A dendrogram of the RAPD results for these samples showed distinct clustering of the isolates from each of the five channels tested, indicating that water transport down drainage channels is an important local dispersal method.

When isolates of *Bryum argenteum* from sites in the Ross Sea region were compared with other isolates of this species from Australia and New Zealand, it was apparent that Antarctic isolates showed less genetic variation (as judged by polymorphic RAPD bands) than those from other regions (Fig. 1b). About 20 isolates from each of the three regions (over distances of about 150 km) showed that the Australian and New Zealand samples clustered together on a dendrogram, whereas the Antarctic samples were less closely related.

3.4 Use of RAPDs for taxonomy

Five or six isolates of each of six species were used to test whether RAPDs could help distinguish the species found in the Antarctic. All moss samples tested in this taxonomic study gave characteristic banding patterns with the four primers used (OP-A1, OP-A17, OP-C4 and OP-P16; Fig. 1). Although a few faint bands were not always reproducible, 64 distinct bands were reproducibly obtained with these four primers. Other primers were also tested, including OP-A8, OP-C7, OP-C18, OP-P1, and OP-P7. These primers all gave banding patterns which distinguished the different species, but generally the RAPD bands obtained were not as easily scorable as those with OP-A1, OP-A17, OP-C4 and OP-P16. Therefore only the bands obtained with these primers were used to produce the dendrogram shown in Fig. 2.

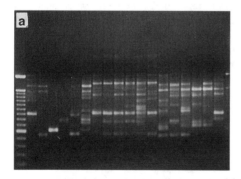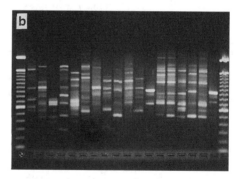

Fig. 1. RAPD banding patterns obtained on an agarose gel for; a - within-population variation for *Bryum argenteum*: 18 isolates of *Bryum argenteum* from the Garwood Valley, Southern Victoria Land with primer OP-C4; b - species identification using primer OPA-17: 6 isolates each of *Sarconeurum glaciale*, *Campylopus pyriformis* and *Bryum argenteum* with primer OP-A17. Far left and right lanes: DNA size standards (BRL 100bp ladder).

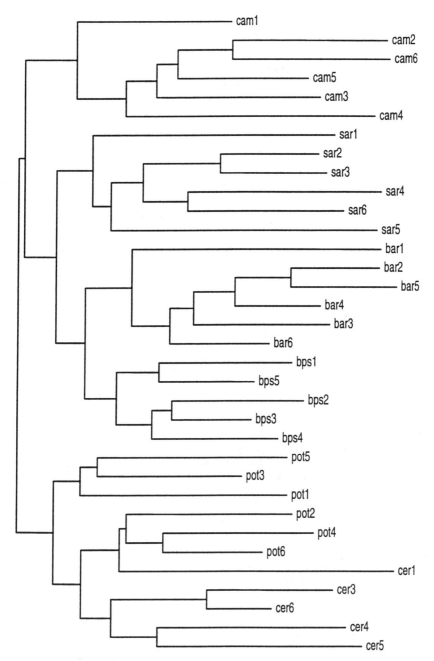

Fig. 2. Neighbour-joining tree obtained using the algorithm of Pearsons Phi in the RAPDistance computer program (Armstrong *et al.* 1995). Five or six isolates of each of six species were used in RAPD reactions with four different primers. The 64 different-sized DNA bands obtained with these primers were scored as present or absent for each isolate, and then analysed by computer. Other algorithms in the program gave similar dendrograms, and a PTP test gave a value of 16.72. cam - *Campylopus pyriformis*; sar - *Sarconeurum glaciale*; bar - *Bryum argenteum*; bps - *Bryum pseudotriquetrum*; pot - *Pottia heimii*; cer - *Ceratodon purpureus*.

Fig. 2 clearly separates four species as distinct subgroups on the phylogenetic tree, with the two *Bryum* species appearing most closely related. *Pottia heimii* and *Ceratodon purpureus* are not completely distinct; other primers appeared to separate these two genera, but did not give clear RAPD bands with some of the other species.

A PTP test on the dendrogram shown in Fig. 2 gave a result of 16.72 standard deviations from random, demonstrating that support for this tree structure is strong. Nine other algorithms from the RAPDistance package also gave very similar dendrograms with these data.

4   DISCUSSION

Although many of the Antarctic species of moss have been known for some years, a lack of genetic information about mosses in general has prevented analysis of how they might have colonised Continental Antarctica, and of whether genetic variation has subsequently occurred. Some studies have also been hindered by the fact that in the extreme climate of the Antarctic, the phenotype of moss species can vary considerably depending on the environmental niche that it occupies, often making positive identification difficult.

The use of RAPDs appears to be a very promising method for assessing the extent of genetic variation in Antarctic mosses, where no other genetic information is available and where small samples must be collected to minimise the impact on the environment. Already, the method has allowed comparison of populations of *Sarconeurum glaciale* from the Ross Sea region and the Vestfold Hills (Selkirk *et al.,* in press), and to populations of *Bryum argenteum* and *Pottia heimii* from Southern Victoria Land (Adam *et al.,* in press; Skotnicki, Selkirk & Dale, unpubl.).

In this communication, the utility of RAPDs as an aid to taxonomic problems is assessed, as phenotypic plasticity of moss species in Antarctica has long been recognised. Several of the isolates tested in this study had originally been tentatively classified in the field as different taxa. The use of RAPDs clearly established that this is a useful method for assisting in the positive identification of species; this has proved especially worthwhile when dealing with mixed clumps of two or more different species. These results were obtained using only four primers from a total of several hundred which are readily available commercially; further experiments may reveal other primers which assist in species identification with both these and other moss species.

Already, results obtained with several moss species have indicated that RAPDs greatly facilitate both population genetics studies, and taxonomic identification of a range of moss species. The results presented here clearly show that RAPDs can be used to confirm morphological identification, and demonstrate the extent of genetic similarity between different isolates and populations.

In future, it should be possible to apply the RAPD technique to more detailed investigations of Antarctic moss populations, and to analysis of local and long-distance vegetative dispersal within Antarctica, as well as determining the origins and colonisation history of these plants.

5   ACKNOWLEDGEMENTS

We thank Dieter Adam, Shawn Walsh and Marie Connett for assistance with collecting the moss isolates, Rod Seppelt for assistance with morphological identification, and Allan Green for helpful discussions. This research was supported by Antarctica New Zealand and the New Zealand Lotteries Board.

REFERENCES

Adam, K.D., P.M. Selkirk, M.B. Connett & S.M. Walsh, in press. Genetic variation in populations of the moss *Bryum argenteum* in East Antarctica. In: Battaglia, B., Valencia, J. & Walton, D.W.H. (Eds.), *Antarctic Communities*. Cambridge University Press, Cambridge.

Armstrong, J., A. Gibbs, R. Peakall & G. Weiller, 1995. *The RAPDistance Package*. WWW:, http://life.anu.edu.au/molecular/software/rapd.html.

Faith, D.P. & P.S. Cranston, 1991. Could a cladogram this short have arisen by chance alone? On permutation tests for cladistic structure. *Cladistics* 7:1-28

Ferguson, J.W.H., 1990. *TDRAW computer program*. Center for Demographic and Population Genetics, University of Texas, Houston.

Jin, L., 1988. *NJTREE computer program*. Center for Demographic and Population Genetics, University of Texas, Houston, Texas, USA.

Longton, R.E., 1988. *Biology of Polar Bryophytes and Lichens*. Cambridge University Press, Cambridge. 391 p.

Rafalski, J.A. & S.V. Tingey, 1993. Genetic diagnostics in plant breeding: RAPDs, microsatellites and machines. *Trends in Genetics* 9:275-280.

Saitou, N. & M. Nei, 1987. The neighbor-joining method: A new method for reconstructing phylogenetic trees. *Molecular Biology and Evolution* 4:406-425.

Selkirk, P.M., K.D. Adam, M.B. Connett, T. Dale, T.W. Joe, J. Armstrong & M.L. Skotnicki, in press. Genetic variation in Antarctic populations of the moss *Sarconeurum glaciale*. *Polar Biology*.

Seppelt, R.D. & P.M. Selkirk, 1984. Effects of submersion on morphology and the implications of induced environmental modifications on the taxonomic interpretation of selected Antarctic moss species. *Journal of the Hattori Botanical Laboratory 55:*273-279

Smith, R.I. Lewis, 1984. Terrestrial plant biology of the sub-Antarctic and Antarctic. In: Laws, R.M. (Ed.), *Antarctic Ecology Vol. 1*. Academic Press, London. pp. 61-162.

Studier, J. & Keppler, K., 1988. A note on the neighbor-joining metric of Saitou and Nei. *Molecular Biology and Evolution* 5:729-731.

Williams, J.G.K., A.R. Kubelik, K.J. Livak, J.A. Rafalski & S.V. Tingey, 1990. DNA polymorphisms amplified by arbitrary primers are useful as genetic markers. *Nuclear Acids Research* 18:6531-6535.

Ecosystem Processes in Antarctic Ice-free Landscapes, Lyons, Howard-Williams & Hawes (eds)
© *1997 Balkema, Rotterdam, ISBN 90 5410 925 4*

Ionic migration in soils of the Dry Valley region

G.G.Claridge & I.B.Campbell
Land and Soil Consultancy Services, Nelson, New Zealand

M.R.Balks
Department of Earth Sciences, University of Waikato, Hamilton, New Zealand

ABSTRACT: The rate at which soluble contaminants migrate in Antarctic soils has been investigated by applying lithium chloride solution to soil surfaces and following its movement in subsequent years. Movement of lithium ions was found to be related to the moisture available at the site. In wet soils, with moisture contents between 3 and 12%, lithium moved up to 5 m in three years. On the other hand, in dry soils, with moisture contents less than 1%, movement was limited to 15 cm depth and 1 m down slope. Therefore considerable movement of ionic solutes may be anticipated where sufficient moisture is available, but generally little or no movement will take place.

1 INTRODUCTION

As human activity increases in Antarctica, the possibility of soil contamination arises, especially around sites where human activity is concentrated. These may range from bases, with most activity and likely contamination, to isolated campsites, where contamination is avoided wherever possible and should only involve the occasional accidental spills.

In general, Antarctic soils are arid and saline. Salts accumulate at or near the surface, as a result of moisture movement from the ice-cemented permafrost beneath to the surface, and from moisture moving into the soil from snow melting on the surface (Campbell & Claridge 1982). The salts differ widely from place to place. Most of the phases that can be formed from the cations sodium, potassium, magnesium and calcium with the anions chloride, sulphate and nitrate have been identified in the soils of the dry valleys (Keys & Williams 1981; Claridge & Campbell 1968, 1977). The cations originate largely from sea water, as indicated by the dominance of sodium. However there are important contributions from weathering, indicated by correlation of the salt composition with geology (Claridge & Campbell 1977).

Salts migrate in Antarctic soils over time, as indicated by the accumulation of salt horizons, the presence of nodules of salt under some surface stones, and the transient appearance of salt efflorescences. Salts move as a result of moisture dynamics within the soil and previous history. Ions migrate through soils in thin films of saturated salt solution that can exist between grains or within cracks in particles (Murrmann 1973).

Moisture is added to the soil surface mainly from melting snow, and moves through the soil by capillary flow. Even at low temperatures, thin films of liquid will remain unfrozen if they consist of extremely concentrated salt solutions (Ugolini & Anderson 1973). There are also considerable humidity gradients within the soil. Air within the soil pores will be fully saturated with moisture at the prevailing temperature ($0°C$ at the warmest part of the summer) at the contact with ice-cemented permafrost and at the prevailing atmospheric humidity at the surface. The soil air is pumped out of the soil as a result of diurnal changes in temperature of up to 10 to 12°C

(Black & Berg 1963), albeit with lesser amplitudes to the contact with ice-cement.

Thus, dry air is taken into the soil as it cools, and air, in hygroscopic equilibrium with ice, is expelled as it warms, moving moisture from the ice cement to the surface, where it is lost to the atmosphere. Some of this moisture moves in the liquid phase, as thin films, carrying salts with it. The salts may accumulate at an equilibrium zone where moisture films finally evaporate. In some soils salt horizons form while in others salts concentrate in nodules under flat stones in the desert pavement.

To complement the evidence from the natural distribution of salt, we obtained some experimental data on the movement of salts in soils by following the movement of lithium chloride added to soil plots in three differing soils moisture regimes. Lithium chloride was used as a surrogate for other salts because lithium is a natural constituent of soil salts, but normally present only in extremely low concentrations, and is easily measurable by simple techniques when present in concentrations similar to those of the other soluble constituents of Antarctic soils.

2   MATERIALS AND METHODS

2.1   Sites

Three sites with differing moisture availability were chosen for this experiment, one close to Lake

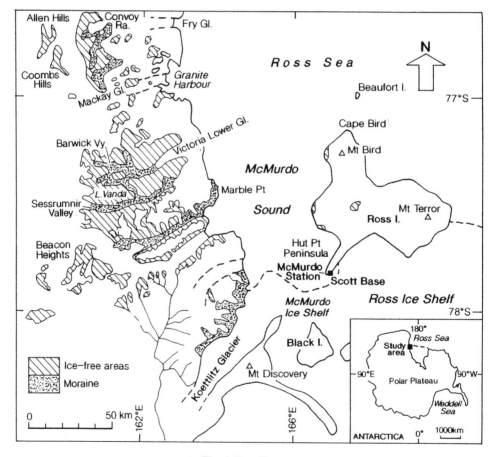

Fig. 1. Locality map.

138

Vanda in the Wright Valley and the other two close to Scott Base on Ross Island (Fig. 1).

The site near Lake Vanda was on an even, broadly sloping surface on the western side of a fan surface. The site had a slope of 9°. Precipitation at the site is about 15 mm and snow does not accumulate on the surface. The soils are very dry, with an average moisture content over the period of the experiment of 0.60%.

The first site at Scott Base was on a small, relatively level plateau, on which the infrequent snowfalls did not persist, and drifting snow did not accumulate. Snowfall is probably around 60 mm. The soils at this site are generally dry, with moisture contents in the active layer averaging 4.4% over the period of the experiment.

The second site was about 30 m from the first, in the head of a small gully, which accumulated drift snow during the winter, and frequently remained snow covered until late in the summer, when the snow eventually melted. The soils at this site have a much greater moisture supply and contain more moisture, averaging 8.5% over the period of the experiment. Further down the gully, below the experimental plot, the soils become wet and drain into the ephemeral streams which flow down the slopes in late summer.

2.2   Application of lithium chloride

At each site, four plots, each 1 m x 1 m were marked out along the contour. Three of the plots were irrigated with 10 l of a solution containing 8 g lithium chloride l^{-1}. The solution was applied using a sprinkler so that the whole of each plot was wetted evenly and the surface material not disturbed by the falling water drops. The fourth plot at each site was irrigated in the same manner with 10 l of water (control plot).

At the site near Lake Vanda, there was sufficient space to lay out a replicate experiment, but this was not possible at the Scott Base sites.

2.3 Sampling

At the Vanda sites, because the soils were very loose and incoherent, pits could only dug to 50 cm. Test pits elsewhere in the vicinity showed that ice-cemented permafrost was not encountered even in pits 1 m deep. At the sites near Scott Base, pits were dug to the ice-cement (35 to 40 cm deep) in the centre of one of the plots and at 1 m intervals downslope from the plot. The ice-cemented material was also sampled by drilling, using a diamond-tipped coring barrel. At the Vanda site, samples were collected 8 days, 12 months and 23 months after application from both sets of plots. At Scott Base, sampling took place seven days after application, 23 months after application and 36 months after application. The control plots were sampled for moisture content seven or eight days after application. No replicate samples were taken, but approximately 1 kg of sample collected was considered sufficient to account for minor variations throughout the sampling zone.

2.4   Analytical

The samples were weighed at field conditions and again after drying at 105°C, and moisture content calculated by difference. The fine earth material (<2 mm) was separated and extracted with 0.1M sodium chloride solution (1:5 soil:extractant ratio) which completely removes all lithium content from exchange sites. The lithium content of the extracts was determined by flame emission spectroscopy, using the 670.8 nm emission line of lithium.

3   RESULTS

Moisture content measurements of the soils of the blank plots at Scott Base revealed that the small amount of water added to the soil had evaporated in the seven days after application. At the Vanda sites, moisture content of the control plots was between 1 and 2% eight days after application, but had returned to baseline values 12 months later, indicating that the experimental procedure had no

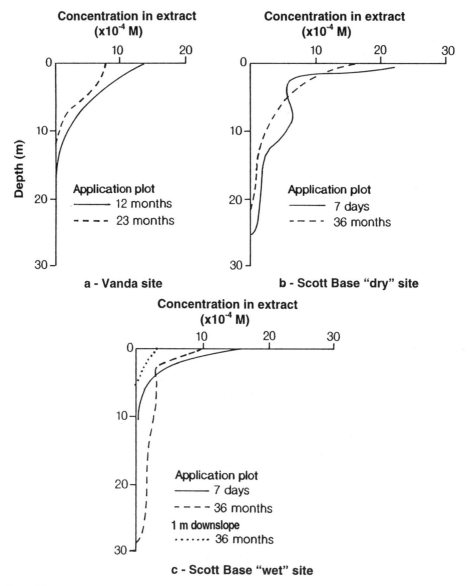

Fig. 2. Plots of lithium content in 1:5 NaCl extracts of soils from the experimental sites near Lake Vanda and Scott Base.
In each of the diagrams, the concentration of lithium immediately after application for the Scott Base site, and 223 months after application for the Vanda site, and that found at the last sampling (36 months later for the Scott Base sites and 23 months for the Vanda site) is plotted.

permanent effect on the soil moisture status and therefore irrigation with lithium chloride could be considered a suitable method of adding salt to the soil surface.

The lithium concentrations of the soils at the three sites are shown in Fig. 2[1]. At the application plots, very little difference in concentration, or a slight decrease with time, was found. Most of the lithium was concentrated at the surface, and only small amounts of solution had penetrated deeper. At the Vanda site, where the soil is very dry, the lithium solution had only penetrated to about 15 cm, whereas at the "dry" site at Scott Base, where the moisture content is higher, the soil was wetted to 25 cm. At the "wet" Scott Base site, which contained even more moisture, penetration was about 15 cm.

At the Vanda site and the Scott Base "dry" site there was a slight decrease in lithium concentration with time but no evidence of migration downwards in the soil. At the Scott Base "dry" site traces of lithium were only found close to the level of the ice-cemented permafrost in plot 1 and 2 m downslope, indicating that there had been minor movement of lithium along the surface of the ice cement where some moisture was available.

At the Scott Base "wet" site, on the other hand, lithium had moved further down the profile and even into the ice cement at the application plot, while appreciable amounts were present 1, and even 2 m, downslope. Traces of lithium could be detected up to 10 m downslope.

4   DISCUSSION

At the Scott Base "dry" site, which has more moisture available than the Vanda site, there has been a significant loss of lithium from the application plot, and movement of small amounts down to the ice-cemented layer and down the slope, presumably by ionic diffusion along thin moisture films in an almost dry soil. The presence of detectable lithium within the permanently frozen layer implies that it may contain cracks and pores through which concentrated salt solutions may move, especially where the moisture content is below saturation.

At the very dry Vanda site very little movement of lithium has taken place in the duration of the experiment. So far, it appears that the lithium applied to the surface has remained largely where it was when the lithium solution dried in the few days after application. The Vanda site is typical of most of the inland valley floors of the Dry Valley region: therefore we expect that movement of salts will be very slow.

We expect that only very small amounts of soluble ions will penetrate further, drawn down by capillary action, until the soil temperature decreases to freezing point. In soils of coastal regions, most of which are underlain by ice-cemented permafrost, pure water will not penetrate further, but salt solutions with depressed freezing points may well infiltrate the permafrost, especially if they become more concentrated by evaporation and freezing out of excess water. Even if the contaminant solution is not sufficiently saline to depress the freezing point appreciably, water added to the soil surface may well dissolve salts already contained within the soil, so that the contaminants may move in a naturally saline solution.

The data presented showed that, where sufficient moisture is available, such as at the Scott Base "wet" site, salts applied to the soil surface move downwards and laterally through the soil, while some of the saline solutions are moving laterally along the surface of the ice-cemented permafrost. At the wet sites where water is available for transport, there has been a measurable loss from the application plots after three years. There does not appear to be any evidence for the upward migration of added salts as no surface efflorescences of lithium chloride were noted in this experiment. This may well be because lithium chloride is slightly deliquescent, absorbing moisture from a humid atmosphere. Efflorescences of less deliquescent salts, such as sodium chloride or sodium sulphate, frequently appear on the soil surface in the vicinity of the trial site at

---

[1] The original analytical data may be obtained from the authors on request.

Scott Base, especially where the surface has been disturbed and melting of subsurface ice has taken place.

5   CONCLUSIONS

Most accidental spills, (e.g. battery acid, urine) are generally of low volume, much less than the 10 1 m^{-2} applied in this trial. Based on our results, any such spill would not be expected to penetrate further into the soil than the lithium addition did at the dry sites. Contaminants could be expected to spread more slowly at inland sites with drier climates, although soils with high salt concentrations might behave differently. This has been borne out by Claridge *et al.* (1995) in their studies of the movement of pollutants at Marble Point, where, even thirty years after a spill, heavy metals were only barely detectable 20 cm from the spill site. Such spills could easily be dealt with by removing a thin layer, a few cm thick, soon after the spill had occurred. Even after several years had elapsed, removal of soil covering a few square metres would be all that would be required to clear the area of contamination.

On wet sites, on the other hand, the conditions are different, and the soluble constituents of spills will move, at rates dependent on the amount of water available. Movement will be downslope to a drainage sink such as the ocean, to an undrained hollow, such as a saline pond, or to a large lake, such as Lake Vanda, where the contaminants will remain. Movement will not necessarily be rapid, as shown in the wet site studied here, but any site where windblown snow can accumulate, concentrating the very low overall precipitation in a limited area, can be considered wet.

This has implications for the fate of material accidentally spilt on the soil surface. At dry sites, soluble ions will remain close to the spill site and would be easily removed in a clean up operation. However, in wet sites, soluble material can be expected to move gradually to the drainage sink, where it will accumulate. Thus greater attention needs to be given to cleaning up contamination in such sites if they drain into an enclosed system. If such spill sites are close to the sea, natural processes will eventually remove any contamination and transfer it to the ocean, where it will be diluted to insignificant concentrations.

6   ACKNOWLEDGEMENTS

The authors acknowledge the Foundation for Research, Science and Technology and the New Zealand Lottery Grants Board for financial assistance and Antarctica New Zealand for logistic support.

REFERENCES

Balks, M.R., D.I. Campbell, I.B. Campbell & G.G.C. Claridge, 1995. Interim results of 1993/94 Soil climate, active layer and permafrost investigations at Scott Base, Vanda and Beacon Heights, Antarctica. *University of Waikato, Antarctic Research Unit, Special Report* 1. 64 p.

Black, T. & T. Berg, 1963. Hydrothermal regimen of patterned ground, Victoria Land, Antarctica. *IASH Commission of Snow and Ice, Publication No.* 61. pp. 121-127.

Campbell, I.B. & G.G.C. Claridge, 1982. The influence of moisture on the development of soils of the cold deserts of Antarctica. *Geoderma* 28:212-238.

Campbell, I.B. & G.G.C. Claridge, 1987. *Antarctica: soils, weathering processes and environment.* Elsevier, Amsterdam. 368 p.

Claridge, G.G.C. & I.B. Campbell, 1968. Origin of nitrate deposits. *Nature* 217:428-430.

Claridge, G.G.C, & I.B. Campbell, 1977. The salts in Antarctic soils, their distribution and relationship to soil processes. *Soil Science* 123:377-384.

Claridge, G.G.C., I.B. Campbell, H.K.J. Powell, Z.H. Amin & M.R. Balks, 1995. Heavy metal contamination in some soils of the McMurdo Sound region, Antarctica. *Antarctic Science* 7:9-14.

Keys, J.R. & K. Williams, 1981. Origin of crystalline, cold desert salts in the McMurdo Sound region, Antarctica. *Geochimica et Cosmochimica Acta* 45:2299-2309.

Murrmann, R.P., 1973. Ionic mobility in permafrost. *Proceedings of the Second International Conference on permafrost, Yakutsk, 1973.* National Academy of Sciences, Washington. pp. 352-359.

Nelson, K.H. & T.G. Thomson, 1954. Crystallisation of salts from sea water by frigid concentration. *Journal of Marine Research* 13:166-182.

Ugolini, F.C. & D.M. Anderson, 1973. Ionic migration and weathering in frozen Antarctic soil. *Soil Science* 115:461-470.

3 Aquatic environments

Ecosystem Processes in Antarctic Ice-free Landscapes, Lyons, Howard-Williams & Hawes (eds)
© *1997 Balkema, Rotterdam, ISBN 90 5410 925 4*

Chemical weathering rates and reactions in the Lake Fryxell Basin, Taylor Valley: Comparison to temperate river basins

W.B. Lyons, K.A. Welch, C.A. Nezat, K. Crick, J.K. Toxey & J.A. Mastrine
Department of Geology, University of Alabama, Tuscaloosa, Ala., USA

D.M. McKnight
INSTAAR, University of Colorado, Boulder, Colo., USA

ABSTRACT: Much has been written recently about the influence of climate on chemical denudation rates. Some authors have argued that climate has little impact on chemical weathering unless coupled with high rates of physical weathering. Others have stated that physical erosion rates do not have a critical influence on chemical weathering but climate does. We present a comparison of chemical weathering rates, as determined by H_4SiO_4 and HCO_3^- fluxes in streams and rivers, between warm-temperate systems in Alabama and desert-polar systems in Taylor Valley. Chemical weathering rates, as represented as H_4SiO_4 fluxes in Alabama, range from 55×10^3 moles km^{-2} yr^{-1} for the Cahaba River basin to 29×10^3 mol km^{-2} yr^{-1} for the Tallapoosa River Basin. These values are slightly lower than the world average of 64×10^3 km^{-2} yr^{-1} calculated by Ming-hui *et al.* (1982). Our preliminary calculations indicate that chemical weathering rates in the Antarctic streams are equal to or greater than the Alabama rivers with rates ranging from 4 to 193×10^3 moles km^{-2} yr^{-1}. These high chemical weathering rates suggest that temperature itself may not play the principal role in determining weathering rates. Geochemical modelling suggests a number of potential reactions occur within the stream reaches in Taylor Valley, Antarctica.

1 INTRODUCTION

The role of climate on chemical weathering, and the rate at which it occurs, has garnered much attention recently. This is due, in great part, to an important global aspect of chemical weathering, the consumption of CO_2 as weathering occurs. Hence, the role of weathering is extremely important in the long-term geological control of CO_2 in the atmosphere. In general, CO_2 is consumed creating HCO_3^- during the process of silicate hydrolysis (Berner 1995). In theory, the faster the rate of weathering, the faster CO_2 is removed from the atmosphere, and the faster the climate cools. As the climate cools, weathering decreases and CO_2 accumulates, thus providing a feedback controlling global temperatures through long periods of geologic time (Berner 1995). It is clear then that understanding weathering rates and reactions in differing temperature regimes will aid in our overall understanding of, not only the evolution of the surface of the earth, but the atmosphere and climate as well.

There are essentially two major "schools" of thought on what controls watershed weathering rates. These two "schools" have been argued in two recent seminal papers on the subject (Bluth & Kump 1994; White & Blum 1995). Bluth & Kump (1994) state that chemical denudation, and increased solute fluxes (i.e. chemical weathering), are governed by a balance between chemical and physical processes. The corollary of this is that warm, wet climate and/or abundant vegetation ALONE cannot guarantee high solute fluxes. The White & Blum (1995) argument implies that as temperature and precipitation increase, so do chemical weathering rates, as evidenced by increased H_4SiO_4 and Na^+ fluxes. The corollary of this is that there is no correlation between chemical fluxes and topographic relief or extent of recent glaciation, implying that physical weathering has no critical influence on chemical denudation rates.

If one opens most geomorphology textbooks, one observes that the most commonly held view regarding the role of temperature and precipitation on chemical weathering is that as temperature decreases the role of physical weathering increases, while chemical weathering diminishes. This implies that physical weathering becomes much more important than chemical weathering in cold, dry or polar desert-type environments.

The polar regions, in general, and the Antarctic Dry Valleys, in particular, are an excellent place to compare and contrast the points of view of these two "schools" of thought. The McMurdo Dry Valleys (MDV), southern Victoria Land, Antarctica, are unusually suited for this because of their general lack of terrestrial vegetation, low mean annual temperature (~20°C) and low precipitation rates of <10 cm yr^{-1} (Clow et al. 1988). Because the debate over what controls chemical weathering rates is fundamental to our overall understanding of earth surficial processes, by measuring weathering rates in this environment and comparing them to those from more temperate and humid environments, it was hoped a better understanding of the importance of climate on weathering could be established.

Our measurements and calculations will show that, indeed, chemical weathering does occur in polar desert environments, such as the MDV. We are not the first to point this out, however, as earlier work by Jones & Faure (1978), Green et al. (1988), and Lyons & Mayewski (1993) demonstrated it quite unequivocally. What is new regarding our work is that we show that the rate of chemical weathering is as high as those in temperate, humid climates, such as Alabama, south-eastern USA. The important caveat that needs to be made here, however, and will be repeated throughout the text, is that in order to have chemical weathering in polar deserts, such as MDVs, liquid water is needed. In the majority of the MDV, liquid water only exists in very specific locations for any period of time - the stream beds and floodplains of the streams and rivers draining the alpine and Piedmont glaciers - and this only happens for a few weeks during the austral summer.

2   STUDY AREA

The Lake Fryxell basin is the easternmost basin in the Taylor Valley with stream inflows from the Canada and Commonwealth Glacier from the south. The landscape is uniformly unconsolidated alluvium. The alluvium is composed mostly of sand-sized particles interbedded with cobbles and boulders. The four stream "watersheds" utilised in this study include: Canada Stream, Lost Seal Stream, Delta Stream and Von Guerard Stream that have recently been described in detail in Alger et al. (in press) and Conovitz et al. (in press), and will not be repeated here. Streamflow is highly variable with large diel variations. The source of water to the streams is solely meltwater from the glaciers.

3   METHODS

3.1   Sampling and analysis

Samples were collected in pre-cleaned polyethylene bottles. During collection, bottles were first rinsed in stream water (as many as three times). The sampler wore polyethylene gloves during collection to minimise any potential contamination. Samples were returned to the laboratory in the field (either at Lakes Bonney, Hoare or Fryxell) and filtered through 0.4-μm Nucleopore filters using pre-cleaned, plastic filter holders and towers that were designated for stream filtration only. One aliquot of filtered samples was placed in an acid-washed, polyethylene bottle for later major cation analysis, while another aliquot was placed in a distilled-deionized washed bottle for later anion analysis. The cations and anions were measured using a Dionex Ion Chromatograph using methods outlined in Welch *et al.* (1996). The cation and anion data are discussed in Lyons *et al.* (in press).

A third aliquot was placed into a water-rinsed bottle for alkalinity determinations. pH were measured immediately in the field, where possible, and in the field laboratory at a later time. H_4SiO_4 (reactive silicate) was also determined on unacidified samples. Alkalinity was determined via titration with a precision of variation of less than ±3%. H_4SiO_4 was determined colourimetrically using the technique of Mullin & Riley (1955). The precision of these measurements was less than ±3%.

Samples from the Cahaba, Tallapoosa and Alabama Rivers in central Alabama were collected in June 1994, using the same protocol as above, and the same analytical methods. The Cahaba and Tallapoosa Rivers are major tributaries of the Alabama/Mobile River System (AMRS), which is the fourth largest riverine system in discharge in the continental United States (Ward *et al.* 1992). The Cahaba drains both Appalachian Plateau and Valley and Ridge geological provinces, whereas the Tallapoosa drains primarily the metamorphosed Piedmont province of the southernmost Appalachians. These river systems are obviously much larger in drainage basin and in discharge than the Lake Fryxell basin ones in Taylor Valley; however we will be comparing and contrasting the H_4SiO_4 and HCO_3^- fluxes between these systems. White & Blum (1995) found no differences between upper basin watershed (first and second order streams) fluxes of H_4SiO_4 between upper basin watershed (first- and second-order streams) fluxes of H_4SiO_4 and those from the larger river systems within the same systems. Therefore, we assume that the H_4SiO_4 fluxes calculated by us for the larger Alabama rivers would be similar to those of the smaller streams (similar in size to the Fryxell basin streams) feeding them.

3.2   Method of determining the "watershed"

The Taylor Valley watersheds are unlike the ones most of us are accustomed to in temperate and tropical climates. In the valley, there is essentially no overland flow and little to no seepage of water into the ground beyond the floodplain. The cold, very dry climate allows for little water influx as sublimation is the major process affecting the soil zone. With this in mind, one can limit the "watershed" in these areas to locations where running water exists for at least a few days to a few weeks of the year. In the Taylor Valley, these areas are, in fact, the stream beds and floodplains themselves where, depending on the year, streams can flow from ~4 to 10 weeks per year.

Four "watersheds" in the Fryxell basin were delineated in this manner (Canada, Lost Seal, Delta and Von Guerard). The streams' length and width were carefully determined by Alger *et al.* (in press). Their dimensions are shown in Table 1. The watershed area was simply calculated by multiplying the width of the stream beds by their length (Table 1). Measurements of the streams were made in January 1994 (Alger *et al.*, in press). The topography of the stream channel and stream banks were determined using a Top Con (a mapping instrument), and the data were used to create small-scale stream contour maps. Lengths were determined using GIS. Our estimate of area in the length measurements is ±0.1 to 0.2 km for the longer streams and ±0.1 km for the shorter streams. The error in the width measurements is ±0.5 to 1.0 m.

3.3   Method of determining H_4SiO_4 and HCO_3^- fluxes

We have used the method set forth in Ming-hui *et al.* (1982) to calculate chemical weathering.

Table 1. Fryxell Basin watersheds.

Name	Length* (km)	Width* (m)	Discharge (1994-95) (m^3)
Canada	1.5	~4	4.17x10^4
Lost Seal	2.2	~4	2.06x10^4
Delta	11.2	~4	3.28x10^3
Von Guerard	4.9	~4	7.22x10^3

*From: Alger et al. 1995

This is done by determining the H_4SiO_4 and HCO_3^- fluxes in each watershed. The mean H_4SiO_4 and HCO_3^- concentrations from each of the Antarctic streams (and rivers in Alabama) were multiplied by the yearly discharge (Table 1) and then divided by the watershed area yielding values in Mol km^{-2} yr^{-1}. This has been done for the major river systems of the world by Minghui et al. (1982) as well as for other river systems by other investigators (e.g. Lyons et al. 1992).

4   RESULTS AND DISCUSSION

4.1   Weathering in the Taylor Valley

Previous workers have demonstrated that chemical weathering does occur within the stream reaches of the MDV. There is strontium isotopic evidence (Jones & Faure 1978), as well as cation and H_4SiO_4 data (Green & Canfield 1984) from the Onyx River in Wright Valley (the valley to the north of Taylor Valley), and major cation and anion data from the Fryxell Basin (Green et al. 1988) that strongly supports the notion of chemical weathering occurring in these systems. The most compelling evidence is, in our opinion, the H_4SiO_4 data from the various streams. The analysis of glacier ice and snow of the alpine glaciers that enter both Taylor and Wright Valleys indicate, in general, very low H_4SiO_4 concentrations (<10 μmol l^{-1}) (Mayewski & Lyons 1982; Lyons et al., in press). The particularly high H_4SiO_4 values (>100 μmol l^{-1}) observed in streams draining the Howard Glacier to the south of Lake Fryxell yare comparable to values from both temperate and tropical rivers (Rose 1994; Edmond et al. 1995). The streams draining the Howard Glacier are generally longer than the other Taylor Valley streams, and, hence, have longer contact time with the stream bed. The *only* source of this H_4SiO_4, above ~5 to 10 μmol l^{-1} of the glacial ice/snow, is from chemical weathering, or silicate hydrolysis of primary materials in the stream beds.

4.2   Rates of chemical weathering in the Fryxell Basin

The rates of silicate mineral weathering within the small, well-defined watersheds of the Fryxell Basin are quite high; surprisingly high, in fact. We note, however, that these should be considered preliminary data for the following reasons. First, they represent only one year of discharge data and we know the discharge can vary dramatically from year-to-year (Lyons et al., in press). Secondly, the H_4SiO_4 and HCO_3^- measurements were not taken in a systematic way, in that they were "spot" measurements taken at one location in the stream, usually at the gauging station. Third and most importantly, we have operationally defined the watershed in a very specific way. In spite of this, the rates determined in this way are within the same order of magnitude as those from more temperate systems (compare Tables 2 & Table 3). In fact, more recent work, using more sophisticated techniques, has shown that in Von Guerard Stream, the

150

Table 2. Chemical weathering rates as determined by H_4SiO_4 and HCO_3^- fluxes in 10^3 mol km^{-2} yr^{-1}.

Stream name	H_4SiO_4	HCO_3^-
Canada	193	1453
Lost Seal	109	1697
Delta	4	403
Von Guerard	26	67

Table 3. Weathering rates from production of H_4SiO_4 (10^3 m km^{-2} yr^{-1}).

World average[1]	64
Tisa, Eastern Europe[2]	90[•]
Mekong[1]	100[+]
Amazon[1]	130[*]
Cahaba[3]	55[•]
Cahaba[3]	45[•]
Cahaba[3]	54[•]
Tallapoosa[3]	29
Alabama[3]	51

[•] Carbonate-dominated drainage

[+] Topographically-elevated drainages, young orogenic belts

[*] Extensive lowlands

[1] Carbonate-dominated drainage

[2] Topographically-elevated drainages, young orogenic belts

[3] Extensive lowlands

silicate weathering rates are greater than any reported watershed study or experimental dissolution rates for plagioclase, and similar to the most rapid experimental dissolution rates for hornblende and augite (Blum et al. 1996).

The highest HCO_3^- weathering rates shown in Table 4 reflect the influence of carbonate rocks, such as limestones and dolomites within the drainage basins of these rivers/streams. The lowest rates shown, the Blue Ridge streams and the Tallapoosa River, Alabama, are in watersheds dominated by metamorphic rocks with little carbonate rocks present. There is abundant $CaCO_3$ in the Taylor Valley soils (Keys & Williams 1981). In addition, there is apparently $CaCO_3$ deposited or blown in via aeolian transport in the Taylor Valley streams during the austral winter, as many of the streams have both high HCO_3^- and Ca^{2+} conentrations (Green et al. 1988; Lyons et al. in press). The Fryxell Basin streams have, for the most part, very high HCO_3^- fluxes (Table 2) reflecting the abundance of $CaCO_3$ within the system. The Canada Stream and Lost Seal Stream fluxes are a factor of ~2 greater than rivers partially draining limestone terrain such as the Cahaba in Alabama and the Tisa in Eastern Europe (Table 4). However, rivers draining solely limestone would have higher fluxes (Meybeck 1987).

4.3   Geochemical modelling

In order to better quantify, as well as qualify the reactions that are controlling the solute composition in one of the Fryxell Basin streams (i.e. Canada Stream), we used the US Geological Survey computer code NETPATH. NETPATH is an inverse model based on the original work of Garrels & Mackenzie (1967) for Sierra Nevadan (California) waters. In our use

151

Table 4. Weathering rates based on HOC_3^- production (10^3 mol km^{-2} yr^{-1})

World average[1]	300
Mekong[1]	680
Amazon[1]	300
Georgia Blue Ridge streams[2]	64
Alabama rivers[3]	138 (Tallapoosa) to 736 (Cahaba)
Tisa[4]	640

[1] Carbonate-dominated drainage

[2] Topographically-elevated drainages, young orogenic belts

[3] Extensive lowlands

[4] Lyons *et al.* 1992

of the model, we have used the mean composition of Canada Glacier (Lyons *et al.*, in press) as the "starting water" and the Canada Stream water at the gauging station as the "final water" composition. Essentially, the code subtracts the two waters and provides the needed geochemical reactions to explain the difference in the two (i.e. what minerals have been dissolved and/or precipitated in order to produce the gain in solute from the starting water to the final water composition). The code provides no unique solution, although the number of model solutions can be constrained by prescribing what mineral phases are reacting. The model was run using the mineral phases listed in Table 5. Seventy different model fits were compatible with the data. These 70 models included some, or all of the reactions shown in Table 5. It should be emphasised that no one solution involved the dissolution of a simple salt alone; that is, the incongruent weathering or hydrolysis of primary silicate minerals always took place within each of the models that fit the stream chemistry data.

Table 5. Geochemical modeling using the USGS inverse model NETPATH

- 70 model fits
- They include some or all of the following:

1.	Dissolution of:	$CaCO_3$
		$NaSO_4 \cdot H_2O$
		$NaCl$
		$CaSO_4 \cdot 2H_2O$
2.	Precipitation of:	Ca-montmorillinite
		chlorite
		illite
		SiO_2
3.	Dissolution of:	K-feldspar, K-mica
		SiO_2
		Albite
4.	Caton Exchange:	Ca^{2+} for Na^+

No one solution involves dissolution of salt ALONE.

5 CONCLUSIONS

This work, along with even more recent, more sophisticated work (Blum *et al.* 1996; Lyons *et al.*, in press), demonstrates clearly that chemical weathering of both silicate and carbonate minerals occurs within streams/floodplains in the MDV. Moreover, the rates of this weathering are quite high and, when compared to rivers/streams in both temperate and humid environments, are similar. The reason for this is not totally clear, but the role of the hyporheic zone, as well as the concurrent process of physical weathering, both may be of extreme importance. Blum *et al.* (1996) have argued that water circulating through the hyporheic zone of Von Guerard Stream (i.e. longer storage) greatly enhances silicate hydrolysis. On the other hand, freezing and thawing, steep temperature gradients within the stream beds, and wetting and drying cycles may also greatly enhance chemical weathering rates in these environments (Ugolini 1986). The continual exposure of fresh mineral surfaces through physical processes may be extremely important in this regard. Peters (1984) has suggested that weathering increases in subzero temperatures due to an increase in such things as frost cracking and wedging. Hopefully, future work will help clarify this. In the true soil zone of the MDV where there is a paucity of liquid water, weathering is indeed slow (Ugolini 1986). However, our work in Antarctica, and that by others in the Arctic (Huh *et al.*, in press) indicates that the notion of low to non-existent chemical weathering occurring in environments where liquid water does exist is incorrect. To the contrary, when there is flowing, liquid water present, chemical weathering rates appear to be equal to those in warmer, wetter climates. These preliminary data also suggest that the role of higher plants on the net chemical weathering rates (Cochran & Berner 1996) may not be a major influence on the watershed scale.

6 ACKNOWLEDGEMENTS

This work was supported by NSF grant OPP-9211773. J.K. Toxey was supported by an REU supplement to the original grant. S. Wilder was supported by an ACS SEED grant to W.B. Lyons. We are grateful to our LTER colleagues, especially H. House, for their help in the collection of these samples. We thank Jordan Hastings for his help with the stream dimension measurements. Fruitful discussions with Dr. A.E. Blum are also gratefully acknowledged. We thank P. Smith for typing the manuscript. We especially thank Jim Saunders and Bill Green who critically reviewed the initial manuscript.

REFERENCES

Alger, A.S., D.M. McKnight, S.A. Spaulding, C.M. Tate, G.H. Shupe, K.A. Welch, R. Edwards, E.D. Andrews & H.R. House, in press. *Ecological Processes in a Cold Desert Ecosystem: the abundance and species distribution of algal mats in glacial meltwater streams in Taylor Valley, Antarctica.* USGS Water Resources Investigative Report.

Berner, R.A., 1995. Chemical weathering and its effect on atmospheric CO_2 and climate. In: White, A.F. & S.L. Brantley (Eds.), *Chemical Weathering Rates of Silicate Minerals.* Elsevier, New York. pp. 565-583.

Blum, A.E., D.M. McKnight & W.B. Lyons, 1996. *Silicate weathering rates along a stream channel drainage into Lake Fryxell, Taylor Valley, Antarctica.* Abstract with Programs GSA meeting, Denver.

Bluth, G.J.S. & L.R. Kump, 1994. Lithologic and climatologic controls of river chemistry. *Geochimica et Cosmochimica Acta* 58:2341-2359.

Clow, G.D., C.P. McKay, G.M. Simmons, Jr. & R.A. Wharton, Jr., 1988. Climatological observations and predicted sublimation rates at Lake Hoare, Antarctica. *Journal of Climate* 1:1-14.

Cochran, M.F. & R.A. Berner, 1996. Promotion of chemical weathering by higher plants: field observations on Hawaiian basalts. *Chemical Geology* 132:71-77.

Conovitz, P.A., D.M. McKnight, L.M. Macdonald, A. Fountain & H. House, in press. Hydrologic processes influencing streamflow variation in Fryxell Basin, Antarctica. In: J.C. Priscu (Ed.), *The*

McMurdo Dry Valleys, Antarctica: a cold desert ecosystem. American Geophysical Union, Washington, D.C.

Edmond, J.M., M.R. Palmer, C.I. Measures, B. Grant & R.F. Stullard, 1995. The fluvial geochemistry and denudation rate of the Guayana Shield in Venezuela, Columbia and Brazil. *Geochimica et Cosmochimica Acta* 59:3301-3325.

Garrels, R.M. & F.T. Mackenzie, 1967. Origin of the chemical compositions of some springs and lakes. In: Sturam, W. (Ed.), *Equilibrium Concepts in Natural Water Systems. Advances in Chemistry Series* 67. American Chemical Society, Washington, D.C. pp. 222-242.

Green, W.J. & D.E. Canfield, 1984. Geochemistry of the Onyx River (Wright Valley, Antarctica) and its role in the chemical evolution of Lake Vanda. *Geochimica et Cosmochimica Acta* 98:2457-2467.

Green, W.J., M.P. Angle & K.E. Chave, 1988. The geochemistry of Antarctic streams and their role in the evolution of four lakes in the McMurdo Dry Valley. *Geochimica et Cosmochimica Acta* 52:1265-1274.

Huh, Y., J.M. Edmond & A. Zartser, in press. Weathering processes and fluxes in the Archean and Proterozoic basement and metamorphic terranes of the Upper Lena: a comparison with the tropics. *Geochimica et Cosmochimica Acta.*

Jones, L.M. & G. Faure, 1978. A study of strontium isotopes in lakes and surficial sediments of the ice-free valleys, southern Victoria Land, Antarctica. *Chemical Geology* 22:107-120.

Keys, J.R. & K. Williams, 1981. Origin of crystalline, cold desert salts in the McMurdo region, Antarctica. *Geochimica et Cosmochimica Acta* 45:2299-2309.

Lyons, W.B. & P.A. Mayewski, 1993. The geochemical evolution of terrestrial waters in the Antarctic: the role of rock-water interactions. In: Green, W.J. & E.I. Friedman (Eds.), *Physical and Biogeochemical Processes in Antarctic Lakes. Antarctic Research Series* 59. American Geophysical Union, Washington D.C. pp. 135-143.

Lyons, W.B., R.M. Lent, N. Djukic, S. Maletin, V. Pujin & A.E. Carey, 1992. Geochemistry of surface waters Vojrodina, Yugoslavia. *Journal of Hydrology* 137:33-35.

Lyons, W.B., K.A. Welch, K. Neumann, J.K. Toxey, R. McArthur, C. Williams, D.M. McKnight & D. Moorhead, in press. Geochemical linkages among glaciers, streams and lakes within Taylor Valley, Antarctica. In: Priscu, J.C. (Ed.), *The McMurdo Dry Valleys, Antarctica: a cold desert ecosystem.* American Geophysical Union, Washington D.C.

Mayewski, P.A. & W.B. Lyons, 1982. Merserve Glacier ice core: reactive iron and reactive silicate concentrations. *Geophysical Research Letters* 9:190-192.

Meybeck, M., 1987. Global chemical weathering of surficial rocks estimated from river dissolved loads. *Aerican. Journal of Sci*ence 287:401-428.

Ming-hui, R., R.F. Stallard & J.M. Edmond, 1982. Major ion chemistry of some large Chinese rivers. *Nature* 298:550-553.

Mullin, J.B. & J.P. Riley, 1955. The colorimetric determination of silicate with special reference to sea and natural waters. *Analytical Chimica Acta* 12:162-175.

Peters, N.E., 1984. Evaluation of environmental factors affecting yields of major dissolved ions of streams in the United States. *USGS Water-Supply Paper* 2228.

Plummer, L.N., E.C. Prestemon & D.L. Parkhurst, 1991. An interactive code (NETHPATH) for modeling NET geochemical reactions along a flow PATH. *USGS Water Resources Investigation Report* 91-4078.

Rose, S., 1994. Major ion variation and efflux related to discharge in a mafic Piedmont province watershed. *Hydrological Processes* 8:481-496.

Ugolini, F.C., 1987. Processes and rates of weathering in cold and polar desert environments. In: Colman, S.M. & D.P. Dethier (Eds.), *Rates of Chemical Weathering of Rocks and Minerals.* Academic Press, Orlando. pp. 193-238.

Velbel, M.A., 1992. Geochemical mass balances and weathering rates in forested watersheds of the southern Blue Ridge. III. Cation budgets and the weathering rate of amphibole. *Americal Journal of Science* 292:58-78.

Ward, A. K., M. Ward & S. C. Harris, 1992. Water quality and biological communities of the Mobile River drainage, eastern Gulf of Mexico region. In: Becker, C.D. & D.A. Neitzel (Eds.), *Water Quality in North American River Systems.* Battelle Press, Columbus. pp. 277-304.

Welch, K.A., W.B. Lyons, E. Graham, K. Neumann, J.M. Thomas & D. Mikesell, 1996. Determination of major element chemistry in terrestrial waters from Antarctica by ion chromatography. *Journal of Chromatography* A 739:257-263.

White, A.F. & A.E. Blum, 1995. Effects of climate on chemical weathering in watersheds. *Geochimica et Cosmochimica Acta* 59:1729-1747.

Ecosystem Processes in Antarctic Ice-free Landscapes, Lyons, Howard-Williams & Hawes (eds)
© *1997 Balkema, Rotterdam, ISBN 90 5410 925 4*

Sources and sinks of nutrients in a polar desert stream, the Onyx River, Antarctica

C. Howard-Williams, I. Hawes & A.-M. Schwarz
National Institute of Water and Atmospheric Research Ltd, Christchurch, New Zealand

J. A. Hall
National Institute of Water and Atmospheric Research Ltd, Hamilton, New Zealand

ABSTRACT: The Onyx River flows for over 30 km from its source in the Greenwood Valley, to discharge into endorheic Lake Vanda. It is the only significant source of meltwater, and hence allochthonous nutrients, to the lake. In recent years localised climatic change has resulted in an increase in the annual discharge in the river from 2.5×10^6 m^3 (1970s and early 1980s) to 5.5×10^6 m^3 (late 1980s and early 1990s). In order to better understand the potential consequences of increased river flow to nutrient loading to the lake, we undertook a series of studies in 1993/94 to identify the sources and sinks of nutrients within the Onyx River system. Concentrations of dissolved inorganic and organic nutrients varied both with time and with distance down the river. During early flows the river water was rich in nutrients, with nitrate nitrogen reaching over 1500 mg m^{-3} and dissolved reactive phosphorous 27 mg m^{-3}. This appeared to reflect two sources, freeze concentration of nutrients in the stream bed over winter, and selective enrichment with nitrate, probably via atmospheric deposition. Another source of nutrients during early flows was leaching from freeze damaged cells in microbial mats. However during these early flows much of this water did not reach Lake Vanda, but rather was lost to lateral groundwater recharge. During the main flow period, nutrients came predominantly from glacial melt, though nitrate enrichment persisted for several weeks. Water reaching Lake Vanda during the main flow period always had low concentrations of nutrients relative to source water. Two microbially active zones were identified which appeared to be responsible for nutrient stripping. These were Lake Brownworth at the head of the river and the "Boulder Pavement", some 4 km upstream of Lake Vanda. There was no relationship between discharge and nutrient concentrations and for most of the year a linear discharge:loading curve was found. Increased flows into Lake Vanda will therefore increase the nutrient loading, but have little effect on concentrations of nutrients in lake water, at least over the range of discharge seen in this study.

1 INTRODUCTION

The Onyx River is Antarctica's longest known water course, flowing for over 30 km from its source at the Lower Wright Glacier, to discharge into endorheic Lake Vanda (Fig. 1). The Onyx River is the only significant source of meltwater to Lake Vanda and in the absence of groundwater flows, the only significant inflow. In recent years, localised climatic change has resulted in an increase in annual discharge in the river, from 2.5×10^6 m^3 in 1970 to 5.5×10^6 m^3 in 1992 (Chinn 1993). This increase has resulted in a rise in the lake's level of 10 m since 1973.

Lake Vanda is extremely oligotrophic and the upper layers of this lake are among the clearest natural waters on earth (Vincent 1981; Howard-Williams *et al.*, in press). Phytoplankton growth in the upper waters of the lake is P-limited (Vincent 1981; Priscu 1995),

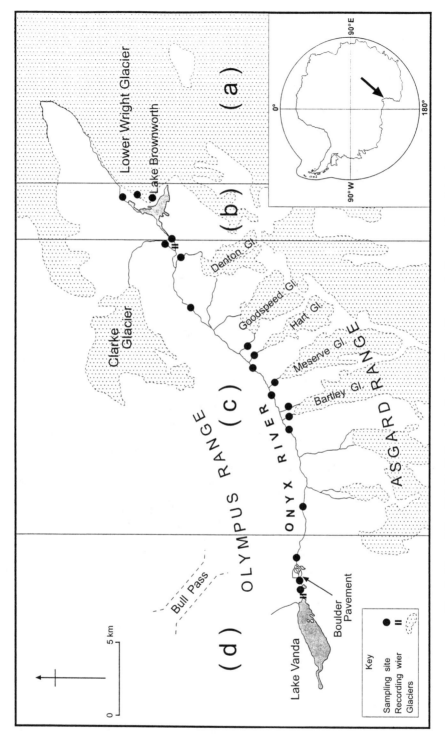

Fig. 1. Map of the Wright Valley showing the Onyx River and sites mentioned in the text. a - Lower Wright Glacier, b - Lake Brownworth and Greenwood Valley, c - Region of sands and tributary glaciers, d - Boulder Pavement, pools and rocky channels.

and consequently any increase in P loading to the lake from the Onyx River may have a significant impact on the productivity and ecology of the lake. In order to refine our understanding of how river flow affects nutrient loading, we carried out a series of experiments during 1993-94 to investigate the link between nutrient-related processes in the river and its headwaters, and the nutrient loading to Lake Vanda. In particular, we examined the role of microbial communities as nutrient sinks.

Biological uptake and transformation of nutrients occur in a number of Antarctic stream systems (Howard-Williams *et al.* 1989), and uptake of nutrients by microbial mats has previously been suggested as a major control on nutrient concentrations in the Onyx River (Canfield & Green 1985; Howard-Williams *et al.* 1986). We found that, while nutrient concentrations could be quite high in snow melt, glacier ice and in the first flows of the Onyx river over its dry river riverbed, the concentrations in river water entering Lake Vanda during the two month flow period were almost always very low. This paper describes the processes which contribute to nutrient stripping from the Onyx River, and discusses how these may be affected by ongoing changes in river discharge.

2   STUDY SITE

The main section of the Onyx River arises from Lake Brownworth, (277 m above sea level) which is in contact with the Lower Wright Glacier. The river normally flows during mid-late December through January down the 30 km channel from Lake Brownworth to Lake Vanda. Towards the end of summer in early February the river freezes, the channel drains and during winter the remaining ice in the channel ablates leaving a dry stream bed for the next year's flow.

Meltwater runs into Lake Brownworth directly from the Lower Wright Glacier, and from the Upper Onyx River, which carries melt from the Greenwood Valley to the north (Howard-Williams *et al.* 1986). For the first 1.5 km below Lake Brownworth the river has a gradient of 0.0037 m m^{-1}, and flows over areas of stable, stony substrate where there is epilithic algal growth. The river then increases in slope (to 0.0075 m m^{-1}) as it cuts a channel through the Trilogy Moraine complex. From 6 to 26 km below Lake Brownworth, the river meanders across the relatively flat, sandy floor of the valley in a series of braided channels. Mosley (1988) describes the mobility of sediments in this area of the river, and little algal material was visible on this unstable substrate, although a limited epilithon occurs where channel edges are armoured with large stable pebbles and stones.

There are two areas of conspicuous biological activity in the system. The first of these is Lake Brownworth (*ca.* 1.5 km^2 in area) which has thick cyanobacterial mats on the bed of the lake. The second is 4 km upstream of Lake Vanda, where the river broadens and flows through the so-called "Boulder Pavement". This comprises an area approximately 0.8 km wide and 1.5 km long, of large, flat boulders, through which the Onyx breaks up into a diffuse network of small streams and ponds with no clear channel. Within the streams and ponds of the Boulder Pavement there are abundant mats of algae and cyanobacteria (Howard-Williams *et al.* 1986). We recognised three visibly distinguishable mat types, all dominated by Oscillatoriaceae. These were a yellow-brown mat with abundant diatoms ("diatom mat"); a cohesive "brown mat" and a thicker cohesive "orange mat".

Below the boulder pavement, the river coalesces to fill a well-defined and relatively steep channel, which flows through shallow Lake Bull before entering Lake Vanda. The large, stable rocks of this lower section of the river bed have crusts and films of cyanobacteria, including brown mat and abundant thin, encrustations of *Gloeocapsa* (Hawes & Howard-Williams, in press).

At intervals along its length, the Onyx River receives episodic flows of meltwater from the alpine glaciers of the valley walls. During our study period, only the Clarke, Meserve and

157

Bartley Glaciers contributed to flows. Chinn (1981) estimated that tributary glacier streams contributed on average 5%, but up to 10%, of daily river flow. High variability in flows both within and between seasons is a feature of the tributaries and the Onyx River. This is a feature of other Dry Valley streams (McKnight *et al.* 1995) and may reflect their complete dependency on ice melt and the close proximity of ambient temperatures to 0°C.

The morphology of the Onyx River is discussed by Shaw & Healy (1980) and the hydrology by Chinn (1981, 1993). There are continuously-recorded flow gauging sites immediately downstream of Lake Brownworth and at Lake Vanda which have been operating since 1969 (Chinn 1981). Recordings from the Lower Wright gauge during the study periods were used in our analysis as the Lake Vanda gauge was not operating. Chinn (1981) found that flow at the Lake Vanda end of the river was 96% of flow at the Lower Wright gauge 24 h previously and we have exploited this relationship in estimating flows into Lake Vanda

3   METHODS

3.1   Water sampling

Water samples were collected at a series of sites down the river (Fig. 1) on a number of occasions between 22 November 1993 and 24 January 1994. Samples were collected into acid washed polythene bottles, and filtered (Whatman GF/F) as soon as possible after collection. They were then frozen and returned to New Zealand for analysis. Analyses for dissolved reactive phosphorus (DRP), nitrate-N (NO_3-N), ammonium-N (NH_4-N), dissolved organic N (DON) and dissolved organic P (DOP) were carried out using a Technicon autoanalyser (Downes 1988).

During the early period of flow, the gradually advancing front of water (termed the "melthead") moving down the channel was sampled on four occasions. Once river flow began to reach Lake Vanda (28 December 1993), samples were collected daily, as close as possible to local noon, at a site approximately 400 m above the lake.

Water from the hyporheic zone below the main channel, and the lateral groundwater from the sandy floor of the Wright Valley on either side of the river was sampled in a transect perpendicular to the river bank 5 km upstream of Lake Vanda. Sampling was carried out using a perforated stainless steel tube pushed into the substrate and connected to a flask which could be evacuated using a hand pump. Water was collected by suction, filtered and analysed as described above.

3.2   Nutrients in river sediments

To determine the potential contribution of nutrients from sediments, samples of bed sediments were sealed into plastic bags and frozen prior to river flow commencing. Sediment was collected as an integrated sample down to the frozen layer. On return to New Zealand, these sediments were thawed and a subsample taken for determination of water content by weighing before and after oven drying (105°C). A known volume of distilled water was then added to a second weighed subsample. These samples were mixed and after 24 h this water was recovered, filtered and analysed for dissolved nutrients. Concentrations of nutrients in the original interstitial water were then calculated as well as the nutrient content per unit dry weight of sediment.

3.3   Mat biomass and nutrient content

There are numerous measurements of nett carbon production by mat communities (Vincent &

Howard-Williams 1986; Hawes & Howard-Williams, in press). In order to estimate the nutrient requirements of these we attempted to measure the C:N:P ratio of the cyanobacteria. Live cyanobacterial trichomes were carefully extracted from the mats by using their phototactic response under low radiation, and samples of this material analysed for C, N, and P (Howard-Williams et al. 1993).

To calculate the extent of microbial mats in the Boulder Pavement, percent cover was estimated using a series of randomly oriented transects. At 10 cm intervals along these 4 m long transects, the presence of absence of mats, and the presence or absence of stream channels was recorded. A total of 750 points were recorded to calculate percent cover of cyanobacterial mats, and percent occurrence of channels.

3.4 Nutrient flux between microbial mats and overlying water

Two experiments were carried out to investigate the potential of cyanobacterial mats from the Boulder Pavement as sources, sinks or transformers of nutrients. The first experiment examined the effect of freeze-thaw cycles on nutrient exchange. The second was designed to measure their maximal uptake rates for DRP and NH_4-N

(i) Freeze-thaw
Five replicated cores of wetted mat were covered with 10 ml river water, and allowed to freeze overnight. The following morning, 50 ml water was added, and a sample taken for nutrient analysis. This was repeated, using the same mat samples, for a total of 5 cycles. Controls comprised mats which were not allowed to freeze. Cumulative release was calculated as the increase in concentration in overlying water relative to the control, and was again normalised to mat surface area.

(ii) Nutrient uptake
Five replicated cores of each mat type were placed in beakers containing 500 ml river water spiked with DRP (final concentration = 50 mg m^{-3}) and NH_4-N (final concentration = 500 mg m^3). This does not provide true estimates of field uptake rates, but gives estimates of uptake under nutrient saturated, still water conditions. Samples (50 ml) were removed after 1, 4, 8, and 12 h and uptake was determined from the rate of nutrient depletion as described in Vincent & Howard-Williams (1986).

4 RESULTS

4.1 River flow

In the 1993/94 summer, flowing water from Lake Brownworth was first recorded on 22 November (Fig. 2). Following a short peak in flows in late November, discharge declined to low values until late December. Flows increased rapidly in late December and maximum flows were recorded in mid January. Flow into Lake Vanda was not recorded until 29 December 1993. Integration of the discharge curve shows that 1.3×10^5 m^3 of water had flowed over the Lake Brownworth weir prior to flow commencing at Lake Vanda. Since this water did not reach Lake Vanda, it may be reasonable to assume that it was lost to evaporation, to the hyporheos and to recharge of lateral ground water. This will involve spreading through the sands above the permafrost on either side of the river channel (c.f. Conovitz et al., in press). A total of 2×10^6 m^3 of water flowed over the Lake Brownworth gauge after water began to flow into Lake Vanda, representing 96% of the year's flow. Maximum flows were recorded in mid January. This was a year of low flows both in the Wright Valley (c.f. Chinn 1993) and the Taylor Valley (Conovitz et al., in press).

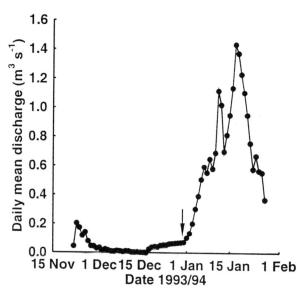

Fig. 2. Mean daily discharge for the Onyx River at the Lower Wright weir, 1993/94. Arrow marks the time when first flows occurred over the Vanda weir.

Table 1. Concentrations of dissolved nutrients in water at the melthead (final 10 cm) of the Onyx River at first flows, in tributaries joining the Onyx River, in a snow bank on the river edge and the river itself on 19 January 1994. Each value is the mean of two replicates. Tributaries are in downstream order. nd - not detectable. Lower Wright - streams running off the glacier into the Onyx River above Lake Brownworth (early and late season). - no data.

		Date	DRP mg m^{-3}	NH$_4$-N mg m^{-3}	NO$_3$-N mg m^{-3}	DOP mg m^{-3}	DON mg m^{-3}
Melthead		24 November 1993	6	26	1668	3	202
		2 January 1993	27	156	967	21	375
		2 January 1993	5	34	362	6	443
		3 December 1993	3	50	533	6	632
Tributary	Lower Wright	12 December 1993	88	7	282	nd	48
	Lower Wright	18 January 1994	1.6	2	45	nd	23
	Clarke		12	5	190	1.2	2.4
	Hart		1.7	6	188	0.4	17
	Meserve		3.2	6	328	0.45	nd
	Bartley West		2.6	4	308	0.5	15
	Bartley East		2.9	7	163	0.5	16
Snowbank	Snowbank	12 December 1994	4	-	647	-	-
Main river	Onyx River	19 January 1994	1.9	4	31	0.2	12

4.2   Nutrient concentrations

(i)   First flows

Longitudinal profiles of nutrient concentrations down the river during the early period of river flow (29 November 1993 and 12 December 1993) showed a downstream accumulation of nutrients and conductivity, particularly close to the melthead as it moved down the channel. The high concentration of nutrients in the water close to the melthead (Table 1) was paralleled by relatively high conductivity (120 to 140 µS cm^{-1}). The melthead was sampled four times and

160

on each occasion high concentrations of organic and inorganic N and P were found (Table 1). For instance, DRP was 2 to 3 times and NO₃-N up to ten times that of mid-season flows.

(ii)   Mid-season flows
A profile of nutrient concentrations down the river in mid-January showed elevated concentrations within the reach of the river influenced by the tributary glaciers (Fig. 3). Concentrations of inorganic nitrogen increased five-fold, and DRP by a factor of six. DON doubled in concentration over this reach. These tributaries were particularly rich in nitrate-N, with concentrations much higher than in Onyx River water at the same time (Table 1). Fewer analyses of glacial melt from the Lower Wright Glacier are available from later in the season but these show that concentrations had declined markedly by 18 January (Table 1). Nutrients in a residual snow bank adjacent to the river in December 1993 were also high (Table 1). In the upper reaches of the river (*ca.* 30 km from Lake Vanda) where it left Lake Brownworth, and at the lower reaches of the river, nutrient concentrations declined (Fig. 3).

Changing processes in the river over time are illustrated by the ratios of DRP: conductivity and NO₃-N:conductivity between early and mid-season flows (Fig. 4). In the early season downstream profile there was a consistent decrease in the ratios of both NO₃-N and DRP to conductivity with distance from Lake Vanda (Fig. 4a). In the second profile in mid-January (Fig. 4b) the downstream patterns were similar until *ca.* 12 km above Lake Vanda. Part of this accumulation of nutrients was the result of the inflows of the small but relatively nutrient-rich tributaries (Table 1). The ratios of both NO₃-N and DRP to conductivity decreased markedly below Lake Brownworth and the section of the river near Lake Vanda, indicating a selective removal of these ions from solution, relative to the total salt at this time.

(iii)   Contributions from stream beds and lateral groundwater
Concentrations of nutrients, particularly NO₃-N, were very high in the interstitial water in the Onyx River sediments prior to first flows (Table 2). The extent to which this was due to atmospheric deposition or freeze-out concentration as a result of ice formation and then sublimation is unknown. However, they clearly represented a potential source of nutrients and

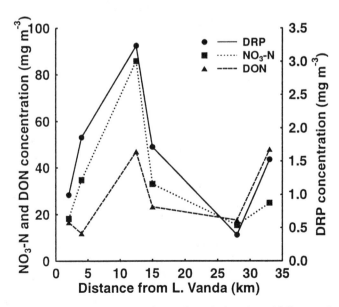

Fig. 3 Profiles of NO₃-N, DON and DRP down river during the mid-flow period (18 January 1994).

Table 2. Concentrations of Nitrate-N and DRP in interstitial water in samples from the bed of the Onyx River taken prior to the commencement of flow. Most samples were taken at the confluence of the river and a tributary stream (see Fig. 1). Each value is the mean of two replicates, all values are mg m^{-3}.

Location	NO$_3$-N	DRP
Bull Pass	1914	49
Bartley confluence	554	17
Meserve confluence	1729	53
Hart confluence	4915	14
Goodspeed confluence	1088	6
Denton confluence	243	17

help to explain the high nutrient concentrations seen at the melthead in early flows (Table 1).

The first flows down the river were unlikely to reach Lake Vanda since most will have been lost to recharge of the hyporheos and lateral groundwater. Our samples of the hyporheos and lateral groundwater show a strong gradient of nutrients with increasing distance from the river (Fig. 5). Conductivity followed the same pattern. Nutrient concentrations in the hyporheic water underlying the river were higher for both DRP and NO$_3$-N than the river water itself. The

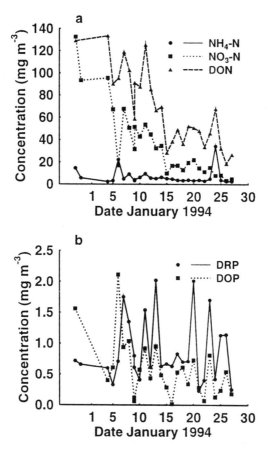

Fig. 4. Concentrations of NO$_3$-N and DRP normalised to conductivity. a - early season (22 November 1993) before the Boulder Pavement mats were fully re-established, b - mid-season (18 January 1994) after development of the Boulder Pavement mats.

162

relationship between the concentration of NO$_3$-N and conductivity in the hyporheos samples was approximately 4:1, similar to that seen in the early river profiles (see Fig. 4a) at this site.

(iv)   Water entering Lake Vanda
Water entering Lake Vanda showed a seasonal pattern in dissolved nutrients consistent with the above observations (Figs. 6a & 6b). Early flows contained high concentrations of NO$_3$-N, which declined gradually over the summer. DON was slightly elevated during early season flows, but was relatively constant during the second half. Ammonium, DRP and DOP concentrations were consistently low (Figs. 6a & 6b), with DRP not exceeding 2 mg m^{-3}.

4.3   Nutrient loadings

There was no clear relationship between discharge and concentration of NH$_4$-N, DRP, DON or DOP but there was a tendency for consistently high NO$_3$-N concentrations to occur during early flow. This resulted in a tendency for NO$_3$-N concentration to decline with increasing discharge. A consequence of this was that the total loading of DRP and NH$_4$-N into Lake Vanda increased steadily with increasing discharge (Figs. 7a & 7b), whereas NO$_3$-N loading showed a more complex pattern with discharge (Fig. 7c). Loading of NO$_3$-N showed two phases within the January flow period, an early phase of high loading between 1 and 15 January when concentrations were relatively high, and a second phase after 15 January when discharge was higher, but loading lower due to low NO$_3$-N concentrations.

From integration of nutrient load *vs.* time curves, we estimated loading of inorganic N and P to the lake over the measured period to be approximately 51 kg and 0.9 kg respectively, of which early season flows (1 to 15 January) comprised 51% and 44%.

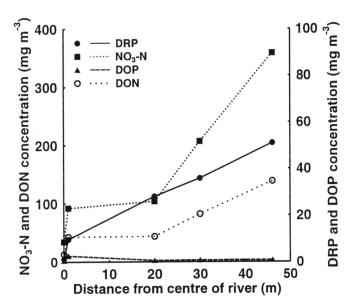

Fig. 5. Fifty metre transect perpendicular to the Onyx River channel showing concentrations of nutrients and conductivity in the groundwaters and channel hyporheos.

163

The Boulder Pavement is an area of active nutrient transformation (Howard-Williams *et al.* 1986). Transects across the Boulder Pavement showed that approximately 33% of the total area comprised wetted channels, and 36% was covered with microbial mats dominated by cyanobacteria. Almost all of this material was categorised as Brown mat (*c.f.* Hawes &

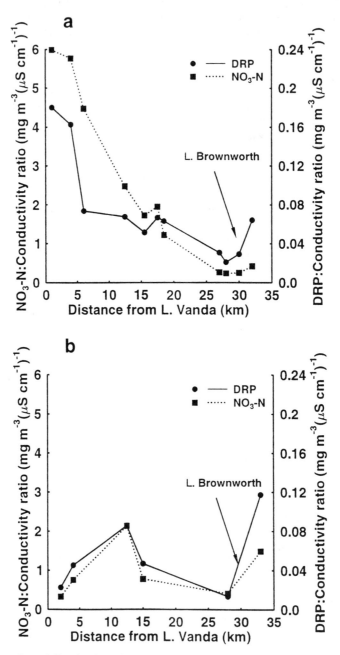

Fig. 6. Concentration of dissolved nutrients in Onyx River water entering Lake Vanda between 26 December and 24 January. a - Nitrogen species, b - Phosphorus species

Howard-Williams, in press), and occurred in the channels. The overall area of the boulder pavement was estimated at 4.95×10^5 m^2, giving a mat area of 1.8×10^5 m^2. Our data from above and below this area indicated that during early flows (30 December 1993 was the second day of flow at this site), it was a source of nutrients, with the exceptions of DRP (Table 3). Subsequently, the Boulder Pavement consistently removed DRP and NO$_3$-N from the river, though was neutral with respect to NH$_4$-N and DOP (Table 3; 9 January 1994 and 18 January 1994). In contrast, concentrations of DON increased across the pavement.

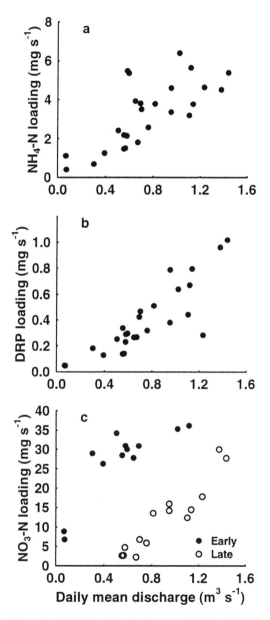

Fig. 7. Nutrient loading plots for Lake Vanda, January 1994. a - NH$_4$-N; b - DRP; c - NO$_3$-N. In c; closed circles - 1 to 15 January, open circles - 16 to 31 January.

165

Table 3. Concentrations of dissolved nutrients upstream and downstream of the Boulder Pavement.
nd - not detectable

Date	Position	DRP	NH$_4$-N	NO$_3$-N	DOP	DON	
30 December 1993	above	2.0	4.3	95.9	nd	1.3	early flow
	below	0.7	9.0	314.5	0.9	nd	
9 January 1994	above	1.3	2.5	45.6	0.2	12.9	mid-flow
	below	0.8	3.2	31.6	0.1	23.8	
18 January 1994	above	1.9	3.8	31.0	0.2	11.9	mid-flow
	below	1.0	3.8	14.4	0.1	16.6	

4.5 Nutrient release and uptake

(i) Freeze-thaw release

Freeze-thaw cycles caused a release of nutrients from the cyanobacterial mats (Fig. 8). Very little DOP and NO$_3$-N was released. However, rates of approximately 0.1 µg N or P cm^{-2} cycle^{-1} were released as NH$_4$-N and DRP, while over 10 times this amount of DON was released. Similar amounts of nutrient were released with each successive cycle for all nutrients except DON. DON release increased with each cycle. Freeze-thaw cycles in microbial mats therefore contribute significant amounts of DON, and provide a limited source of NH$_4$-N and DRP to the river.

(ii) Nutrient uptake

A consistent range of nutrient saturated uptake rates for each mat type (Table 4) show that DRP uptake (µg cm^{-2} h^{-1}) ranged from 0.07 (orange mat) to 0.12 (brown mat) and NH$_4$-N uptake ranged from 0.23 (brown and diatom mats) to 0.46 (orange mat). The NH$_4$-N uptake rates were very similar to NO$_3$-N uptake rates recorded for nearby Canada Stream mats by Howard-Williams et al. (1989) of 0.37 µg cm^{-2} h^{-1}.

5 DISCUSSION

The sources of nutrients to Dry Valley streams were identified in Howard-Williams et al. (1986) as nutrients in glacier ice and nutrients in streambed sediments. In the Onyx River there are several sources of nutrients which are illustrated in this present study. These include glacier melt, snow melt, stream bed sediment release, stream bed weathering, groundwater and hyporheic exchanges, and freeze-thaw effects on microbial mats. Early season glacier melt from the Lower Wright Glacier provides six times the concentration of NO$_3$-N and DRP, and twice the concentration of DON than late season melt (Table 1).

Howard-Williams et al. (1986) speculated that high early season nutrient flows from glaciers may be caused by concentration of nutrients on glacier surfaces by freeze-thaw cycles

Table 4. Substrate-saturated rates of uptake of DRP and NH$_4$-N by algal and cyanobacterial mats from the Onyx River. All values are means ± sd of three replicates, in µg cm^{-2} h^{-1}

Date	Mat type	DRP	NH$_4$-N
12 December 1993	brown	0.12±0.1	0.38±0.02
12 January 1994	brown	0.10±0.01	0.23±0.17
12 January 1994	orange	0.07±0.03	0.46±0.21
12 January 1994	diatom	0.8±0.04	0.23±0.22

early in the season, and from wind-blown particulates from adjacent land surfaces and from the sea. It is noteworthy that snow drifts have relatively high nutrient concentrations as shown in Table 1. Nitrate in elevated concentrations has been measured in snow at several sites in the Antarctic interior (Legrand *et al.* 1984). The chloride to nitrate ratio of 5.45 in inland nitrate-rich melt waters (Vincent & Howard-Williams 1994) is the same as that in the snow suggesting snow as the source of nitrate.

The complex pattern of the nitrate loading *vs* discharge plot (Fig. 7c) can be explained by melt taking place in two phases: an early season phase when the winter snow melts, providing a high nutrient pulse on the glacier surfaces; and a later season phase when the main core of the glacier provides lower nutrient meltwater. Concentrations of nutrients in glacier ice itself have been shown to be substantially lower than those in early season melting icicles (Howard-Williams *et al.* 1986) and the difference shown in early season and late season melt in the Lower Wright glacier (Table 1) is consistent with this. Another potential source of nitrate that needs further investigation is that of dry atmospheric deposition of nitrate derived from stratospheric denitrification or electrostatic effects in blowing snow (Legrand & Kirchner 1990, M. Harvey, pers. comm.).

Streambed sediments provide a source of nutrients to early season flow (Tables 2 & 4). These are probably derived from residual salts frozen out in late summer from the previous year's flow and from rock weathering during winter. Green & Canfield (1984) showed how major ions increased downstream in the Onyx River due to the weathering of valley rocks and saline soils. The river chemistry showed a downstream trend from atmospheric dominance of salts in the upper reaches to a rock dominance in the lower reaches. Weathering, particularly during the winter months, may provide nitrogen and phosphorus to the river from stream bed sediments and also from the lateral groundwaters. The presence of two flow gauges on the Onyx River providing the lag time between flows over the 30 km stretch of river, allows a first approximation to be made of the amount of water supplied by the river to the lateral groundwater pool. In addition to this there may well be a contribution to the groundwaters from melting permafrost on the sides of the valley. The lateral groundwaters between the permafrost and the surface have elevated solute concentrations including nitrate and DRP with a gradient

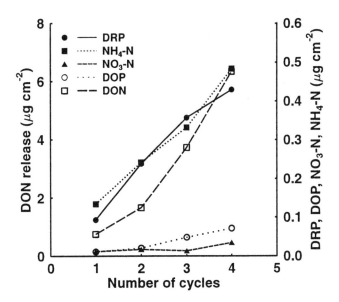

Fig. 8. Cumulative release of nutrients from Onyx River cyanobacterial mat during successive freeze-thaw cycles.

from the lateral sands to the river (Fig. 5). This reflects an exchange with the waters of the river channel (McKnight & Andrews 1993). The exchange will be enhanced by daily freeze-thaw cycles when the river level drops and rises again. In flat wide streams such as the reaches of the Onyx River above the Boulder Pavement, the interaction between the hyporheic zone and lateral groundwaters will be significant particularly where diel variations in water level occur (Conovitz et al., in press).

The increase in nutrient concentrations relative to conductivity between lake Brownworth and the Boulder Pavement (Fig. 4) is therefore likely to be due to a combination of weathering sources enhanced by groundwater and hyporheic exchanges, and flows from the alpine tributary glaciers. The tributary glacier flows maintained high nutrient concentrations throughout the season (Table 1). As the flows were small they were probably derived from snowmelt rather than the glacier ice core. The tributary glaciers provide less than 10% of total flow at Vanda weir (Chinn 1981), but concentrations of nitrate in their melt streams were up to 10 times those in the main river channel and DRP and DON concentrations were more than double the channel values. For many nutrients, the tributary glaciers were the major source, as indicated in Table 5.

The major sink for nutrients was the lower river, and particularly Boulder Pavement (Table 5). This area of the river with its rich microbial mats acts as a nutrient filter, analogous to wetland filters in temperate latitudes. There appear to be both nutrient removal mechanisms and nutrient transformations. An example of the efficiency of the Boulder Pavement to remove nutrients is shown by a calculation for a day of high flow in the river. Over the course of the two days of maximal flow (17 and 18 January 1994), flow across the Boulder Pavement was estimated at 125 000 m^3. At this time the daily average reduction in nitrate and DRP concentrations was 25 mg m^{-3} and 1.5 mg m^{-3} respectively. The nutrient sink on these days was therefore 3.1 kg N and 0.19 kg P . With a total mat area of 1.8×10^5 m^2, this equates to aerial uptake rates of 0.1 μg N and 0.01 μg P cm^{-2} h^{-1}. These values are of similar magnitudes to those calculated from nutrient uptake experiments on different mat types from the Boulder Pavement (Table 4). The data in Table 4 were obtained from nutrient saturated rates of uptake. With the relatively low levels in the river we may expect that actual rates will be less than those shown

Table 5. Sources and sinks for inorganic and organic N and P in the Onyx River in early and late parts of the flow cycle. Fluxes g day^{-1} are calculated from concentration changes and discharge along the river. On 1 December 1993, the river did not reach Lake Vanda and discharge at sampling sites below Lake Brownworth were estimated from cross-sectional area and velocity. + indicates a source to the river, - a sink.

Source or sink	Early flow (1 December 1993)				Peak flow (16 January 1994)			
	DRP	DIN	DOP	DON	DRP	DIN	DOP	DON
Tributaries draining lower Wright Glacier	+50	+2300	+10	+320	+180	+3000	+1200	+5700
Lake Brownworth	-40	-2100	-3	+400	-130	-1600	-1200	-3700
Soils in mid river (upper desert zone)	+1	+440	0	-400	0	0	0	0
Valley wall tributary glaciers	0	0	0	0	+340	+8300	+48	+700
Lower river including Boulder Pavement	+10	+100	+10	+60	-300	-7800	+12	+720
Loss to lateral soils	-21	-740	-3	-380	0	0	0	0
Net flux to Lake Vanda	no flow				90	1900	60	3420

in Table 4, but it is apparent that the mats alone are capable of accounting for the observed loss in N and P. As in a temperate wetland system there are significant transformations of nutrients, the most obvious being the change from the uptake of nitrate to the export of DON. DON is exported during the main flow period (Table 3) and is released from the mat in large amounts during freeze-thaw cycles (Fig. 8). Small releases of DOP were recorded during freeze-thaw cycles (Fig 8). A transformation from nitrate and DRP to particulate N and P and an export in this form to Lake Vanda is also likely. Examples of such transformations for microbially rich Canada Stream (previously referred to as Fryxell Stream) were shown in Howard-Williams *et al.* (1989).

Modelling work by Canfield & Green (1985) suggested that bottom water removal rates of N (by denitrification) and P (by mineral formation) in Lake Vanda were comparable to the input rates from the Onyx River indicating that an overall steady state exists with respect to nutrient levels in the Lake. Their data showed a net annual flux to the lake of 17 and 2.5 kg of N and P respectively. They acknowledged, however, that they had not taken into account potential uptake by the benthic cyanobacterial mats within the lake itself. Our data for the 1993/94 year show that the flux of nitrate was 51 kg, much higher than that reported by Canfield and Green (1983), and that without the Boulder Pavement the fluxes of nitrate and DRP would be of the order of five times these values, considerably more than the calculated removal rates in the lake.

At this stage it is tempting to hypothesise that the low phytoplankton biomass in Lake Vanda, and the consequent high water clarity, is controlled by the large biological wetland filter provided by the Boulder Pavement which significantly reduces loads of inorganic nutrients to the lake.

6   ACKNOWLEDGEMENTS

This work was funded by the Foundation for Research Science and Technology under contract Number CO1601.

We thank Faye Richards for skilled analytical assistance and Carol Whaitiri for word processing. The work benefited from discussions with Dr. Warwick Vincent and Dr. Bill Green, and we thank Dr. Brian Sorrell for critical review of the manuscript.

REFERENCES

Canfield, D.E. & W.J. Green, 1985. The cycling of nutrients in a closed basin antarctic lake: Lake Vanda. *Biogeochemistry* 1:233-256

Chinn, T.J.H., 1981. Hydrology and climate in the Ross Sea area. *Journal of the Royal Society of New Zealand* 11:373-386

Chinn, T.J.H., 1993. Physical hydrology of the Dry Valley lakes. In: Green, W.J. & E.I. Friedmann (Eds.), *Physical and biogeochemical processes in Antarctic lakes. Antarctic Research Series* 59. American Geophysical Union, Washington D.C. pp. 1-51.

Conovitz, P.A., D.M. McKnight, L.M. MacDonald & A. Fountain, in press. Hydrologic processes influencing streamflow variation in Fryxell basin, Antarctica. In: Priscu, J.C. (Ed.), *The McMurdo Dry Valleys, Antarctica: a cold desert ecosystem.* American Geophysical Union, Washington D.C.

Downes, M.T., 1988. Taupo Research Laboratory Chemical Methods Manual. *Department of Scientific and Industrial Research, TRL Report* 102:1-70

Green, W.J. & D.E. Canfield, 1984. Geochemistry of the Onyx River (Wright Valley, Antarctica) and its role in the chemical evolution of Lake Vanda. *Geochimica et Cosmochimica Acta* 48:2457-2467

Hawes, I. & C. Howard-Williams, in press. Primary production processes in the streams of the McMurdo Dry Valleys, Antarctica. In: Priscu, J.C. (Ed.) *The McMurdo Dry Valleys, Antarctica: a cold desert ecosystem.* American Geophysical Union, Washington, D.C.

Howard-Williams, C., J.C. Priscu & W.F. Vincent, 1989. Nitrogen dynamics in two Antarctic streams. *Hydrobiologia* 172:51-61.

Howard-Williams, C, C.L. Vincent, P.A. Broady & W.F. Vincent, 1986. Antarctic stream ecosystems Variability in environmental properties and algal community structure. *Internationale Revue der gesamten Hydrobiologie* 71:511-544.

Howard-Williams, C., R.D. Pridmore, M.T. Downes & W.F. Vincent, 1993. Microbial biomass, photosynthesis and chlorophyll *a* related pigments in the ponds of the McMurdo Ice Shelf, Antarctica. *Antarctic Science* 1:125-131

Howard-Williams, C., A-M. Schwarz, I. Hawes. & J.C. Priscu, in press. Optical properties of the McMurdo Dry Valley lakes, Antarctica. In: Priscu, J.C. (Ed.). *The McMurdo Dry Valleys, Antarctica: a cold desert ecosystem.* American Geophysical Union, Washington D.C.

Legrand, M.R. & S. Kirchner, 1990. Origins and variations of nitrate in South Polar precipitation. *Journal of Geophysical Research* 95 D4: 3493-3507.

Legrand, M.R., R.J. De Angelis & R.J. Delmas, 1984. Ion chromatographic detection of common ions at ultratrace levels in Antarctic snow and ice. *Analytica Chemica Acta* 156:181-192

McKnight, D.M. & N. Andrews, 1993. Hydrologic and geochemical processes at the stream-lake interface in a permanently ice-covered lake in the McMurdo Dry Valleys, Antarctica. *Verhandlungen. Internationale Vereiningung fuer Theoretische und Angewandte Limnologie* 25:957-959

McKnight, D.M., H.R. House & P. von Guerard, in press. Streamflow measurements in Taylor Valley. *United States Antarctic Journal.*

Mosley, M.P., 1988. Bedload transport and sediment yield in the Onyx River, Antarctica. *Earth Surface Processes and Landforms* 13:511-567.

Priscu, J.C., 1995. Phytoplankton nutrient deficiency in lakes of the McMurdo Dry Valleys, Antarctica. *Freshwater Biology* 34:215-227.

Shaw, J. & T.R. Healy, 1980. Morphology of the Onyx River system, McMurdo Sound region, Antarctica. *New Zealand Journal of Geology and Geophysics* 23:223-238

Vincent, W.F., 1981. Production strategies in Antarctic inland waters:phytoplankton ecophysiology in a permanently ice-covered lake. *Ecology* 6:1215-1224

Vincent, W.F. & C. Howard-Williams, 1986. Antarctic stream ecosystems: physiological ecology of a blue-green agal epilithon. *Freshwater Biology* 16:219-233

Vincent, W.F. & C. Howard-Williams, 1994. Nitrate rich inland waters of the Ross Ice Shelf region, Antarctica. *Antarctic Science* 6:39-346

Ecosystem Processes in Antarctic Ice-free Landscapes, Lyons, Howard-Williams & Hawes (eds)
© *1997 Balkema, Rotterdam, ISBN 90 5410 925 4*

Species composition and primary production of algal communities in Dry Valley streams in Antarctica: Examination of the functional role of biodiversity

Dev K. Niyogi, Cathy M. Tate & Diane M. McKnight
US Geological Survey, Denver, Colo., USA

John H. Duff
US Geological Survey, Menlo Park, Calif., USA

Alex S. Alger
Department of Biology, University of Michigan, Ann Arbor, Mich., USA

ABSTRACT: Species composition and primary production of different types of algal mats were quantified in three streams in Lake Fryxell Basin, Taylor Valley, Antarctica. These data can be used to examine relationships between species diversity and ecosystem processes, such as primary production. At the scale of this study, the data support the redundancy hypothesis, in that monospecific communities (algal mats composed of *Nostoc* sp.) are at least as productive as more diverse algal mats. Higher species diversity may, however, be interpreted differently at other spatial and temporal scales. Long-term monitoring of Antarctic streams should provide more information on the link between ecosystem structure and function.

1 INTRODUCTION

The functional role of biodiversity in ecosystems remains largely unknown. Only recently, the link between species diversity and rates of ecosystem processes, such as primary production, has been examined in environmental chambers (Naeem *et al.* 1994) and in natural ecosystems (Tilman *et al.* 1996). In this paper, we examine the link between species diversity and primary production for algal communities in streams of the McMurdo Dry Valleys region of southern Victoria Land, Antarctica.

Lawton (1994) outlined several possible relationships between species richness and the process rate for a given ecosystem function. The first possibility, termed the "rivet" hypothesis, suggests that a given process rate will be higher when more species are present. The main alternate hypothesis, the "redundancy" hypothesis, predicts that only one to a few species are required to perform a process at a certain rate, and that more species do not increase this rate. The other alternatives, the "idiosyncratic" and null hypotheses, predict that the relationship between species number and the process rate is unpredictable or not significant.

The Dry Valleys region is an extreme cold desert environment, caused by low temperatures and low precipitation. In spite of this, microbial life exists in soils, lakes, streams, and even in porous rocks (Vincent 1988). The streams of the Dry Valleys have intermittent flow regimes, draining glaciers and snowfields during the austral summer for periods of up to ten weeks. An advantage of studying extreme environments is that they often present a natural gradient in species diversity, including habitats with only one or a few species. Dry Valley streams have a variety of algal communities, including some with a diverse assemblage of species, and others dominated by one species (Vincent & Howard-Williams 1986; Alger *et al.*, in press; McKnight *et al.*, in press).

This paper presents data on species diversity and primary production from three Dry Valley streams that are used to examine the relationship between ecosystem structure and function. We compare rates of primary production between diverse and monospecific algal communities to determine if species diversity affects this ecological process in these streams. The data presented in this paper are from an ongoing, long-term project designed to document ecological processes at a variety of spatial and temporal scales.

2   SITE DESCRIPTION

The McMurdo Dry Valleys of southern Victoria Land, Antarctica, are the largest of the desert oases found along the coast of Antarctica. Air temperatures range from -60°C in winter to +5°C during summer. Precipitation is less than 10 cm in most years. Taylor Valley is located in this region between the Asgard Range and the Kukri Hills, and has three large perennially ice-covered lakes: Lake Bonney, Lake Hoare, and Lake Fryxell.

This study examined three streams in the Lake Fryxell basin (Fig. 1), located at approximately 77°37'S 163°10'W: Canada Stream, Von Guerard Stream, and the Relict Channel. The Relict Channel is a stream that has been reactivated with water diverted from Von Guerard stream as part of an experiment on stream colonization and geomorphological processes. Canada Stream and Von Guerard Stream are two of the larger streams in Taylor Valley, with maximum flows of about 500 l s^{-1} (Von Guerard et al. 1994). Mean stream temperatures are approximately 4°C, with maximum values about 10°C (Von Guerard et al. 1994).

Several distinct algal communities are observed in the streams, and can be classified based on color (Alger et al., in press). Orange-colored algal mats predominate in the channel center, which has flowing water for most of the period of flow. These communities generally have a diverse assemblage of Oscillatoriales present, and have been called *Phormidium* mats by others (Vincent 1988). Other species, however, often including chlorophytes and diatoms, are present in these communities. Black-colored algal mats are often found in streams along the edge of the channels. These mats are composed almost exclusively of the cyanobacterium genus

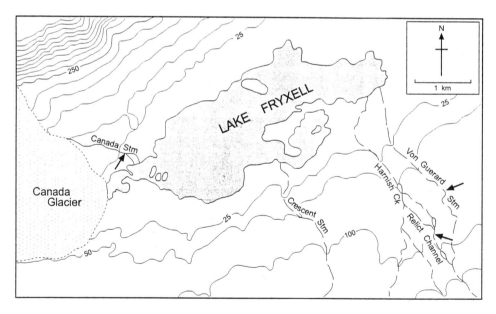

Fig. 1. Lake Fryxell basin showing three streams from this study: Canada Stream, Von Guerard Stream, and Relict Channel. Arrows indicate sampling sites along the three streams. Contour interval is 50 m, with an additional contour at 25 m.

172

Nostoc. Finally, some streams in the region, such as Canada Stream, have filamentous growths of *Prasiola*, a chlorophyte, growing in the main channel; these algae can be classified as green-colored mats.

3   METHODS

Different algal mat communities (orange, black, and, if present, green) were sampled in the three study streams during the austral summer of 1995/96 (December to January). The overall experimental design was to compare rates of primary production between diverse (orange-colored) algal mats and unialgal (black- and green-colored) algal mats in each of the three streams.

Primary production of the algal communities was measured with chambers. Algal samples were collected from areas in the streams where a particular mat type dominated. Usually, orange-colored and green-colored mats were found in the mid-channel while black-colored mats were found along stream edges or in areas where the channel widened. A cork borer was used to obtain a known area of algal mat, which was then placed into a small vial (approximately 25-ml) for incubation. Vials were filled with stream water and sealed prior to incubation. Light and dark replicate vials were run during each experiment. Light was measured with a quantum sensor, and was relatively constant during the primary production measurements. Incubations were only done during sunny weather, and as light usually exceeded 500 µmol photons m^{-2} s^{-1}, we considered algal photosynthesis to be light-saturated. Samples were incubated in the stream or in a water bath at ambient stream temperature for 4 to 8 h. Changes in dissolved oxygen over time were measured and used to calculate rates of net primary production (light vials) and respiration (dark vials).

After incubation, mat samples were frozen for chlorophyll analysis or ash-free dry mass determination. Chlorophyll content of mat samples was estimated with an acetone extraction and acidification step for pheophytin correction (Strickland & Parsons 1972). Additional samples were taken from the different mat communities and fixed in a 5% formalin solution. Algae in these samples were later identified and quantified as percent community composition (in terms of biomass estimates from biovolume calculations). Alger *et al.* (in press) and McKnight *et al.* (in press) present more detailed descriptions of the algal identification procedures.

Means and standard errors of primary production rates were calculated for each algal mat type from each stream. Primary production results are based on 10 to 36 replicates for areal primary production rates and 4 to 19 replicates for chlorophyll-specific production rates. Percent community composition data are based on 3 to 5 replicate samples for each algal mat type from each stream.

4   RESULTS

Figs. 2 through 4 present the species composition data for the three types of algal mats from the three streams of this study. Orange-colored algal mats generally had a diverse assemblage of Oscillatoriales as well as other cyanobacteria, chlorophytes, and diatoms. Diatoms were the dominant taxon in orange-colored mats from Von Guerard Stream (Fig. 2). *Nostoc* sp. was common in orange-colored mats from Von Guerard Stream and the Relict Channel. Species composition was variable among replicate samples, as indicated by the large standard error bars in Fig. 2. Different species were dominant in different replicates (*c.f.* Alger *et al.*, in press). The dominant taxon in the orange-colored mats never exceeded 40% of the mean species biomass composition.

Black-colored algal mats were composed of greater than 90% *Nostoc* sp. in all three

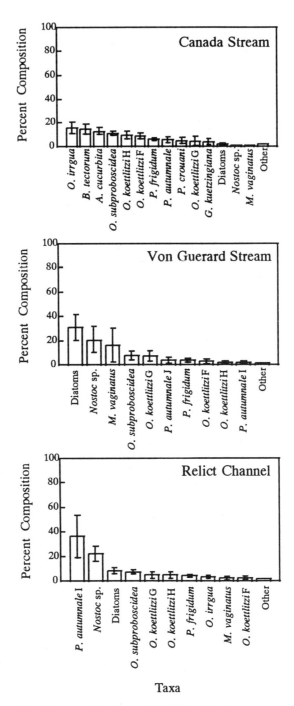

Fig. 2. Composition (percent of community biomass) of orange-coloured algal mats from Canada Stream, Von Guerard Stream, and the Relict Channel. Genus abbreviations are as follows: *O - Oscillatoria, P - Phormidium, M - Microcoleus, G - Gleocapsa, A - Actinotaenium, B - Binuclearia*. Letters after species' name represent different morphotypes (see Alger *et al.*, in press for descriptions of algal identifications). Values are means of 3 to 5 samples. Error bars represent ±1 standard error of the mean.

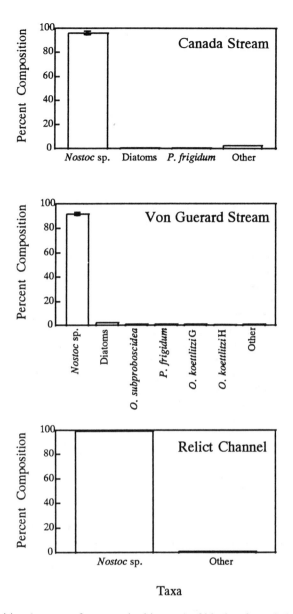

Fig. 3. Composition (percent of community biomass) of black-coloured algal mats from Canada Stream, Von Guerard Stream, and the Relict Channel. Genus abbreviations are as follows: *O* - *Oscillatoria*, *P* - *Phormidium*. Letters after species' name represent different morphotypes (see Alger *et al.*, in press for descriptions of algal identifications). Values are means of 3 to 5 samples. Error bars represent ±1 standard error of the mean.

streams (Fig. 3) . The remaining portion of the community was composed of Oscillatoriales and diatoms, and was similar to the community found in the orange-colored mats. Green-colored algal communities from Canada Stream were similarly composed almost exclusively of *Prasiola crispa* (mean = 97% species composition; Fig. 4). Other species in green-colored mats included cyanobacteria, diatoms, and another chlorophyte, *Actinotaenium cucurbita*.

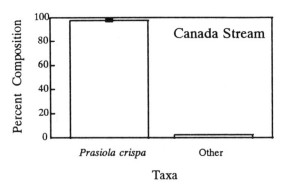

Fig. 4. Composition (percent biomass of green-coloured algal mats from Canada Stream. Values are means of 3 samples. Error bars represent ±1 standard error of the mean.

Primary production data for the algal communities in the three streams are presented in Fig. 5. The black-colored algal mats from the Relict Channel had the highest mean net primary production rates (on an areal basis), which were significantly higher than the orange-colored mats from the same stream. Similarly, black-colored algal mats had higher production rates than orange-colored mats in Canada Stream. Green-colored mats of *Prasiola* had intermediate rates. The black- and orange-colored mats from Von Guerard Stream had similar primary production estimates. Areal respiration rates were not significantly different between the black- and orange-colored mats from any of the three streams. Mean respiration rates varied from 83 mg O_2 m^{-2} h^{-1} for Canada Stream orange-colored algae, to 158 mg O_2 m^{-2} h^{-1} for Von Guerard Stream black-colored algae. Gross primary production rates (not shown) showed a similar trend to the net primary production rates presented in Fig. 5.

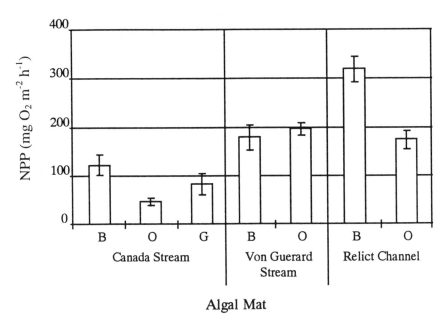

Fig. 5. Areal net primary production (NPP) rates for algal mats from Canada Stream, Von Guerard Stream, and the Relict Channel. mat types are as follows: B - black-coloured, O - orange-coloured, G - green-coloured. Values are means of 10 to 36 samples. Error bars represent ±1 standard error of the mean.

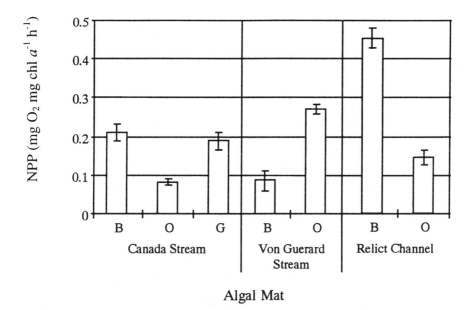

Algal Mat

Fig. 6. Net primary production (NPP) rates per unit chlorophyll values for algal mats from Canada Stream, Von Guerard Stream, and the Relict Channel. Mat types are as follows: B - black-coloured, O - orange-coloured, G - green-coloured. Values are means of 4 to 19 samples. Error bars represent ±1 standard error of the mean.

The black-colored (*Nostoc*) mats had even higher chlorophyll-specific production rates, relative to the orange-colored algal mats, in Canada Stream and the Relict Channel (Fig. 6). The trend was reversed in Von Guerard Stream, where orange-colored algal mats had a higher net primary production rate per unit chlorophyll. Production rates per ash-free dry mass (not shown) yielded similar results to the chlorophyll-specific production rates. Mean values ranged from 0.16 mg O_2 (g afdm)$^{-1}$ h^{-1} for Canada Stream orange-colored algae to 0.84 mg O_2 (g afdm)$^{-1}$ h^{-1} for Relict Channel black-colored algae.

5   DISCUSSION

The primary production rates measured for these three streams are similar to those previously measured in Canada Stream (Howard-Williams & Vincent 1989), other streams in the Taylor Valley (Tate, unpubl.), and other areas of Antarctica (Hawes & Brazier 1991). The fate of this stream algal production remains unresolved, although some probably travels downstream to lakes. High winds may also carry some algal biomass across the landscape, providing a source of carbon for soil communities. Most algal biomass remains in the stream channels at the end of the growing season and subsequently becomes freeze-dried. The biomass becomes active again soon after wetting during the following summer (Vincent & Howard-Williams 1986).

We examined the relationship between species diversity and primary production for different algal communities. The scale of our study was limited to small patches of algal mats (about 1 cm^2) and short duration (incubations of several hours). At this scale, our results generally support the redundancy hypothesis, where only a single species is required to perform an ecosystem process at a given rate. In this study, the monospecific communities (*Nostoc* and *Prasiola*) usually had greater areal primary production rates than the more diverse communities (orange-colored mats), despite usually having lower biomass. Why *Nostoc* mats have higher primary production rates than the more diverse orange-colored mats is unclear. One possible explanation is that the orange-colored mats generally have a flat surface, whereas *Nostoc* mats are more foliose, and thus potentially intercept more light. Additionally, the heterocystous *Nostoc* sp. may have greater nitrogen fixation capabilities, although other studies have shown rates of nitrogen fixation to be low relative to uptake rates of dissolved nitrogen compounds (Howard-Williams *et al.* 1989). The *Nostoc* mats may also simply have a higher rate of primary production because they may be adapted to a shorter growing season, growing along the edges of these streams with a shorter period of submersion.

One way for species diversity to affect process rates such as primary production involves enhanced use of limiting resources by diverse biota, an observation suggested even by Darwin (1859). Tilman *et al.* (1996) reported that increased species diversity in grasslands may increase nitrogen retention, and levels of this limiting nutrient may then affect primary production. In Dry Valley streams, it is unclear if increased species diversity allows for more efficient use of a limiting resource, such as nutrients or light.

Understanding the role of species diversity in controlling ecosystem process rates may require a broader spatial or temporal scale of study. Comparing algal mat communities within one stream, and at one time of year, shows that species diversity may not affect primary production. Certain streams without *Nostoc* or *Prasiola*, however, may have lower overall production rates. Recent algal surveys (Alger *et al.*, in press), though, suggest that most streams have a similar species list, and that impediments to colonization do not appear to limit species diversity as much as the physical characteristics of the streams. At broader temporal scales, it is unclear if our instantaneous rates of primary production lead to seasonal changes in biomass accumulation for all sites. The Relict Channel experiment has seen large increases in biomass levels of *Nostoc* mats under newly wetted conditions (Tate, unpubl.), so the high production rates we measured do appear to lead to increases in biomass for this one site.

Another possible role of species diversity in affecting process rates involves disturbances, which occur at time scales outside the scope of this and many other studies. Recently, Tilman & Downing (1994) found higher resilience of primary production in more diverse grassland communities recovering from a severe drought. Similarly, Niyogi *et al.* (1996) found that a flood from a summer thunderstorm decreased primary production in a monospecific algal community affected by acid mine drainage to a greater extent than a more diverse community in a pristine reference stream. The *Nostoc* communities in Antarctic streams may also be more vulnerable to high flows or other disturbances, such as freeze-thaw cycles, when compared to the more diverse communities, where different species may help compensate for biomass losses due to sloughing.

Overall, the link between species diversity and ecosystem function continues to be an important area of study. Possible relationships between biodiversity and function (Lawton 1994) need to be tested at a variety of temporal and spatial scales, because different relationships may be apparent at different scales. Long-term monitoring of Antarctic streams and other ecosystems will help resolve some of these questions regarding species diversity and ecosystem function.

6   ACKNOWLEDGMENTS

We wish to thank Antarctic Support Associates and the US Navy for logistical support in the field throughout the study. Funding was provided by the US National Science Foundation's Long-Term Ecological Research Program (OPP 92-11773). Reviews of the manuscript by Drs. William M. Lewis, Jr., Paul Brooks, and two anonymous reviewers are appreciated.

REFERENCES

Alger, A.S., D.M. McKnight, S.A. Spaulding, C.M. Tate, G.H. Shupe, K.A. Welch, R. Edwards, E.D. Andrews & H.R. House, in press. Ecological processes in a cold desert ecosystem: the abundance and species distribution of algal mats in glacial meltwater streams in Taylor Valley, Antarctica. *INSTAAR Occasional Paper No.* 51.

Darwin, C., 1859. *On The Origin of Species by Means of Natural Selection.* John Murray, London. 426 p.

Hawes, I. & P. Brazier, 1991. Freshwater stream ecosystems of James Ross Island, Antarctica. *Antarctic Science* 3:265-271.

Howard-Williams, C. & W.F. Vincent, 1989. Microbial communities in southern Victoria Land streams (Antarctica). I. Photosynthesis. *Hydrobiology* 172:27-38.

Howard-Williams, C., J.C. Priscu & W.F. Vincent, 1989. Nitrogen dynamics in two Antarctic streams. *Hydrobiology* 172:51-61.

Lawton, J.H., 1994. What do species do in ecosystems? *Oikos.* 71:367-374.

McKnight, D.M., A. Alger, C.M. Tate, G. Shupe & S. Spaulding, in press. Longitudinal patterns in algal abundance and species distribution in meltwater streams in Taylor Valley, South Victoria Land, Antarctica. In: Priscu, J.C. (Ed.*), The McMurdo Dry Valleys Antarctica: a cold desert ecosystem.* American Geophysical Union, Washington D.C.

Naeem, S., L.J. Thompson, S.P. Lawler, J.H. Lawton & R.M. Woodfin, 1994. Declining biodiversity can alter the performance of ecosystems. *Nature* 368:734-737.

Niyogi, D.K., W.M. Lewis, Jr. & D.M. McKnight, 1996. Relationships between species diversity and ecosystem function in streams affected by acid mine drainage. *North American Benthological Society, Annual Meeting.*

Strickland, J.D.H. & T.R. Parsons, 1972. *A Practical Handbook of Seawater Analysis, 2nd edition.* Fisheries Research Board of Canada, Ottawa, Bulletin 167.

Tilman, D. & J.A. Downing, 1994. Biodiversity and stability in grasslands. *Nature* 367:363-365.

Tilman, D., D. Wedin & J. Knops, 1996. Productivity and sustainability influenced by biodiversity in grassland ecosystems. *Nature* 379:718-720.

Vincent, W.F., 1988. *Microbial Systems of Antarctica.* Cambridge University Press, Cambridge. 304 p.

Vincent, W.F. & C. Howard-Williams, 1986. Antarctic stream ecosystems: physiological ecology of a blue-green algal epilithon. *Freshwater Biology* 16:219-233.

Vincent, W.F. & C. Howard-Williams, 1989. Microbial communities in southern Victoria Land streams (Antarctica). II. The effects of low temperature. *Hydrobiology* 172:39-49.

Von Guerard, P., D.M. McKnight, R.A. Harnish, J.W. Gartner & E.D. Andrews, 1994. Streamflow, water-temperature, and specific-conductance data for selected streams draining into Lake Fryxell, Lower Taylor Valley, Victoria Land, Antarctica, 1990-92. *US Geological Survey Open-File Report* 94-545.

Ecosystem Processes in Antarctic Ice-free Landscapes, Lyons, Howard-Williams & Hawes (eds)
© 1997 Balkema, Rotterdam, ISBN 90 5410 925 4

Carbon dynamics of aquatic microbial mats in the Antarctic dry valleys: A modelling synthesis

Daryl L. Moorhead & W. Shane Davis
Texas Tech University, Lubbock, Tex., USA

Robert A. Wharton
Desert Research Institute, Reno, Nev., USA

ABSTRACT: Algae associated with microbial mats in the McMurdo Dry Valleys are responsible for the bulk of primary production in streams, as well as a fraction of the primary production in ice-covered lakes. Reported photosynthetic characteristics and respiratory coefficients of various mat types were used to develop a net primary production (NPP) model. Recorded sunlight regimes in Taylor Valley, Antarctica, adjusted for transmission through lake ice where necessary, were used to drive this model for lake and stream ecosystems. In initial simulations, annual NPP for lake mats was consistently negative; stream mats avoid this paradox because they are frozen during the dark, winter months. There seem to be two reasons why simulated lake mat production was negative: sunlight intensities seldom exceed 5% ambient intensity beneath lake ice and respiratory rates are high relative to photosynthesis. Thus, winter respiration exceeded net production at other times. Additional simulations were able to maintain lake mats when respiratory rates were reduced during periods of darkness and photosynthetic responses were adjusted for spectral differences between ambient and incident radiation. Stream mats maintain positive annual NPP, with production limited by the saturation of photosynthesis at low light intensities (*ca.* 20 to 50 µmol photons m^{-2} s^{-1}), regardless of location.

1 INTRODUCTION

In polar deserts, low temperatures, limited precipitation, high winds and long periods with little or no sunlight, provide a challenging environment for biota. Biological communities are generally dominated by microorganisms. Nevertheless, a unique combination of regional and local features support surprisingly productive lake and stream ecosystems in the Antarctic Dry Valleys. In fact, aeolian transport of organic carbon from these aquatic systems may provide an important source of organic matter for adjacent terrestrial food webs. For these reasons, understanding the patterns of primary production in dry valley streams and lakes is critical to understanding the functioning of valley ecosystems and assessing likely impacts of environmental changes.

The term "Dry Valley" refers to the relatively ice-free valleys located along the coast of Antarctica. The McMurdo Dry Valleys are the largest of these ice-free areas (*ca.* 4800 km^2), and are located on the western coast of the Ross Sea, approximately 100 km west of the USA scientific base at McMurdo Station. The lakes in these valleys are permanently ice-covered and receive seasonal inputs of water from glacial meltwater streams. Their existence as ice-water equilibria results from two conditions: mean summer temperatures are too low for winter accumulations of ice to melt completely, and meltwater from local glaciers replace ablation losses from the lake surface (Wilson 1982; Clow *et al.* 1988). Water level is established by the balance between summer glacial melting (inputs) and annual ablation (outputs) (Chinn 1985; Clow *et al.* 1988), while thickness of ice cover is determined by the energy balance of the lake (McKay *et al.* 1985). Streams in the McMurdo Dry Valleys are fed primarily by glacial melt, with virtually no inputs of water from the adjacent, terrestrial landscape. Thus, streams are usually short (<3 km) and flow more-or-less directly from glacial sources into the dry valley lakes.

Microbial mats are common in many of the dry valley streams and lakes. The algal communities of these mats are composed primarily of cyanobacteria (e.g., *Phormidium*, *Oscillatoria*, and *Lyngbya*), pennate diatoms, and eubacteria (Wharton *et al.* 1983; Vincent 1988; Howard-Williams & Vincent 1989; Alger *et al.* 1996). Previous studies have examined primary production of mat communities in Canada Stream, a representative meltwater stream in the McMurdo Dry Valleys (Vincent & Howard-Williams 1986; Howard-Williams & Vincent 1989). More recently, investigations have examined photosynthesis and respiration of mats at various depths in Lake Hoare, also located in the McMurdo Dry Valleys (Hawes & Schwarz 1996). However, to date there has been no attempt to simulate net annual productivities of mat communities in lakes or streams, based on diurnal and seasonal variations in light regimes and the reported responses of mats to light intensity. In this paper, we develop models of primary production for benthic mat communities in Lake Hoare and Canada Stream (Taylor Valley, Antarctica), which we then use to explore net primary productivity in response to patterns of light availability.

2   BACKGROUND

2.1   *Site description*

The McMurdo Dry Valley Long Term Ecological Research (LTER) site is located in Taylor Valley, southern Victoria Land, Antarctica (77°00'S, 162°52'E). Taylor Valley is approximately 33 km long by 12 km wide and contains three major lakes (Bonney, Hoare and Fryxell), fed by 15 glaciers. The valley bottoms are predominantly glacial till and the higher slopes consist of granites, dolerites, sandstones and occasional volcanics. Outlet, piedmont and alpine glaciers drain from the Polar Plateau and cirques onto the valley floors. The Dry Valleys have been ice-free for most of the past 4 million years and the lakes in Taylor Valley are remnants of the much larger Glacial Lake Washburn, that existed 10 000 to 24 000 years ago (Denton *et al.* 1989; Lawrence & Hendy 1989).

The Taylor Valley climate is cold and dry. Precipitation is received as snow and averages <10 cm y^{-1} (Keys 1980). Mean annual temperature is -20°C, average wind speeds are 5.0 m s^{-1}, and relative humidities average <50% (Clow *et al.* 1988). Low precipitation, low surface albedo, and dry föhn winds descending from the Polar Plateau produce extremely arid conditions (Clow *et al.* 1988). Furthermore, this region of Antarctica has approximately four months each of continuous sunlight, twilight and darkness.

2.2   *Primary productivities of stream and lake mats*

The primary productivity of mats in antarctic lakes varies considerably, with estimated annual carbon fixation ranging from 5 to 1400 g C m^{-2} yr^{-1} (Table 1). In general, photosynthesis rates are low, varying with mat type and species composition. A pair of studies conducted on mats in seasonally ice-free lakes on Signy Island (a maritime, subantarctic island) are among the most detailed investigations of benthic mats available and report net primary productivity rates of 1.5 to 4.0 µg C mg^{-1} ash-free dry weight d^{-1} (Priddle 1980a; 1980b). More recently, Hawes & Schwarz (1996) reported maximum net photosynthesis rates ranging from 3.04 to 0.85 µg O_2 cm^{-2} h^{-1} for mats between 8 and 23 m depth, respectively, in Lake Hoare (Taylor Valley).

Similar investigations performed on algal mats in several seasonally ice-free ponds lakes located on the Ross Ice Shelf found gross photosynthetic rates of 9 to 21 mg C m^{-2} h^{-1} (Howard-Williams et al. 1989). Differences in methodologies, environments and units of measure of these studies make comparisons difficult, but rates of primary productivity are consistently low.

Microbial mats also occur in dry valley streams, are composed primarily of filamentous cyanobacteria and, as in lake mats, exhibit low photosynthetic rates (Table 2). Given the apparent lack of light and nutrient limitation for stream mat communities (Howard-Williams & Vincent 1989), low assimilation numbers may result from the persistence of photosynthetically inactive chlorophyll a associated with inactive or senescing cells (Vincent & Howard-Williams 1989). For example, in Canada Stream, which flows into Lake Fryxell (Taylor Valley), both the amount of measured chlorophyll a and standing stock carbon are extremely high relative to the amount of carbon fixed (>1 µg chl a and >500 µg biomass-C exist per 1 µg C fixed h^{-1}, respectively; Vincent & Howard-Williams 1989). Apparently, cyanobacterial mats maintain low rates of productivity regardless of light or nutrient regimes, both in Antarctic lakes and Dry Valley streams.

2.3   Light regimes

In addition to temporal pecularities in ambient light regimes of polar regions, the permanent ice cover (3 to 4 m thick) on Dry Valley lakes affects the quantity and spectral distribution of radiation reaching the underlying water (Ragotzkie & Likens 1964; Wharton et al. 1989; Lizotte & Priscu 1992; Howard-Williams et al., in press). In general, only a small amount of the incident solar flux (about 1 to 3%) is transmitted through the ice, depending on albedo, sediment content, thickness, and scattering properties (e.g. McKay et al. 1994). For example, spectral downwelling measurements in Lake Hoare show full wavelength PAR beneath the ice at <5% of maximum incident surface PAR (Palmisano & Simmons 1987; McKay et al. 1994). Moreover, McKay et al. (1994) show that transmitted light has greatly reduced intensities of longer wavelengths (>600 nm). After light penetrates the overlying ice, intensity then diminishes with depth in the water column. In the stable, highly stratified Dry Valley lakes, attenuation varies more with depth than is common in well-mixed bodies of water, with considerable differences also existing between Dry Valley lakes.

Table 1. Production of benthic algal mats in Antarctic ponds and lakes (g C m^{-2} y^{-1}).

Primary production	Lake and location	Reference
11	Changing Lake; Signy Island	Priddle 1980a
45	Sombre Lake; Signy Island	Priddle 1980a
88	Skua Lake; McMurdo Ice Shelf	Goldman 1964
161	Algal Lake; McMurdo Ice Shelf	Goldman 1964
37	Fresh Pond; McMurdo Ice Shelf	Howard-Williams et al. 1989
57	Skua Lake; McMurdo Ice Shelf	Howard-Williams et al. 1989
60	Ice Ridge; McMurdo Ice Shelf	Howard-Williams et al. 1989
36	P-70 Lake; McMurdo Ice Shelf	Howard-Williams et al. 1989
39	Brack Pond; McMurdo Ice Shelf	Howard-Williams et al. 1989
26	Salt Pond; McMurdo Ice Shelf	Howard-Williams et al. 1989
140-230	Skua Lake; McMurdo Ice Shelf	Goldman et al. 1972
172-327	Algal Lake; McMurdo Ice Shelf	Goldman et al. 1972
730	Skua Lake; McMurdo Ice Shelf	Goldman et al. 1967
1,388	Algal Lake; McMurdo Ice Shelf	Goldman et al. 1967
5.5	Watts Lake; Vestfold Hills	Heath 1988
0-113	Lake Bonney; Taylor Valley	Parker & Wharton 1985

Table 2. Production and respiration rates for benthic microbial mats in antarctic streams.

Photosynthesis	Respiration	Algal Community	Location	Reference
0.004 mg C mg C^{-1} h^{-1}		Prasiola crispa	?	Davey 1989
4.004 mg C mg C^{-1} h^{-1}		Phormidium autumnale	?	Davey 1989
0.3–0.7 mg C mg C^{-1} h^{-1}	0.67 mg C mg C^{-1} h^{-1}	Phormidium	Livingston Island	Davey 1993
0.5–1.4 mg C mg C^{-1} h^{-1}	0.67 mg C mg C^{-1} h^{-1}	Phormidium	Livingston Island	Davey 1993
0.4 µg C g^{-1} AFDW h^{-1}		Phormidium	Signy Island	Hawes 1993
3.0 µg C g^{-1} AFDW h^{-1}		Phormidium	Signy Island	Hawes 1993
3.81 mg C g^{-1} AFDW h^{-1}	2.29 mg C g^{-1} AFDW h^{-1}	Klebsormidium	James Ross Island	Hawes & Brazier 1991
0.47 mg C g^{-1} AFDW h^{-1}	0.23 mg C g^{-1} AFDW h^{-1}	Phormidium (thin mat)	James Ross Island	Hawes & Brazier 1991
0.20 mg C g^{-1} AFDW h^{-1}	0.15 mg C g^{-1} AFDW h^{-1}	Phormidium (thick mat)	James Ross Island	Hawes & Brazier 1991
2.6 µg C cm^{-2} h^{-1}		Nostoc	McMurdo Region	Howard-Williams & Vincent 1989
3.6 µg C cm^{-2} h^{-1}		Prasiola	McMurdo Region	Howard-Williams & Vincent 1989
2.9 µg C cm^{-2} h^{-1}		Binuclearia	McMurdo Region	Howard-Williams & Vincent 1989
2.0 µg C cm^{-2} h^{-1}		Phormidium	McMurdo Region	Howard-Williams & Vincent 1989
1.48 µg C cm^{-2} h^{-1}	0.97 µg C cm^{-2} h^{-1}	Nostoc @ A1	McMurdo Region	Vincent & Howard-Williams 1986
0.00 µg C cm^{-2} h^{-1}	0.24 µg C cm^{-2} h^{-1}	Nostoc @ A2	McMurdo Region	Vincent & Howard-Williams 1986
0.31 µg C cm^{-2} h^{-1}	0.68 µg C cm^{-2} h^{-1}	Nostoc @ A3	McMurdo Region	Vincent & Howard-Williams 1986
0.33 µg C cm^{-2} h^{-1}	1.22 µg C cm^{-2} h^{-1}	Nostoc @ F1	McMurdo Region	Vincent & Howard-Williams 1986
2.15 µg C cm^{-2} h^{-1}	1.22 µg C cm^{-2} h^{-1}	Nostoc @ F2	McMurdo Region	Vincent & Howard-Williams 1986
1.65 µg C cm^{-2} h^{-1}	0.57 µg C cm^{-2} h^{-1}	Nostoc @ F3	McMurdo Region	Vincent & Howard-Williams 1986
0.91 µg C cm^{-2} h^{-1}	0.11 µg C cm^{-2} h^{-1}	Phormidium @ A4	McMurdo Region	Vincent & Howard-Williams 1986
0.61 µg C cm^{-2} h^{-1}	0.54 µg C cm^{-2} h^{-1}	Phormidium @ F4	McMurdo Region	Vincent & Howard-Williams 1986
1.41 µg C cm^{-2} h^{-1}	1.23 µg C cm^{-2} h^{-1}	Phormidium @ F5	McMurdo Region	Vincent & Howard-Williams 1986
0.88 µg C cm^{-2} h^{-1}	0.07 µg C cm^{-2} h^{-1}	Phormidium @ F6	McMurdo Region	Vincent & Howard-Williams 1986
0.39 µg C cm^{-2} h^{-1}	0.10 µg C cm^{-2} h^{-1}	Phormidium @ F7	McMurdo Region	Vincent & Howard-Williams 1986
1.2–3.1 µg C cm^{-2} h^{-1}		Phormidium	McMurdo Region	Vincent & Howard-Williams 1989
9.7–23.7 µg C cm^{-2} h^{-1}		Nostoc	McMurdo Region	Vincent & Howard-Williams 1989
7.6–14.0 µg C cm^{-2} h^{-1}		Binuclearia	McMurdo Region	Vincent & Howard-Williams 1989
2.0–5.4 µg C cm^{-2} h^{-1}		Prasiola	McMurdo Region	Vincent & Howard-Williams 1989
0.53 mg C mg Chl a^{-1} h^{-1}	0.75 mg C mg Chl-a^{-1} h^{-1}	Phormidium	McMurdo Region	Vincent & Howard-Williams 1989
0.94 mg C mg Chl a^{-1} h^{-1}	0.80 mg C mg Chl-a^{-1} h^{-1}	Phormidium	McMurdo Region	Vincent & Howard-Williams 1989
0.17 mg C mg Chl a^{-1} h^{-1}	0.29 mg C mg Chl-a^{-1} h^{-1}	Nostoc	McMurdo Region	Vincent & Howard-Williams 1989
0.79 mg C mg Chl a^{-1} h^{-1}	0.18 mg C mg Chl-a^{-1} h^{-1}	Binuclearia	McMurdo Region	Vincent & Howard-Williams 1989
0.53 mg C mg Chl a^{-1} h^{-1}	0.08 mg C mg Chl-a^{-1} h^{-1}	Prasiola	McMurdo Region	Vincent & Howard-Williams 1989

In contrast to the microbial mats in Dry Valley lakes, stream mats apparently receive near-ambient levels of sunlight. Most streams are shallow (<10 cm), with little or no ice and snow cover, have low concentrations of light absorbing compounds (e.g. fulvic and humic acids), and those with significant mat communities, carry little suspended particulate matter. However, stream mats are frozen or dessicated throughout most of the year, with stream flow limited to the short, austral summer when ambient light intensities are near maximum. Clearly, the location of a mat influences the light regime it experiences as well as the period of physiological activity.

3   MODELLING APPROACH

In the present study, we assumed that the biomass of microbial mats was increased by photosynthetic gains and decreased by respiratory losses:

$$dB/dt = P - R \qquad\qquad [1]$$

where B is biomass, P is photosynthesis, and R is respiration. Photosynthesis was assumed to be driven by hourly light intensity. Light intensity was simulated to include daily, seasonal and site differences, particularly with respect to depth in lakes. Photosynthetic and respiratory characteristics for representative microbial mats were obtained from the work of Howard-Williams & Vincent (1989) for use in simulating mat production in Canada Stream, and from Hawes & Schwarz (1996) for mats in Lake Hoare. Values of state variables and parameters used in models are given in Tables 3 and 4. No estimates of advective loss or grazing are included, although both processes are expected to be minimal.

3.1   Photosynthesis

The relationship between light intensity and photosynthesis has often been described with rectangular hyperbolic and hyperbolic tangent models (c.f. Jassby & Platt 1976). Moreover, previous studies of mat production in antarctic streams and lakes demonstrate the utility of

Table 3. Values of parameters describing photosynthesis and respiration activities for lake and stream simulations.

Simulation	[a]P_{max}	[b]β	[c]P_{max}	[d]I_k	[e]m	[f]b	[g]g
Stream (*Nostoc*)	2.25	80	-	-	-	-	0.053
Stream (*Phormidium*)	3.50	110	-	-	-	-	0.019
Lake at 8 m	4.50	9	3.04	18	0.3767	-0.9004	0.670
Lake at 12 m	3.70	11	2.96	12	0.1560	0.0748	0.570
Lake at 15 m	2.27	13	1.12	9	0.0310	0.3035	0.580
Lake at 16 m	1.50	10	1.39	9	0.0748	-0.0531	0.450
Lake at 23 m	0.90	10	0.85	9	0.0291	-0.0499	0.320

[a] used in Eq. 2; $\mu g\ C\ cm^{-2}\ h^{-1}$
[b] used in Eq. 2; $\mu mol\ m^{-2}\ s^{-1}$
[c] used in Eq. 3; $\mu g\ O_2\ cm^{-2}\ h^{-1}$
[d] used in Eq. 3; $\mu mol\ m^{-2}\ s^{-1}$
[e] used in Eq. 4; $(\mu g\ O_2\ cm^{-2}\ h^{-1})/(\mu mol\ m^{-2}\ s^{-1})$
[f] used in Eq. 4; $\mu g\ O_2\ cm^{-2}\ h^{-1}$
[g] used in Eq. 5; $\mu g\ O_2\ cm^{-2}\ h^{-1}$ for lake models and $\mu g\ C\ cm^{-2}\ h^{-1}$ for stream models

these approaches (e.g. Priddle 1980a, b; Howard-Williams & Vincent 1989; Hawes & Schwarz 1996). We first used a rectangular hyperbolic expression of photosynthesis:

$$P=(P_{max}.I)/(ß+I)$$ [2]

where P_{max} is the maximum photosynthetic rate, ß is the half-saturation coefficient, and I is the light intensity (Fig. 1). Data reported by Howard-Williams & Vincent (1989) were used to estimate parameter values for two mat types in Canada Stream; one dominated by *Nostoc* spp. and the other dominated by *Phormidium* spp.

A hyperbolic tangent function also was used to describe photosynthesis:

$$P=P_{max}.tanh(I/I_k)$$ [3]

where I_k is the intensity at which P_{max} is reached. Hawes & Schwarz (1996) provided parameter

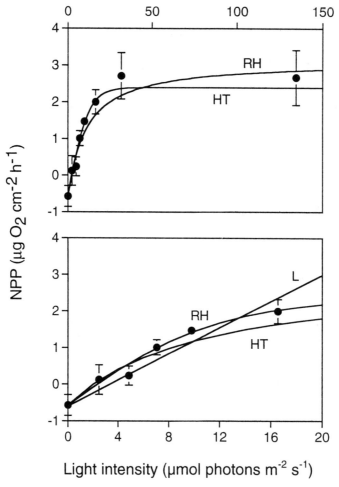

Fig. 1. Observed (dots) and simulated (lines) rates of net primary productivity at various light intensities, for benthic microbal mats taken from 12 m depth in Lake Hoare. Top: response to full range of light intensities; Bottom: response to low light intensities. Lines are labeled RH (Eq. 2; rectangular hyperbola), HT (Eq. 3; hyperbolic tangent) and L (Eq. 4; linear).

values for hyperbolic tangent models of primary production, which we used to simulate primary production of mats in Lake Hoare (Fig. 1).

The photosynthetic responses of mats in Lake Hoare to high light intensities may be irrelevant because light intensities beneath the perennial ice cover is <5% of ambient values (Palmisano & Simmons 1987; McKay *et al.* 1994). Thus, mat communities at significant depths normally do not experience light intensities much greater than 20 µmol photons m^{-2} s^{-1}. At such low light levels, photosynthesis approaches a linear function of intensity:

$$P = m \cdot I + b \qquad \qquad [4]$$

where m and b are the slope and intercept of the relationship, respectively. All three equations were used to estimate annual production patterns for mats at various depths in Lake Hoare to evaluate the ramifications of the different formulations.

3.2   Respiration

Respiration rates for mat communities are often determined by measuring net production in the absence of light and expressed as a function of standing biomass, for example:

$$R = B \cdot \gamma \qquad \qquad [5]$$

where γ is a respiratory coefficient. Vincent & Howard-Williams (1986) report respiratory coefficients for mats in Canada Stream. However, Hawes & Schwarz (1996) provide estimates of respiration on an areal basis rather than per unit biomass (Table 3).

Several studies have shown that algal respiration varies with previous light exposure (Webster & Frenkel 1953; Kratz & Myers 1955; Padan *et al.* 1971; Gibson 1975; Prezelin & Sweeney 1978; Markager & Sand-Jensen 1989); in general, a rapid decrease in respiration often occurs at the beginning of dark incubation with a gradual stabilization at 10 to 40% of the initial rate after 10 to 12 hours. This may be a very important consideration in evaluating productivity patterns for mat communities of Dry Valley lakes, which experience about four months of continuous darkness per year. Hawes & Schwarz (1996) demonstrated that the relative dark respiration rate of mats in Lake Hoare declined with light intensity at depth of collection:

$$R/P_{max} = c \cdot I + f \qquad \qquad [6]$$

where c and f are the slope and intercept of the relationship, respectively. We included this function to simulate responses of respiration rates to seasonal light intensity for lakes.

3.3   Sunlight regimes

The driving variable for model simulations is sunlight intensity. A synthetic light regime was constructed to provide hourly sunlight intensity, based on meterological data obtained from the McMurdo LTER project (Fig. 2). The pattern of sunlight intensity (I) over the course of a day was estimated as a function of time:

$$I = A \cdot \cos(2^1 \cdot (H + \emptyset)/24) + S \qquad \qquad [7]$$

where H is the hour of the day, $\emptyset$ is phase shift and A is the amplitude of the function. The effect of season (S) on daily light regime was calculated as a polymomial function of time (in days):

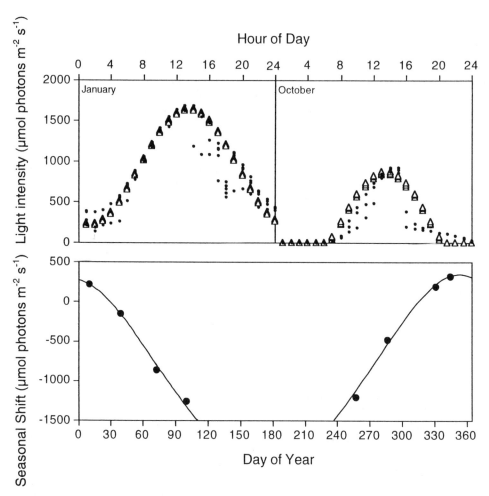

Fig. 2. Daily light regimes and seasonal variations in intensity. Top left: observed (dots) and simulated (triangles) hourly sunlight intensities for January 7-11, 1995; Top right: observed and simulated sunlight intensities for October 12-14, 1995; c - observed (dots) and simulated (line) shifts in average maximum sunlight intensities (clear sky) between November 1994 through November 1995.

$$S = a_0 + a_1 \cdot D + a_2 \cdot D^2 + a_3 \cdot D^3 + a_4 \cdot D^4 \qquad [8]$$

where D is the day of the year (1-365) and values of a_i were best-fit estimates (Table 4) based on measurements of incident PAR at Lake Hoare between November 1993 to November 1994, provided by the McMurdo LTER.

For the stream model, ambient light intensities estimated by this method were used to drive the primary production model, assuming no reduction in light intensity due to albedo or attenuation. For Lake Hoare, previous studies suggest that light transmission through the perennial ice cover is ±5% of ambient intensities (Parker & Wharton 1985; Lizotte & Priscu 1992; McKay et al. 1994). We assumed a constant 5% transmission of ambient light through the ice and then estimated attenuation through the water column according to McKay et al. (1994):

Table 4. Values of state variables and parameters for equations describing sunlight intensity and mat dynamics.

Parameter	Units	Value	Equation
B	mg C cm^{-2} (*Nostoc*)	24.39	5
	mg C cm^{-2} (*Phormidium*)	9.21	5
c	(μg O^2 cm^{-2} h^{-1})/(μmol m^{-2} s^{-1})	1.4772E-3	6
f	μg O^2 cm^{-2} h^{-1}	9.2585E-3	6
A	μmol m^{-2} s^{-1}	710.0	7
Ø	d	10.5	7
a_0	μmol m^{-2} s^{-1}	264.9	8
a_1	μmol m^{-2} s^{-1} d^{-1}	2.1978	8
a_2	μmol m^{-2} s^{-1} d^{-2}	2.7437E-1	8
a_3	μmol m^{-2} s^{-1} d^{-3}	1.6106E-3	8
a_4	μmol m^{-2} s^{-1} d^{-4}	2.3059E-6	8
e	m^{-1}	0.14	9

$$I_d = I_t \cdot e^{(-e \cdot d)} \tag{9}$$

where I_d is light intensity at depth (d), I_t is the intensity of light transmitted through the ice cover (5% of ambient), and e is the attenuation coefficient of the water column.

Not only is light intensity greatly reduced by transmission through the ice cover of dry valley lakes, the spectral composition also differs from ambient light. McKay *et al.* (1994) showed that intensities of longer wavelengths (>600 nm) of light were reduced beneath the ice cover of Lake Hoare. The photosynthetically active layers of benthic microbial mats in Lake Hoare show high absorbance at short and middle wavelengths (<600 nm), but little absorbance at longer wavelengths (Fig. 3; Hawes & Schwarz, unpubl.). Comparisons of the areas under these absorbance and transmission curves suggests that while active layers of mats absorb 60 to 65% of full spectrum light, they may utilize as much as 85 to 90% of the transmitted light. Thus, measures of mat photosynthesis in response to ambient light (full spectrum) used in most experiments (e.g. Hawes & Schwarz 1996) may underestimate photosynthesis beneath ice

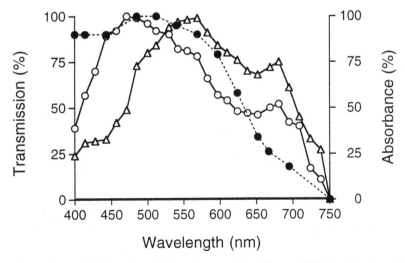

Fig. 3. Transmission of light through the ice at Lake Hoare (filled circles), and light absorbances of photosynthetically active layers of benthic microbial mats taken from Lake Hoare (triangles - surface 0 to 1 mm of mat, open circles - surface 0 to 3 mm of mat).

covers by as much as 20 to 50%. We examined the impact of this higher apparent photosynthetic efficiency of lake mats by increasing light transmission through the ice cover by 50% (see below).

3.4 Simulations

Available information on ambient light regimes, light attenuation, and physiological responses of microbial mats to light intensity, was sufficient to support the development and application of a model for evaluating patterns of net primary production. We chose Lake Hoare and Canada Stream as representatives of Dry Valley lakes and streams because the mats in these systems have been most closely studied to date. The synthetic light regime (discussed previously) was used to drive this model for a period of one year.

In stream simulations, only the rectangular hyperbolic equation (Eq. 2) was used to simulate net primary productivity given characteristics of two mat types (*Phormidium* and *Nostoc*; Vincent & Howard-Williams 1986; Howard-Williams & Vincent 1989). We assumed that mats received ambient light intensitites, were active only when streams were usually flowing (a 90-day period from December through February), and showed no direct respiratory response to light intensity. We also assumed that the mats were metabolically inactive (no photosynthesis or respiration) during the other months of the year when they are usually frozen or dessicated.

In the lake model, net primary productivity was calculated at five depths for which model parameters were available (i.e. 8, 12, 15, 16 and 23 m; Hawes & Schwarz 1996). We then explored the effects of: (1) altered spectral composition of transmitted light on photosynthetic rates, and (2) low light intensities on respiration rates, in conjunction with the rectangular hyperbolic function of photosynthesis (Eq. 2). Subsequently, all three photosynthetic equations were used (Eqs. 2 to 4) in separate simulations, to evaluate the impact of each equation on model behavior. We assumed that lake mats were metabolically active throughout the year.

4 MODEL RESULTS AND DISCUSSION

4.1 Stream mat productivity

In Canada Stream, both mat types showed a substantial increase in simulated biomass over a year (Fig. 4a). The *Nostoc* mat increased from an initial mass of 28.1 mg C cm^{-2} to a final value of 32.6 mg C cm^{-2}, representing an increase of 16%. Similarly, the mass of the *Phormidium* mat increased from 7.9 to 10.7 mg C cm^{-2} (a 35% increase). Therefore, mats have reasonably high rates of net primary production, relative to standing stocks, despite saturation of photosynthesis at light intensities that are much lower than ambient levels throughout much of the austral summer.

Existing knowledge of mat communities in antarctic streams has been summarized by Howard-Williams *et al.* (1986), Vincent (1988) and Vincent *et al.* (1993). Moreover, previous research has included detailed examinations of photosynthetic and respiratory characteristics of mats in streams located on the South Shetland Islands (Davey 1993), James Ross Island (Hawes and Brazier 1991), Signy Island (Hawes 1989, 1993), and in the McMurdo Sound region of Antarctica, including the nearby Dry Valleys (Vincent & Howard-Williams 1986, 1989; Howard-Williams & Vincent 1989). However, the current study is the first to estimate annual patterns of primary production for such systems. Model parameter values for respiratory and photosynthetic attributes of *Nostoc* and *Phormidium* mats are generally representative of these reported values (Tables 2 & 3), and closely approximate the general responses of these mat

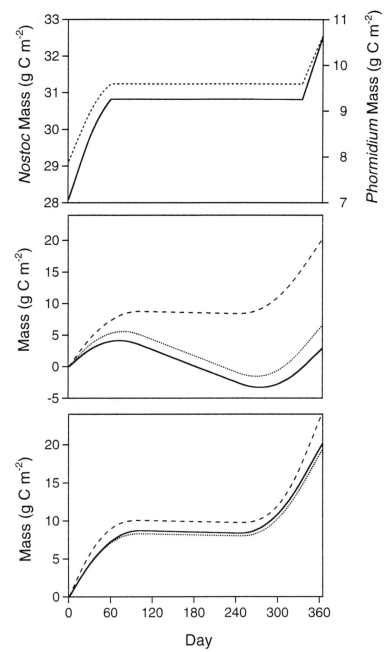

Fig. 4. Simulated biomass dynamics of aquatic microbial mats. Top: *Nostoc* (solid line) and *Phormidium* (dotted line) dominated communities in Canada Stream; Middle: benthic mat at 12 m depth in Lake Hoare using the rectangular hyperbolic equation for photosynthesis (solid line - baseline conditions with 5% light transmission through lake ice, dotted line - baseline conditions with 7.5% light transmission, dashed line - 7.5% light transmission and varying respiration rate with light intensity); Bottom: benthic mat at 12 m depth in Lake Hoare with 7.5% light transmission through lake ice and varying respiration rate with light intensity (solid line - rectangular hyperbolic function of photosynthesis; dotted line - hyperbolic tangent function; dashed line - linear function).

191

types to light intensity. Thus, we believe that simulated net annual production estimates are reasonable. However, these predictions raise the question of whether these systems are accumulating biomass or approximate steady-state dynamics.

The general lack of herbivores in antarctic streams suggest that the distribution and standing stock of mat communities are influenced primarily by abiotic factors. While competition may play a role in determining the location of particular mat types within stream reaches, correlations between general physical characteristics of mats and their locations also suggest strong physical controls (Algers *et al.* 1996). For example, *Nostoc* mats consist of a loose association of thalli that are primarily limited to the wet margins of stream channels, out of the main current. In contrast, *Phormidium* mats are strongly consolidated, firmly affixed to the substrate, and exist in the main channels of streams. It has been hypothesized that current velocity, substrate stability and sediment scouring are primary factors controlling the distributions and accumulations of mat communities (Howard-Williams & Vincent 1989; Algers *et al.* 1996), although these hypotheses have not been tested experimentally. Moreover, frequent freezing-thawing cycles may disrupt mat communities and contribute to export of organic matter. In any case, if benthic communities in stable reaches of Canada Stream have net annual production values of 16 to 35% standing stocks, as our model predicts, then annual losses must be substantial to maintain steady-state biomass values. The mechanisms causing such losses have not been verified.

4.2   Lake mat productivity

For benthic mats in Lake Hoare, the model predicted a decrease in net annual production with depth under all simulation scenarios (Table 5). However, substantial differences existed between scenarios (Fig. 4b). In the first three model runs, photosynthesis was described with a rectangular hyperbolic function (Eq. 2). When light transmission through the ice cover of the lake was held constant at 5% of ambient intensity, to compensate for the spectral effects previously discussed, microbial mats achieved a positive net annual productivity only at 8 and 12 m depths. Increasing the transmission of light to 7.5% of ambient intensity to compensate for the spectral shifts previously discussed, increased the productivity of mats at all depths, but net annual production was still only positive at the two shallower depths. In contrast, net annual production was positive at all depths with greater light transmission through ice and when mat respiration was assumed to repond to light intensity.

Differences in simulated net annual production also resulted from using different equations to describe photosynthesis (Fig. 4c). The rectangular hyperbolic function (Eq. 2) predicted higher values of production than the hyberbolic tangent equation (Eq. 3) at all depths except 16 m. In addition, the linear equation (Eq. 4) estimated higher production than the rectangular hyperbolic function at all depths except 23 m. In general, the incorporation of the linear photosynthesis equation yielded much higher values of net annual production than the other approaches (Table 5).

Insights into the annual light regimes at various depths in Lake Hoare can be gained by

Table 5. Simulated net annual production for Lake Hoare microbial mats ($g\ C\ m^{-2}\ y^{-1}$).

Model	Depth (m)				
	8	12	15	16	23
Rectangular Hyperbolic; 5% Transmitted	14.3	3.0	-7.7	-6.4	-6.4
Rectangular Hyperbolic; 7.5% Transmitted	18.2	6.7	-5.5	-4.9	-5.7
Rectangular Hyperbolic; 7.5% + R/P_{max}	32.8	20.3	9.5	6.7	2.8
Hyperbolic Tangent; 7.5% + R/P_{max}	24.3	19.6	6.4	7.1	2.1
Linear; 7.5% + R/P_{max}	57.9	24.5	8.1	4.9	0.4

evaluating differences in model output with respect to the functional behaviors of the three photosynthesis equations (Fig. 1). For example, the large differences between models at the shallower depths demonstrated substantial differences in their responses to high light intensities. This was particularly evident in the much higher values of production estimated by the linear photosynthesis equation. Differences between output generated by the two nonlinear approaches were closely related to the maximum rates of photosynthesis used in the two formulations (P_{max}). This became apparent when the ratios of net annual primary production were plotted against the ratios of the P_{max} values for the two models (Fig. 5); a strong linear relationship existed for depths other than 23 m. These observations suggested that light regimes at all but the greatest depth were dominated by intensities sufficient to drive both hyperbolic functions at near maximum rates. The divergence from this linear relationship at 23 m depth indicated that both models were not driving photosynthesis at maximum rates (see also Table 5). In contrast, all three equations predicted similar net annual production at 23 m depth, indicating that light intensities were generally low enough that differences between equations had little impact on model output.

Previous studies have reported high variation in net annual production values for benthic communities of Antarctic lakes (Table 1). In Dry Valley lakes, estimates have ranged between 5.5 and 113 g C m^{-2} y^{-1} (Parker & Wharton 1985; Heath 1988), comparable to our simulation results. However, daily and seasonal variations in light availability, as well as attenuation by overlying ice and the water column, were usually not considered in earlier studies. Moreover, simulation results suggest that adaptations of microbial mats to these limiting conditions may permit mats to persist at depths where light availabilities would otherwise preclude their existence. These adaptations may include a spectral absorbance pattern that corresponds to the spectral distribution of light transmitted through overlying ice and reduced respiratory rates during extended periods of darkness. Although model behavior cannot definitively test such hypotheses in the absence of experimental data, model parameters and responses to light were based on empirical data obtained from mats in Lake Hoare (Hawes & Schwarz 1996) and are comparable to values reported in other studies (Tables 1 & 3).

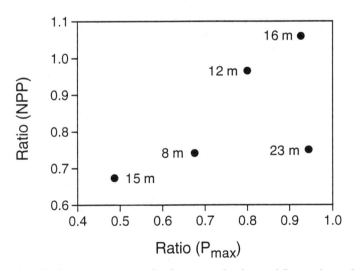

Fig. 5. Relationship between net annual primary production and P_{max} values of models using hyperbolic tangent and rectangular hyperbolic functions of photosynthesis. Ratios are calculated as; 1 - NPP hyperbolic tangent:NPP rectangular hyperbolic model, and 2 - P_{max} hyperbolic tangent:P_{max} rectangular hyperbolic model.

5 CONCLUSIONS

Microbial mat communities are common throughout the streams and lakes of the Antarctic Dry Valleys. Although experimental data is limited, observations suggest substantial levels of net annual production in these aquatic habitats. Results of our modeling studies corroborated these observations and identified some important gaps in knowledge of these systems. For streams, factors controlling biomass accumulations and turnover are uncertain, and loss rates (other than respiration) have not been examined in detail. In lakes, light regimes clearly play a strong role in limiting primary production and influencing the distributions of benthic mats. Simulations suggest that some metabolic adaptations may allow communities to exist where light regimes would otherwise be insufficient. However, experimental data does not yet exist to test these hypotheses.

6 ACKNOWLEDGMENTS

We wish to thank I. Hawes and A-M. Schwarz for providing unpublished data and considerable discussions of this modelling effort. Meterological data were provided by the USA McMurdo Long Term Ecological Research Program. Funding for this work was provided by USA National Science Foundation Office of Polar Programs research grant OPP-9211773 and National Aeronautics and Space Administration Exobiology Program research grant NAGW-1947.

REFERENCES

Alger, A.S., D.M. McKnight, S.A. Spaulding, C.M. Tate, G.H. Shupe, K.A. Welch, R. Edwards, E.D. Andrews & H.R. House, 1996. *Ecological processes in a cold desert ecosystem: The abundance and species distribution of algal mats in glacial meltwater streams in Taylor Valley, Antarctica.* United States Geological Survey. Boulder, Colorado. 102 p.

Chinn, T.J., 1985. Structure and equilibrium of the Dry Valley glaciers, *New Zealand Antarctic Records* 6, Special Supplement:73-88.

Clow, G.D., C.P. McKay, G.M. Simmons Jr. & R.A. Wharton Jr., 1988. Climatological observations and predicted sublimation rates at Lake Hoare, Antarctica. *Journal of Climatology* 1:715-728.

Davey, M.C., 1989. The effects of freezing and desiccation on photosynthesis and survival of terrestrial Antarctic algae and cyanobacteria. *Polar Biology* 10:29-36.

Davey, M.C., 1993. Carbon and nitrogen dynamics in a maritime Antarctic stream. *Freshwater Biology* 30:319-330.

Denton, G.H., S.C. Bockheim & M. Stuiver, 1989. Late Wisconsin and early Holocene glacial history, Inner Ross Embayment, Antarctica. *Quartely Research* 31:151-182.

Gibson, C.E., 1975. A field and laboratory study of oxygen uptake by planktonic blue-green algae. *Journal of Ecology* 63:876-870.

Goldman, C.R., 1964. Primary productivity studies in Antarctic lakes. In: Carrick, R., M.W. Holdgate & M. Prevost (Eds.), *Biologie Antarctique. Premier Symposium Organise par le SCAR, Paris 2-8 September 1962, comptes rendus.* Herman, Paris. pp. 291-299.

Goldman, C.R., D.T. Mason, and J.E. Hobbie, 1967. Two Antarctic desert lakes. *Limnolgy and Oceanography* 12:295-310.

Goldman, C.R., D.T. Mason & B.J.B. Wood, 1972. Comparative study of the limnology of two small lakes on Ross Island, Antarctica. In: Llano, G.A. (Ed.), *Antarctic Terrestrial Biology, Antarctic. Research Series* 20. American Geophysical Union, Washington, D.C. pp. 1-50.

Hawes, I., 1989. Filamentous green algae in freshwater streams on Signy Island, Antarctica. *Hydrobiology* 172:1-18.

Hawes, I., 1993. Photosynthesis in thick cyanobacterial films: a comparison of annual and perennial Antarctic mat communities. *Hydrobiology* 252:203-209.

Hawes, I. & P. Brazier, 1991. Freshwater stream ecosystems of James Ross Island, Antarctica. *Antarctic Science.* 3:365-271.

Hawes, I. & A-M. Schwarz, 1996. Photosynthesis in benthic cyanobacterial mats from Lake Hoare, Antarctica. *Antarctic Journal of the United States.*

Heath, C.W., 1988. Annual primary productivity of an Antarctic continental lake: Phytoplankton and benthic algal mat production strategies. *Hydrobiology.* 165:77-87.

Howard-Williams, C. & W.F. Vincent, 1989. Microbial communities in southern Victoria Land streams (Antarctica) I. Photosynthesis. *Hydrobiology* 172:27-38.

Howard-Williams, C., A-M. Schwarz, I. Hawes & J.C. Priscu, in press. Optical properties of the McMurdo Dry Valley lakes, Antarctic. In: Priscue, J.C. (Ed.), The McMurdo Dry Valleys, Antarctica: a cold desert ecosystem. American Geophysical Union, Washington, D.C.

Howard-Williams, C., R. Pridmore, M.T. Downes & W.F. Vincent, 1989. Microbial biomass, photosynthesis and chlorophyll *a* related pigments in the ponds of the McMurdo Ice Shelf, Antarctica. *Antarctic Science* 1:125-131.

Howard-Williams, C., C.L. Vincent, P.A. Broady & W.F. Vincent, 1986. Antarctic stream ecosystems: Variability in environmental properties and algal community structure. *Internationale Revue der gesamten Hydrobiologie* 71:511-544.

Jassby, A.D. & T. Platt, 1976. Mathematical formulation of the relationship between photosynthesis and light for phytoplankton. *Limnology and Oceanography* 21:541-547.

Keys, J.R., 1980. Air temperature, wind, precipitation, and atmospheric humidity in the McMurdo region. *Department of Geology Publication 17 (Antarctic Data Series No. 9).* Victoria University of Wellington, New Zealand. 52 p.

Kratz, W.A. & J. Myers, 1955. Photosynthesis and respiration of three blue-green algae. *Plant Physiology* 30:275-280.

Lawrence, M.J.F. & C.H. Hendy, 1989. Carbonate deposition and Ross Sea ice advance, Fryxell Basin, Taylor Valley, Antarctica. *New Zealand Journal of Geology Geophysics* 32:267- 277.

Lizotte, M.P. & J.C. Priscu, 1991. Natural fluorescence and photosynthetic quantum yields in vertically stable phytoplankton from perennially ice-covered lakes (dry valleys). *Antarctic Journal of the United States* 26: 226-228.

Lizotte, M.P. & J.C. Priscu, 1992. Spectral irradiance and bio-optical properties in perennially ice-covered lakes of the dry valleys (McMurdo Sound, Antarctica). In: Elliot, D.H. (Ed.), Contributions to Antarctic Research III. *Antarctic Research Series* 57. American Geophysical Union, Washington D.C. pp. 1-50.

Markager, S. & K. Sand-Jensen, 1989. Patterns of night-time respiration in a dense phytoplankton community under a natural light regime. *Journal of Ecology* 77:49-61.

McKay, C.P., G.D. Clow, D.T. Andersen & R.A. Wharton Jr., 1994. Light transmission and reflection in perennially ice-covered Lake Hoare, Antarctica. *Journal of Geophysical Research* 99:427-444.

McKay, C.P., G.D. Clow, R.A. Wharton Jr. & S.W. Squyers, 1985. Thickness of ice on perennially frozen lakes. *Nature.* 313:561-562.

Padan, E., B. Raboy & M. Shilo, 1971. Endogenous dark respiration of the blue-green alga, *Plectonema boryanum. Journal of Bacteriology* 106:45-50.

Palmisano, A.C. & G.M. Simmons Jr., 1987. Spectral downwelling irradiance in an Antarctic lake. *Polar Biology* 7:145-151.

Parker, B.C. & R.A. Wharton Jr., 1985. Physiological ecology of bluegreen algal mats (modern stromatolites) in Antarctic oasis lakes. *Archiv fuer Hydrobiologie,* Algological Supplement 71(1/2) Albological Studies 38/39:331-348.

Prezelin, B.B. & B.M. Sweeney, 1978. Photoadaptation of photosynthesis in *Gonyaulax polyedra. Marine Biology* 48:27-35.

Priddle, J., 1980a. The production ecology of benthic plants in some Antarctic lakes I. *in situ* production studies. *Journal of Ecology* 68:141-153.

Priddle, J., 1980b. The production ecology of benthic plants in some Antarctic lakes II. Laboratory physiology studies. *Journal of Ecology* 68:155-166.

Priscu, J.C., 1992. Particulate organic matter decompostion in the water column of Lake Bonney, Taylor Valley, Antarctica. *Antarctic Journal of the United States.* 260-262.

Ragotzkie, R.A. & G.E. Likens, 1964. The heat balance of two Antarctic lakes. *Limnology and Oceanography.* 9:412-425.

Vincent, W.F., 1988. *Microbial ecosystems of Antarctica.* Cambridge University Press, New York. 304 p.

Vincent, W.F. & C. Howard-Williams, 1986. Antarctic streams ecosystems: Physiological ecology of a blue-green algal epilithon. *Freshwater Bioogy.* 16:219-233.

Vincent, W.F. & C. Howard-Williams, 1989. Microbial communities in southern Victoria Land streams (Antarctica) II. The effects of low temperature. *Hydrobiology* 172:39-49.

Vincent, W.F., M.T. Downes, R.W. Castenholz & C. Howard-Williams, 1993. Community structure and pigment organization of cyanobacteria-dominated microbial mats in Antarctica. *European Journal of Phycology* 28:213-221.

Webster, G.C. & A.W. Frenkel, 1953. Some respiratory characteristics of the blue-green alga, *Anabeana*. *Plant Physiology* 28:61-69.

Wharton Jr., R.A., B.C. Parker & G.M. Simmons Jr., 1983. Distribution, species composition and morphology of algal mats in Antarctic dry valley lakes. *Phycologia.* 22:355-365.

Wharton Jr., R.A., G.M. Simmons Jr. & C.P. McKay 1989. Perennially ice-covered Lake Hoare, Antarctica: physical environment, biology, and sedimentation. *Hydrobiologia* 172:305-320.

Wilson, A.T., 1981. A review of the geochemistry and lake physics of the Antarctic dry areas. In: McGinnis, L. (Ed.), *Dry Valley Drilling Project. Antarctic Research Series* 33. American Geophysical Union, Washington D.C. pp. 185-192.

Ecosystem Processes in Antarctic Ice-free Landscapes, Lyons, Howard-Williams & Hawes (eds)
© 1997 Balkema, Rotterdam, ISBN 90 5410 925 4

A simple temperature-based model for the chemistry of melt-water ponds in the Darwin Glacier area, 80°S

Michael H.Timperley
National Institute of Water and Atmospheric Research, Auckland, New Zealand

ABSTRACT: The total dissolved salts concentrations (TDS) in the stream flowing into Lake Wilson, and in 15 melt water ponds and associated streams of the Darwin Glacier area, ranged from 15 g m^{-3} to 5500 g m^{-3}. Principal components analysis of the standardised major ion and NO$_3$ concentrations identified three components accounting for 89% of the data variance. Component 1 accounting for 61% of the variance, contained positive loadings on Na, Mg, Cl, and NO$_3$, and negative loadings on Ca, K, HCO$_3$ and SO$_4$. This result is consistent with the salts in the pond catchments originating from atmospheric deposition of sea salts, plus NO$_3$ and SO$_4$ and weathering processes in the catchment soils. It is suggested that the loading for SO$_4$, which does not follow from the probable atmospheric origin of this ion, reflects the fact that as the ions move through the catchment soils towards the ponds, the less mobile SO$_4$ salts are retarded together with the CO$_3$ salts from weathering, whereas the other ions of atmospheric origin, Cl and NO$_3$, form mobile salts which move more readily. The relative concentrations of these mobile ions in the pond waters increase with the TDS concentration and a model is proposed to explain both this observation and the wide range of TDS concentrations in ponds within a small area. It is proposed that during particularly warm summers the frozen ground water table lowers sufficiently to allow some ponds to drain into the ponds in the next lower catchments. The draining waters are enriched, relative to the catchment inputs, in the more mobile Cl and NO$_3$ salts, and these ions would accumulate in the non-draining ponds increasing both the TDS concentrations and the ratios of Cl and NO$_3$ relative to CO$_3$ and SO$_4$.

1 INTRODUCTION

One of the most intriguing characteristics of the meltwater ponds that are common features of most Antarctic "dry" areas during summer, is the extremely wide range of dissolved salts concentrations that occur in ponds and their associated streams of relatively small areas (Green *et al.* 1989, Matsumoto *et al.* 1985; Torii *et al.* 1989, Webster *et al.* 1994). It is now generally accepted that the dissolved salts in the catchments of these ponds originate partly from atmospheric inputs of sea salts plus NO$_3$ and SO$_4$ and partly from chemical weathering in the pond catchments (Lyons & Mayewski 1993; Campbell & Claridge 1982; Claridge & Campbell 1968, 1977; Green & Canfield 1984; Green *et al.* 1989; Keys & Williams 1981; Miotke & Von Hodenberg 1983) but the reasons for the often very different TDS concentrations in closely adjacent ponds have not been elucidated. On the north side of the Darwin Glacier at 80°S there is an ice-free area with Lake Wilson as the prominent limnological feature (Webster *et al.* 1996). Within this area there exist a number of melt water ponds (Fig. 1) with widely differing water chemistry. In this paper I discuss this chemistry and the origins of the dissolved salts and propose a mechanism to explain the differences in water chemistry among the ponds.

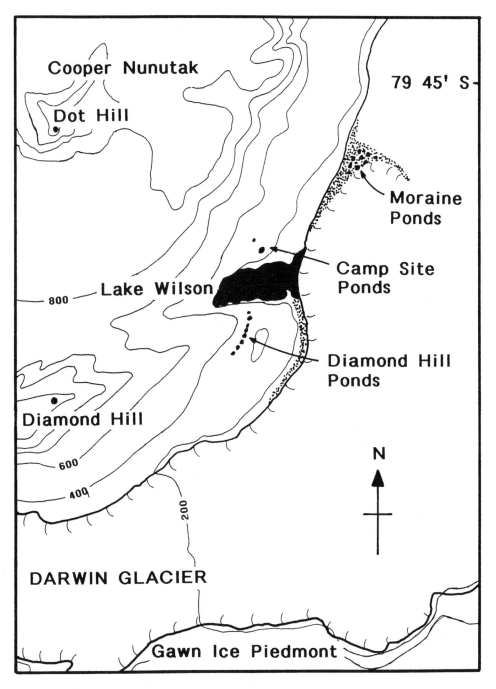

Fig.1. The ice-free area to the north of the Darwin Glacier showing the three groups of melt water ponds sampled in this study.

2   METHODS

Three groups of ponds were studied (Fig. 1). The camp-site ponds included one pond situated immediately below a large area of apparently permanent ice and fed by melt water from the ice. Water flowed from this pond via a small stream approximately 600 m long into a second larger pond with no obvious outflow. Samples were collected from both ponds and from two sites in the interconnecting stream. The moraine ponds comprised several ponds on lateral moraine of the Darwin Glacier. Six of these ponds lying in an area of approximately 200 m x 200 m were sampled. The third series of ponds were situated in a valley on the extended eastern slopes of Diamond Hill. Of the eight ponds in this series, four ponds occurring over a distance of about 800 m were sampled. The melt water stream flowing into Lake Wilson was also sampled. Duplicate samples were collected at all sites, except four samples were collected from the Diamond Hill pond with the highest concentration of TDS. The results from these replicate samples were included in the statistical analyses and, where appropriate, the results for all replicates are shown in the figures.

Water samples from the ponds were collected by hand from depths of 0.2 to 0.5 m into 1 l HDPE bottles which had been cleaned before use by soaking in HCl (5M) for several days, followed by repeated rinsing with distilled, deionised (DD) water and storing full of DD water for several weeks.

Water samples were analysed for Na, K, Ca, Mg, Cl, NO_3 and SO_4 by standard HP Ion Chromatography techniques. Bicarbonate ion was measured by titration with HCl to pH 4.5, sparging with N_2, and back-titration with NaOH, to the original sample pH to correct for interference from other ions. An estimate of the overall precision (sampling and analytical) of the data for each variable was obtained from the results for duplicate samples by expressing the standard deviation of the differences between all duplicates as a percentage of the mean concentration for all samples. The values for Na, K, Ca, Mg, Cl, SO_4, NO_3, HCO_3 and organic-N were 6.2, 2.5, 4.8, 2.5, 2.3, 3.5, 15, and 33 respectively. Analytical error was not determined separately from sampling error for this set of samples. Based on other samples, however, the analytical techniques would have contributed less than 10% of the variance in the differences between duplicates. As noted above, the results for duplicate samples and consequently, both sampling and analytical error, are included in the statistical analyses below. The influence of the cleaning procedure for the sample containers on Cl concentrations can be assessed from the analyses of the two samples collected for the Lake Wilson inlet stream. If it is assumed that the difference between cations and anions in the ion balance is due solely to Cl then one sample was deficient in Cl by 1.8 g m^{-3} and the other contained a surplus of 0.07 g m^{-3}. These results imply that the cleaning procedure contributed negligible amounts of Cl to the samples relative to the total error between duplicate samples.

The Systat package was used for statistical analyses (Wilkinson 1990).

3   RESULTS AND DISCUSSION

The ranges of ion concentrations in the waters sampled are given in Table 1. The Lake Wilson inlet stream was a dilute water dominated by Ca and HCO_3. The moraine pond waters were mixtures of similar amounts of Cl, SO_4 and HCO_3 salts, with TDS concentrations spanning an order of magnitude. The camp site ponds and the waters in the connecting stream contained mostly Cl and SO_4 salts, and the TDS concentrations differed by about a factor of 2. The TDS concentrations in the ponds on Diamond Hill varied by 2 orders of magnitude. The dominant anions in the more dilute waters were Cl and SO_4, whereas in the more concentrated waters Cl and NO_3 were the main anions. A notable feature of all these pond waters was the presence of NO_3 at concentrations typical of a major ion. Substantial amounts of organic-N were found in some of the pond waters with concentrations often, but not always, correlated with those of NO_3.

The ionic compositions of the pond waters were examined by principal components analysis (PCA). To eliminate the influence of the strong correlations among the ionic concentrations because of the wide range of TDS concentrations among the ponds, each ion concentration was standardised by dividing by the TDS concentration of the sample. The PCA therefore factored the relative proportions of the ions.

The results of the PCA are given in Table 2. Component 1 explained 61% of the total variance in the data, with strong positive loadings on Na, Mg, Cl, and NO_3, strong negative loadings on Ca and HCO_3, and weaker negative loadings on K and SO_4. This result implies that the pond water compositions are mostly mixtures of two groups of ions. The division of the pond water ions into these two groups would seem to be consistent with Na, Mg and Cl

Table 1. Concentration ranges (g m^{-3}) for major ions, nitrate and organic-N in melt waters of the Darwin Glacier area.

Ion	Lake Wilson inlet stream	Moraine ponds	Diamond Hill ponds	Camp site ponds
Na	0.72	1.2 - 11	6.5 - 780	17 - 25
K	0.97	1.1 - 4.6	1.3 - 160	2.5 - 4.3
Ca	3.8	1.9 - 19	3.0 - 300	10 - 23
Mg	0.31	0.43 - 4.3	1.2 - 340	4.7 - 11
Cl	2.4	2.3 - 27	9.7 -2000	27 - 59
SO_4	1.8	2.9 - 42	9.0 - 500	24 - 49
HCO_3	7.7	2.0 - 38	3.9 - 60	6.5 - 22
NO_3	0.68	0.05 - 1.0	5.0 -1300	22 - 53
org-N	0.002	0.01 - 0.21	0.05 - 1.1	0.02 - 0.24

Table 2. Principal component loadings for melt water dissolved major constituents

Variable	Component		
	1	2	3
Na	0.872	0.213	0.300
K	- 0.631	0.449	0.575
Ca	- 0.945	0.058	- 0.225
Mg	0.778	0.374	- 0.190
Cl	0.813	- 0.158	0.379
HCO_3	- 0.908	0.260	- 0.192
SO_4	- 0.345	- 0.880	0.199
NO_3	0.787	- 0.098	- 0.480
% variance explained	61.0	15.8	12.0

Table 3. Median composition of the non-sea salt (NSS) component of pond waters

Ion	% NSS equivalent TDS equivalent	Concentrations (g m^{-3}) for TDS = 35 g m^{-3}
Na	0.04	0.009
K	5.1	2.0
Ca	31.5	6.3
Mg	12.0	1.5
HCO_3	11.8	7.2
SO_4	22.1	10.6
NO_3	12.1	7.5

200

originating from sea salts in which they are the dominant ions, and Ca and HCO_3 originating from chemical weathering of minerals in the pond catchments.

The origins of NO_3 and SO_4 are more obscure. It has been suggested that NO_3 in Antarctic soil salt deposits is of marine origin (Claridge & Campbell 1968; Keys & Williams 1981) although fixation by an upper atmosphere ionization process has been proposed as a possible contributor (Wilson & House 1965). Wada *et al.* (1981), on the basis of N isotope abundance, concluded that NO_3 in Antarctic soils is deposited from the atmosphere where it originates from photochemical and/or auroral activity. In the Arctic, however, N and S in recent snow and ice is believed to have partly anthropogenic origins, (Jaffe & Zukowski 1993, and references therein) primarily the burning of fossil fuels. Although the contributions from anthropogenic sources of N and S in Antarctica would probably be less than those in the Arctic, such sources for Antarctic N and S cannot be dismissed.

Keys & Williams (1981) demonstrated a distinct increasing trend in the proportion of NO_3 and SO_4 salts to Cl salts with increasing distance from the sea. If the sea is the origin of these ions then their result would seem to imply that with increasing distance from the sea, fractionation of sea salts in atmosphere aerosols enriches both NO_3 and SO_4 relative to Cl. Alternatively, the trend could indicate that NO_3 and SO_4 have an origin other than the sea. Tomiyama & Kitano (1985) studied S isotopes in the salts from the Wright Valley, and although their results were inconclusive for the Don Juan basin, SO_4 in the Vanda basin seemed to have originated from sea salts formed when the valley was a fiord. Overall, there seems to be general agreement among scientists that atmospheric deposition is the pathway by which a substantial proportion of both NO_3 and SO_4 arrives in Antarctic snow, meltwaters and soils, but there has been no conclusive proof of the original sources of these ions.

The association of NO_3 in component 1 with the other ions of atmospheric origin is, therefore, quite logical, but the association of SO_4 with the ions derived from weathering implies the influence of other processes. Although some SO_4 might originate from the weathering of sulfides in the pond catchments it is likely that most SO_4 is deposited from the atmosphere as discussed above. The association of SO_4 in component 2 with the ions derived from weathering is probably because the concentrations used in the PCA were those in the pond waters rather than in the catchment as a whole. Although all the ions deposited from the atmosphere will be distributed more or less equally over the catchments, Cl and NO_3 which form the most soluble salts will move more readily into the ponds. Sulfate salts are less soluble and, like the CO_3 salts from weathering, are likely to be less mobile in the catchment soils. The two components therefore reflect both the origins of the ions and their mobility in catchment soils. This concept of differential mobility of salts in Antarctic soils was first described by Wilson (1979) who proposed that as the relative humidity in the soils increased during summer the more deliquescent salts, i.e. chlorides, became more mobile. A similar explanation based on the different solubilities of salts was used by Miotke & Von Hodenberg (1983) to explain the vertical migration of salts in Antarctic soils.

Component 2, which explains 15.8% of the data variance, has relatively weak positive loadings on Na, K, Mg, and HCO_3, probably reflecting a minor variation in the weathering process, but has a strong negative loading on SO_4. Overall, this component indicates that the pattern of SO_4 concentrations does not completely match the origins and behaviour of the other ions as indicated by component 1. This result is not surprising because SO_4 is unique among the ions; it is mostly of atmospheric origin but is of low mobility in the catchment soils. Component 3 explains 12 % of the data variance and has a strong negative loading on NO_3, probably reflecting the unique atmospheric origin of NO_3.

When the sea salt contribution (Mason 1966) is subtracted from the pond water compositions the median composition of the residual solution, given in Table 3, is a mixture of HCO_3, SO_4 and NO_3 salts of Ca and Mg. This composition is consistent with the weathering of Ca and Mg carbonates and ferromagnesian silicates by carbonic, sulphuric and nitric acids from the atmosphere.

As noted from the concentration data in Table 1, the influence of salt origins and the mobility of salts in the catchments on pond water chemistry, varies among the different ponds. Fig. 2 shows a plot of component 2 versus component 1, and it is apparent from this figure that with the exception of a small amount of overlap between some camp site and Diamond Hill ponds, the three groups of ponds are distinguishable according to their component 1 scores. The moraine ponds and the water in the stream flowing into Lake Wilson are dominated by the products of weathering, whereas the ponds of Diamond Hill are more influenced by the mobile ions of atmospheric origin. The camp site ponds and stream fit between the moraine and Diamond Hill ponds.

Negative scores on component 2 indicate a relatively strong SO_4 influence, and all samples from the camp site ponds and stream were strongly influenced by SO_4. The influence of SO_4 in the waters of the moraine and Diamond Hill ponds ranged from weak to strong. The reason for the strong SO_4 influence is believed to be related to the proximity of the ponds to melting ice. The ponds with the most negative scores (particularly the camp site ponds) all received melt water directly (i.e. without passing through soil) from ice on the edge of the ponds and in this situation SO_4 deposited from the atmosphere onto the ice would move into the pond waters just as readily as would NO_3 and Cl. In these pond catchments, therefore, part of the SO_4 was behaving as a mobile ion like Cl and NO_3, rather than as a relatively immobile ion characteristic of SO_4 deposited in catchment soils.

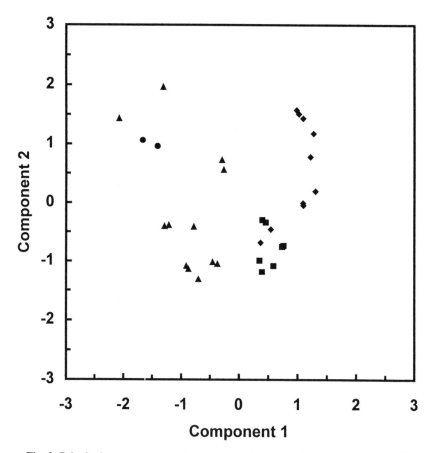

Fig. 2. Principal components analysis pond water scores for components 1 and 2.
◆ Diamond Hill ponds, ● Lake Wilson inlet, ▲ moraine ponds, ■ camp site ponds.

It would be expected that over the relatively small area in which these ponds exist both the atmospheric deposition of NO_3, SO_4 and sea salts, and the weathering processes in the catchments, would occur at almost constant rates. This would result in the same relative proportions of ions in all catchments, and even allowing for the influence of the ion transport processes identified in relation to component 1 from the PCA, the relative ion proportions in the pond waters would be expected to be reasonably similar. This was not the case. One possible explanation is that the rates of either, or both, atmospheric deposition and weathering vary over the area of the ponds. Although these rates are likely to vary over short distances and for short periods of time in response to changing climatic conditions, it is extremely unlikely that long term changes large enough to cause differences in salt accumulations in the catchments would occur between ponds only a few metres apart.

An alternative, more likely explanation is that the deposition and formation of salts occur at reasonably constant rates in all pond catchments, but the movement of ions through the catchment soils and into the ponds occurs at different rates in different catchments. A clue to the explanation of this behaviour is given by the plot of component 1 scores against the concentrations of TDS in Fig. 3. This figure shows that the combined effect on pond and stream water composition of atmospheric deposition, weathering and ion transport processes in the pond catchments, is related to the concentrations of TDS in the waters. At low concentrations of TDS the water chemistry is dominated by the immobile salts from weathering and SO_4, but

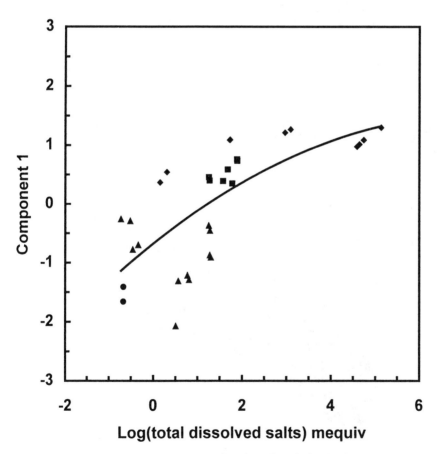

Fig. 3. Pond water component 1 scores as a function of total dissolved salts concentrations. ◆ Diamond Hill ponds, ● Lake Wilson inlet, ▲ moraine ponds, ■ camp site ponds.

as the TDS concentrations increase, the water chemistry becomes more influenced by the mobile salts from atmospheric deposition. If it is accepted that the rates of atmospheric deposition and weathering are similar in all pond catchments, then the plot in Fig. 3 implies that the transfer rates of the mobile ions Cl and NO_3 from the catchments into the ponds relative to the immobile CO_3 and SO_4 ions, is higher in the catchments of ponds with high concentrations of TDS than it is in the catchments of ponds with low concentrations of TDS.

One of the features of the ponds in the area we studied is their "nested" character, with many of the ponds having other higher altitude ponds lying within their catchments. The other feature of the ponds is that they appear to exist in depressions in either frozen ground water (camp site and Diamond Hill ponds), or in glacier ice (the moraine ponds). Although there are basement rock outcrops on the higher slopes of the valleys occupied by the Diamond Hill ponds, the water seeping through the lower walls of some of these ponds indicates that even if they are formed directly on basement rock, they are not completely retained by it.

To explain how the ionic composition of the pond waters might vary in the manner observed, consider two ponds, A and B, with pond A in the catchment of pond B as shown in Fig. 4. The possible progression of salt accumulation in the catchments and waters of the two ponds is shown in Fig. 5. Starting from some point in time, salts from atmospheric deposition and weathering accumulate in the catchments of the two ponds in amounts proportional to their catchment areas. It is assumed for the purposes of this explanation that the rates of salt accumulation in the catchments are the same irrespective of season. During summer, the more mobile Cl and NO_3 salts move rapidly into the ponds relative to the less mobile CO_3 and SO_4 salts which are largely retained in the catchment soils. This process continues until in one

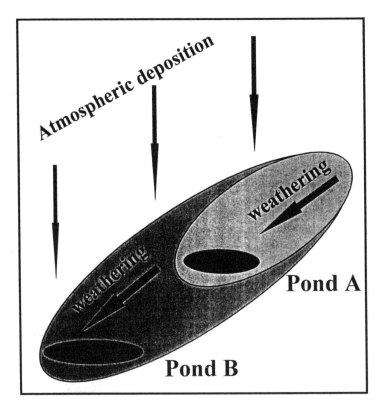

Fig 4. Atmospheric deposition and weathering in the catchments of two "nested" ponds; pond A in the catchment of pond B.

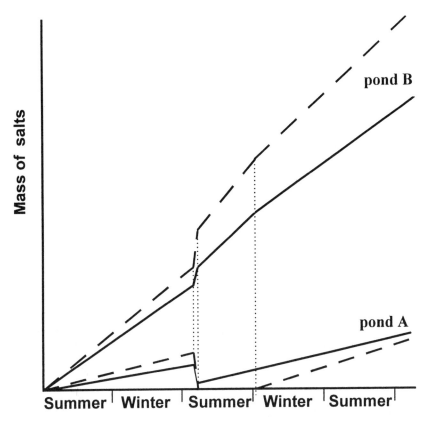

Fig. 5. The proposed progression of salt accumulation in the catchments of pond A and pond B (see Fig. 4). — immobile CO_3 and SO_4 salts, --- mobile Cl and NO_3 salts.

particularly warm summer the frozen ground water, or the glacier ice retaining the ponds, melts sufficiently to allow pond A to drain partially or completely into pond B. Similar losses of pond water because of the failure of ice dams has been noted for ponds on the McMurdo ice shelf near Bratina Island (Hawes *et al.* 1992) Of the total amount of salts accumulated in catchment A, the waters of pond A contain a greater proportion of the mobile ions than of the less mobile ions. When pond A drains these salts are transferred into the permanent catchment of pond B, and the water in this pond becomes enriched with the mobile Cl and NO_3 salts originally from pond A. The catchment of pond A reforms the following autumn but now this catchment is deficient in the mobile salts relative to the less mobile CO_3 and SO_4 salts. During the following summer when pond A again contains liquid water, it has a low concentrations of TDS dominated by Ca, CO_3 and SO_4. This model explains both the wide variation in TDS concentrations in ponds within small areas and also why the ionic compositions of the melt water changes from HCO_3 and SO_4 dominance at low TDS concentrations to, Cl and NO_3 dominance at high TDS concentrations.

This model implies that the increasing temperatures observed over the past two decades in the Darwin Glacier area and elsewhere (Webster *et al.* 1996) would have increased the transfer of salts, particularly the mobile Cl and NO_3 salts, into the "terminal", i.e. lower altitude ponds, of the nested series. As a result more melt water ponds would now have dilute waters dominated by Ca, HCO_3 and SO_4 ions, whereas those ponds with high concentrations of TDS would now be even more concentrated. The influence of pond water major ion chemistry on the often extensive cyanobacterial communities in these ponds is presently unclear. If my proposed model is correct, then these communities not only survive freezing, seasonal desiccation (Hawes *et al.* 1992) and the short term change in pond water ionic strength over the seasonal freeze/thaw cycle, but when a pond drains, they must also adapt to the sudden change in their long term aqueous environment, from a relatively concentrated salt solution to a solution which is much more dilute.

4   ACKNOWLEDGEMENTS

I am grateful to Clive Howard-Williams and Ian Hawes who studied the biology of these ponds, for their help with the sampling programme, and I thank them and also Jenny Webster and Malcolm Downes, the other members of our team, for their company and assistance during our field expedition.

REFERENCES

Campbell, I. B. & G.G.C. Claridge, 1982. The influence of moisture on the development of soils in the cold deserts of Antarctica. *Geoderma* 28:221-238.

Claridge, G.G.C. & I.B. Campbell, 1968. Origin of nitrate deposits. *Nature* 217:428-430.

Claridge, G.G.C. & I.B. Campbell, 1977. The salts in Antarctic soils, their distribution and relationship to soils processes. *Soil Science* 123:428-430.

Green, W.J. & D.E. Canfield, 1984. Geochemistry of the Onyx River (Wright Valley, Antarctica) and its role in the chemical evolution of Lake Vanda. *Geochimica et Cosmochimica Acta* 48:2457-2467.

Green, W.J., T.J. Gardner, T.G. Ferdelman, M.P. Angle, L.C. Varner & P. Nixon, 1989. Geochemical processes in the Lake Fryxell Basin (Victoria Land, Antarctica). *Hydrobiologia* 172:129-148.

Hawes, I., C. Howard-Williams & W.F. Vincent, 1992. Desiccation and recovery of antarctic cyanobacterial mats. *Polar Biology* 12:587-594.

Jaffe, D.A., & M.D. Zukowski, 1993. Nitrate deposition to the Alaskan snowpack. *Atmospheric Environment* 27A(17/18):2935-2941.

Keys, J.R. & K. Williams, 1981. Origins of crystalline, cold desert salts in the McMurdo regions, Antarctica. *Geochimica et Cosmochimica Acta* 45:2299-2309.

Lyons, W.B. & P.A. Mayewski, 1993. Geochemical evolution of terrestrial waters in the Antarctic: the role of water-rock interactions. In: Green, W. J. & E.I. Friedmann (Eds.), *Physical and biogeochemical processes in Antarctic lakes. Antarctic Research Series* 59. American Geophysical Union, Washington D.C. pp. 135-143.

Mason, B., 1966. *Principles of Geochemistry*. John Wiley & Sons, Inc., New York.

Matsumoto, G.I., K. Watanuki, & T. Torii, 1987. Total organic carbon in pond waters from the Labyrinth of southern Victoria Land in the Antarctic. *Antarctic Record* 31:171-185

Miotke, F-D. & R. Von Hodenberg, 1983. Salt fretting and chemical weathering in the Darwin Mountains and the dry valleys, Victoria Land, Antarctica. *Polar Geography and Geology* 7(2):83-122.

Torii, T., S. Nakaya, O. Matsubaya, G.I. Matsumoto, N. Masuda, T. Kawano & H. Murayama, 1989. Chemical characteristics of pond waters in the Labyrinth of southern Victoria Land, Antarctica. *Hydrobiologia* 172:255-264.

Tomiyama, C. & Y. Kitano, 1985. Salt origin in the Wright Valley, Antarctica. *Antarctic Record* 86:17-27.

Wada, E., R. Shibata & T. Torii, 1981. [15]N abundance in Antarctica: origin of soils nitrogen and ecological implications. *Nature* 292:327-329

Webster, J.G., K.L. Brown & W.F. Vincent, 1994. Geochemical processes affecting meltwater chemistry and the formation of saline ponds in the Victoria Valley and Bull Pass region, Antarctica. *Hydrobiologia* 281:171-186.

Webster, J., I. Hawes, M. Downes, M. Timperley & C. Howard-Williams, 1996. Evidence for regional climate change in the recent evolution of a high latitude pro-glacial lake. *Antarctic Science* 8(1):9-59.

Wilkinson, L., 1990. *Systat: the system for statistics*. Systat. Inc., Evanston, Ilinois. 4 volumes.

Wilson, A.T., 1979. Geochemical problems of the Antarctic dry areas. *Nature* 280:205-208.

Wilson, A.T. & D.A. House, 1965. Chemical composition of South Polar snow. *Journal of Geophysical Research* 70:5515-5518.

Ecosystem Processes in Antarctic Ice-free Landscapes, Lyons, Howard-Williams & Hawes (eds)
© *1997 Balkema, Rotterdam, ISBN 90 5410 925 4*

Palaeosalinity reconstruction from saline lake diatom assemblages in the Vestfold Hills, Antarctica

D. Roberts & A. McMinn
Antarctic CRC and Institute of Antarctic and Southern Ocean Studies, University of Tasmania, Hobart, Tas., Australia

ABSTRACT: The relationship between surface sediment diatom assemblages and measured limnological variables in thirty-three coastal Antarctic lakes from the Vestfold Hills was examined by constructing a diatom-water chemistry dataset. Canonical correspondence analysis of this dataset revealed that salinity accounted for a significant amount of the variation in the distribution of the diatom assemblages in this set of lakes. Weighted averaging regression and calibration of this diatom-salinity relationship established a transfer function for the reconstruction of past lakewater salinity from fossil diatom assemblages in this coastal Antarctic region. Application of this palaeosalinity reconstruction tool to a sediment core from Anderson Lake revealed distinct patterns of lakewater salinity change from the changes in species assemblages throughout the lake's history.

1   INTRODUCTION

Lakes in arid and semi-arid regions of the world respond rapidly to climate-driven hydrological change (Fritz *et al.* 1993), and because of the sensitivity of salt lake chemistry and biology to even small changes in the climate of the region, salt lakes are particularly well suited as sources of palaeoclimatic information (Williams 1981; Last & Schweyen 1983; Evans 1993).

Diatoms are sensitive ecological indicators and many researchers have demonstrated their use in reconstructing a number of limnological variables(e.g. Gasse & Tekaia 1983; Hall & Smol 1992; Bennion 1994), including salinity (e.g. Fritz *et al.* 1991, 1993, 1994). Their remains in dated lake sediment cores in temperate and tropical regions have been used extensively for reconstructing past changes in water chemistry and show great potential for studying environmental change in the Antarctic (Bjorck *et al.* 1991; Wasell & Håkansson 1992; Jones *et al.* 1993; Jones & Juggins 1995).

The variety of lakes found in the Vestfold Hills, an ice free landscape on the coast of east Antarctica, provide a unique environment in which to investigate the response of diatoms to changing water chemistries. As the ice sheet retreated from the Vestfold Hills, melt water and sea water filled the pre-existing low areas to form the lakes which are present today. Their current water chemistries are a product of their history and their present water balance. As salinity (and ionic composition) reflects effective precipitation (i.e. precipitation minus evaporation) on a broad scale, the relationship between salinity and diatom distribution and abundance enables fossil diatom assemblages to indicate past precipitation/evaporation gradients. This method allows past salinity to be inferred for saline lakes in the Vestfold Hills and, consequently lake level and climatic history to be reconstructed.

2   SITE DESCRIPTION

The Vestfold Hills form a 400 km^2 ice-free area on the coast of east Antarctica (68°25'S-68°40S, 77°50'E-78°35'E). The regional geology of the Prydz Bay coastline has been described by Collerson & Sheraton (1986) and the local geology of the Vestfold Hills described in detail by Pickard (1986). These hills have a low-lying, rugged topography with a maximum elevation of 158 m a.s.l. (Gore 1992). Glacial land-forms dominate the geomorphology of the Vestfold Hills (Adamson & Pickard 1986) and prominent relict marine terraces, approximately 6000 years BP in age and now at 6 m a.s.l., testify to the marine conditions prevailing in the Vestfold Hills following the retreat of the ice-sheet in the early Holocene (Bird et al. 1991; Peterson et al. 1988).

The climate of this area is discussed in detail by Streten (1986). Owing largely to the low temperature of the Antarctic atmosphere, water vapour content is an order of magnitude less than that of temperate latitudes and consequently precipitation is very low (Horowitz et al. 1971).

A prominent feature of the Vestfold Hills are the numerous lakes which range in size, depth, salinity and history. There are approximately 300 lakes, at least 30 of which are meromictic, in which water balance and salinity history has been recorded in fossil diatom assemblages. Thirty-three lakes spanning a large gradient in salinity, ranging from fresh through to hypersaline (summer epilimnion salinity values range from 0.5 to 165‰) were selected for a training dataset, i.e. a collection of modern lake surface sediment diatom assemblages representative of an integrated sample of the various living diatom communities in the lake related to contemporary water chemistry (Jones et al. 1993) (Table 1). These lakes, which were mostly derived from marine inlets by isolation, uplift and evaporation (Pickard 1986), are representative of the water chemistry environments encountered within the Vestfold Hills lake district, and they provide a useful selection from which to study the response of diatoms to environmental variability.

3   METHODS

Surface sediment diatom samples were collected during November-December 1992 and November-December 1994 from the training set lakes. The sediment/water interface was collected from the deepest part of each lake with a Glew Corer (Glew 1989) or Eckman Grab. A sediment core was collected from Anderson Lake during November 1991 by Dr Michael Bird of the Research School of Earth Sciences, Australian National University. A piston corer was used to collect the core from the northern arm of the lake through a 1.4 m thick ice cover. The core was frozen and divided into subsamples for diatom analysis. Radiocarbon dating of selected sections of the core were also supplied by Dr. Bird.

Water samples were collected with a 2 l Kemmerer bottle during November and December 1994. All samples were collected from the epilimnion (0 to 2 m depth). Samples were stored in 125-ml polyethylene bottles, frozen at -20°C and later analysed for major ionic concentrations following the methodology of Mackereth et al. (1978). Major nutrient concentrations were analysed with an ALPKEM autoanalyser. Field measurements of salinity were recorded with a CTD meter (Platypus Engineers, Hobart, Tasmania). Measurements were taken at 1 m intervals in all lakes. Salinity of each water sample was calculated from individual CTD profiles using a conductivity-salinity conversion equation (Fofonoff & Millard 1983). Salinity of the markedly hypersaline lakes sampled (>100‰ at 2 m) was converted in the same manner although conversion equations are not designed for such high salinity concentrations. These markedly hypersaline lakes were nevertheless included in all analyses, although actual salinity for these lakes could be higher than that estimated here. Each lake is categorised by the salinity range encountered: a - from surface to bottom if the lake is of holomictic or uncertain mixing status; or b - within the mixolimnion where meromixis occurs. The salinity categories used herein are fresh (<3‰), hyposaline (3 to 30‰), marine (30 to 35‰) and hypersaline (>35‰) (Table 1).

All samples (surface sediment and core sediments respectively) were analysed following the methodology of Battarbee (1986). Diatom species are expressed as relative abundances (% total diatoms) of the 1200 frustules counted per surface sediment sample and the 400 frustules counted per core sample. Identification and taxonomy of the diatom species enumerated were based principally on: Round et $al.$ (1990), Medlin & Priddle (1990), Priddle & Fryxell (1985) and Wasell (1993). Diatoms that occurred in at least one lake at ≥2% abundance (47 in total) were used in the numerical analyses. Taxa are listed with reference number and author citation in Table 2.

The first step in a transfer function is the collection of species data and current water chemistry parameters; a training dataset. Canonical correspondence analysis (CCA), a direct gradient ordination technique (Ter Braak 1986), was used to identify possible relationships between diatom distributions and abundances and the initial 11 measured environmental variables (salinity, Na, Ca, K, Mg, Cl, SO_4, alkalinity, NO_3, PO_4, and SiO_2) (Roberts & McMinn 1996). All ordinations were performed using CANOCO version 3.12 (Ter Braak 1988, 1990). Forward selection identified salinity and silicate as variables that contributed significantly and independently ($P<0.05$) to the explanation of the variation in the species data. Salinity independently and significantly explained 38% ($P = 0.01$) of the variance explained by all variables. The more important an environmental variable is in explaining the species data, the larger the first constrained axis will be in comparison with the second unconstrained axis (Bennion 1994). Ratios greater than 0.50 suggest that there is an independent diatom signal for that variable (Dixit et $al.$ 1991). A ratio of 0.85 for salinity indicates this variable is a viable and useful environmental variable for a diatom-based transfer function.

Weighted averaging regression and calibration (Ter Braak & Barendregt 1986; Ter Braak & Prentice 1988) was applied to the 33 lakes and 47 selected diatom taxa in the training dataset, to generate a transfer function for inferring lakewater salinity from fossil diatom assemblages in the Vestfold Hills. Weighted averaging (WA) estimates salinity optima for species in current lake environments in the Vestfold Hills and infers past lakewater salinity from the relative abundance of these species in lake sediment cores. Weighted averaging with tolerance downweighting (WA(tol)) takes into account species salinity tolerances by downweighting each species by its variance for salinity. Analysis of the dataset with both WA and WA(tol) resulted in the selection of WA(tol) for palaeosalinity reconstructions as it gave lower error estimates for both the root mean square of the error (RMSE) (0.32 and 0.25 respectively) and the jackknifed RMSE (0.36 and 0.34 respectively), and higher squared correlation between observed and inferred salinity values (r^2) (0.78 and 0.86 respectively). The RMSE and r^2 give a measure of the "apparent" error in the model, while the jackknife RMSE is a more reliable indicator of the true predictive ability of a transfer function (Dixon 1993; Jones & Juggins 1995).

In WA reconstructions, averages are taken twice, once in WA regression and once in WA calibration. The resulting shrinkage of the inferred environment variable is corrected for by using inverse or classical deshrinking regression. Following analyses using both methods, classical deshrinking was chosen as the more reliable estimate method for palaeosalinity in the Vestfold Hills.

The resulting WA(tol) classical deshrinked transfer function was applied to fossil diatom assemblages enumerated from the Anderson Lake core.

WA regression and calibration (both with and without tolerance downweighting) were performed using CALIBRATE (Juggins & Ter Braak 1994).

Table 1. Relevant physico-chemical details for lakes sampled. Note: "Admin", "LP1", "LP2" and "Pointed" are informal names not approved by the Antarctic Names and Polar Medal Committee. All entries are collated from 1994 information.

Lake Name	Location (°S)	Location (°E)	Elevation (m)	Maximum Recorded Depth (m)	Salinity (‰ at 2m)	Salinity Range (min - max ‰)	Mixing Status	Lake category
Abraxas	68 29.5	78 17	13.11	23	15.8	15.36 - 22.85	meromictic	hyposaline
Ace	68 28.4	78 11.1	8.91	25	16.2	16.19 - 40.35	meromictic	hypo - marine
"Admin"	68 27.2	78 16.5	0.95	6	14.8	14.84 - 17.53	holomictic ?	hyposaline
Anderson	68 36.0	78 10.0	3.50	21	62.9	57.24 - 144.05	meromictic	hypersaline
Burch	68 27.3	78 16.0	-0.07	7	138	135.02 - 167.83	meromictic ?	hypersaline
Burton	68 37.5	78 06.0	0.11	16.2	41.6	41.52 - 42.62	meromictic	hypersaline
Camp	68 32.5	78 04.5	–	7.4	16.4	15.56 - 18.58	uncertain	hyposaline
Clear	68 39.0	78 00.0	-8.28	60.5	10.7	8.73 - 13.84	meromictic	hyposaline
Collerson	68 35.0	78 11.0	4.99	8.2	8.6	7.74 - 9.33	holomictic	hyposaline
Ekho	68 31.0	78 15.5	-1.405	39	52.0	46.13 - 149.14	meromictic	hypersaline
Fletcher	68 27.0	78 16.0	0.36	12	65.3	65.30 - 100.01	meromictic	hypersaline
Franzmann	68 29.0	78 14.9	–	8.5	71.4	71.16 - 96.02	meromictic	hypersaline
Grace	68 25.3	78 27.5	–	3	1.1	0.57 - 1.17	holomictic	fresh
Hand	68 33.2	78 19.0	9.55	29	5.5	4.90 - 5.59	meromictic ?	hyposaline
Highway	68 27.9	78 11.3	8.30	17.4	4.7	4.70 - 5.11	holomictic	hyposaline
Johnstone	68 30.0	78 25.0	–	9.8	157	155.80 - 167.42	meromictic	hypersaline
Lichen	68 28.8	78 26.0	–	26	0.5	0.50 - 0.61	holomictic	fresh
"LP 1"	68 28.6	78 16.0	7.51	4.9	74.1	73.94 - 127.38	meromictic	hypersaline
"LP 2"	68 28.6	78 15.5	7.51	1.8	140	132.77 - 170.22	holomictic	hypersaline
McCallum	68 38.0	78 01.0	-1.71	32	14.5	10.05 - 23.92	meromictic	hyposaline
McNeil	68 35.6	78 22.0	27.30	3.8	8.8	8.35 - 11.05	holomictic?	hyposaline
Oblong	68 37.5	78 14.2	-2.89	14.8	165	148.43 - 178.05	meromictic	hypersaline
Organic	68 27.5	78 11.5	2.75	7	142	138.78 - 177.07	meromictic	hypersaline
Oval	68 32.0	78 16.0	-28.44	16	143	142.86 - 175.31	meromictic	hypersaline
Pendant	68 27.7	78 14.5	3.045	18.4	13.6	13.53 - 36.60	meromictic	hyposaline
"Pointed"	68 31.5	78 19.75	5.52	5	5.1	5.04 - 5.14	holomictic	hyposaline
Scale	68 35.0	78 10.0	–	10.6	16.3	16.29 - 32.39	meromictic	hypo - marine
Shield	68 32.0	78 15.0	-6.915	33	77.5	71.31 - 154.14	meromictic	hypersaline
South Angle	68 37.5	77 55.0	-0.385	20	138	104.64 - 181.54	meromictic	hypersaline
Vereteno	68 31.2	78 24.0	0.96	25	3.7	3.65 - 3.75	holomictic	hyposaline
Watts	68 36.0	78 11.0	–	29.5	2.3	2.24 - 2.40	holomictic	fresh
Weddell	68 33.2	78 06.5	–	6	59.4	58.91 - 72.61	holomictic	hypersaline
Williams	68 29.0	78 09.5	1.165	7	46.9	46.56 - 134.31	meromictic	hypersaline

4 RESULTS

4.1 Training data set

Canonical correspondence analysis revealed that diatom distribution and abundance in the Vestfold Hills lakes is clearly related to measured chemical gradients (Roberts & McMinn 1996). On the basis of the significant fraction of the variance in the diatom data accounted for by salinity (38%) and its high inference ratio (0.85), salinity was chosen as an appropriate variable for a diatom-based transfer function for the Vestfold lakes.

4.2 Weighted averaging regression and calibration

Weighted averaging with tolerance downweighting and classical deshrinking regression was applied to the full training dataset of 33 lakes and 47 diatoms, with salinity as the environmental variable to be inferred. Salinity was log-transformed (base-ten) to reduce the skewed distribution (Zar 1984).

Statistical results for both simple and tolerance downweighted WA indicated that the predictive ability of the WA(tol) model, in terms of the RMSE, r^2 and the jackknifed RMSE, is better than the simple WA model. Therefore, salinity inferences were calculated from both the optima and tolerance of the 47 diatom taxa in the 33 surface sediment samples. The majority of the diatom taxa had salinity optima in the hyposaline and hypersaline ranges, with 4 freshwater taxa, 22 hyposaline taxa, 2 marine taxa and 19 hypersaline taxa revealed (Fig. 1).

Diatom-inferred salinity concentrations ($\log_{10}$ ppt) were then derived from the regression equation: Final x_i =(Initial x_i-0.53)/0.61, where x_i is the inferred value of the environmental variable x for the sample i (Ter Braak & Van Dam 1989; Birks et al. 1990).

4.3 Application of the transfer function

Anderson Lake is currently a closed basin meromictic, hypersaline lake situated on Mule Peninsula, the southernmost of the three Vestfold Peninsulas, at an elevation of 3.5 m a.s.l. Lewis (1994) recorded a surface area and volume of ~0.12 km^2 and ~1.35 Mm^3 respectively. The current summer mixolimnion (i.e. 1994) salinities range from 57 to 75‰. The mixolimnion freezes seasonally to ~1.8 m depth and the salty ice tends not to thaw completely (Pickard et al. 1986). The monimolimnion is an anoxic environment in which no diatom growth takes place. Water temperatures are currently as high as 11°C below the ~3 to 4 m oxycline. Surveys by Seppelt et al. (1988) revealed no mosses or algae, and a low abundance of lichens, near Anderson Lake. Therefore, negligible amounts of allochthonous organic matter are probably ending up in the lake, indicating that organic material preserved in the sediments is from in situ biological production (Lewis 1994).

The WA(tol) classical deshrinked transfer function was applied to the fossil diatom assemblages enumerated from the sediment core taken from Anderson Lake. The core extended from the surface (0 cm) to 135.50 cm, although below 109.75 cm dissolution excluded the identification of any diatom species. These highly dissolved basal sediments are glacial moraine of undetermined age (Lewis 1994). Diatom stratigraphy of the remaining sediments reveal that Anderson Lake has been previously exposed to freshwater, marine and polar desert influences.

Looking at individual species optima as derived from the WA(tol) model (Fig. 1), and taking into account relative abundances throughout the training lakes, certain species reveal themselves to be excellent indicator taxa for specific salinity categories.

Table 2. List of diatom species included in numerical analyses, reference code names and authorities.

	Pennales
ACde	*Achnanthes* cf. *delicatula* (Kützing) Grunow
ACpe	*Achnanthes* cf. *petersenii* Hustedt
ACsa	*Amphora* species a
ACve	*Amphora veneta* Kützing
Acsb	*Amphora* species b
ACsc	*Amphora* species c
ACsd	*Amphora* species d
COco	*Cocconeis costata* Gregory
COfa	*Cocconeis fasciolata* (Ehrenberg) Brown
COpi	*Cocconeis pinnata* Gregory
FRcv	*Fragilaria construens* var. *venter* Ehrenberg
FRpi	*Fragilaria pinnata* Ehrenberg
HAvi	*Hantzschia virgata* (Roper) Grunow
NAdi	*Navicula directa* W. Smith
NAgl	*Navicula glaciei* Van Heurck
Namu	*Navicula mutica/Navicula muticopsis* Kützing/Van Heurck
NAtr	*Navicula tripunctata* Muller
NAsa	*Navicula* species a
NAum	*Navicula* cf. *seminulum* Grunow
NAsb	*Navicula* species b
NAsd	*Navicula* species d
NApe	*Navicula perminuta* Grunow
NAse	*Navicula* species e
NAgr	*Navicula* cf. *gregaria* Donkin
NAsf	*Navicula* species f
NIcu	*Fragilariopsis curta* (Van Heurck) Hustedt
NIcy	*Fragilariopsis cylindrus* Grunow
NIle	*Nitzschia lecointei* Van Heurck
NIsa	*Nitzschia* species a
NIsb	*Nitzschia* species b
SYsa	*Synedra* species a
PIcy	*Pinnularia cymatopleura* West & West
PIlu	*Pinnularia lundii* Hustedt
PImm	*Pinnularia microstauron* var. *microstauron* (Ehrenberg) Cleve
PImi	*Pinnularia microstauron* (Ehrenberg) Cleve
PIqa	*Pinnularia quadratera* var. a (A. Schmidt) Cleve
PIvi	*Pinnularia viridis* Nitzsch
PIvc	*Pinnularia viridis* var. "*constricta*" Ehrenberg
STsa	*Stauroneis* species a
STan	*Stauroneis* cf. *anceps* Ehrenberg
STsl	*Stauroneis* cf. *salina* Smith
TRas	*Trachyneis aspera* Ehrenberg
TRma	*Tryblionella marginulata* Grunow
UNsa	Genus indetermined species a
	Centrales
CHce	*Chaetoceros* species
CHsp	*Chaetoceros* spores
THan	*Thalassiosira antarctica* Comber

Optimum Salinity (log10 ppt)

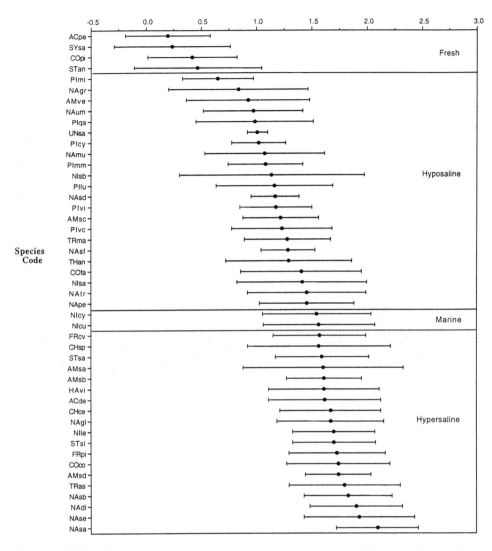

Fig. 1. Species optima and tolerances for salinity as determined by WA(tol) diatom-salinity model. Species are arranged in order of increasing optima for lakewater salinity and assigned a salinity category. Species codes are those given in Table 2.

Navicula directa, Fragilariopsis cylindrus and *Pinnularia microstauron* have been selected as taxa representative of hypersaline, marine, and hyposaline to fresh conditions respectively. Distinct changes in the abundance of these indicator diatom species occur throughout the core (Fig. 2). Based solely on diatom distribution and abundance the core can be divided into 3 zones or units with representative taxa (Figs. 2 & 3):

• Zone 1 (109.75 to 102 cm): The base of the core is almost completely dominated by *Pinnularia microstauron*. This hyposaline-freshwater taxon has an estimated salinity optimum of 3.48‰ with a tolerance of 1.10‰. Below 102 cm this species accounts for 9-73% of the total diatoms enumerated.

213

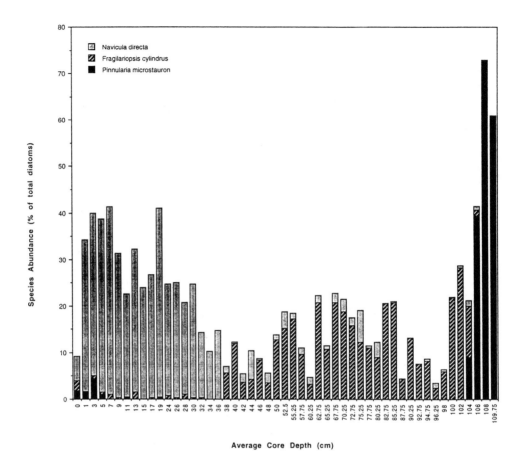

Fig. 2. Selected indicator taxa throughout Anderson Lake core sediments.

- Zone 2 (102 to 36 cm): Between 38 and 102 cm inclusive *Nitzschia* and *Fragilariopsis* species, particularly *F. cylindrus*, and *Chaetoceros* cells dominate, reaching abundances of up to 28 and 68% respectively. These marine diatoms have estimated optimas of 33.87 and 45.36‰ respectively. *Pinnularia microstauron* was not found in this section of the core.
- Zone 3 (36 cm to 0 cm): From 36 cm to the top of the core *Navicula directa* dominates the sediment, becoming markedly abundant (5 to 40 %). This zone is also dominated by *Navicula* sp. e and *Navicula* sp. a. All 3 species have high estimated salinity optima's (77.89, 83.45 and 122.42‰ respectively). *Fragilariopsis cylindrus* and *Chaetoceros* cells still occur in this core section but decline in abundance from a mean average abundance of 4.40 and 37.35% respectively in zone 2, to a mean average abundance 0.51 and 7.07% respectively in this section..

An interesting observation in the very top section of this zone (7 to 0 cm) is the re-appearance of *Pinnularia microstauron*. In this sub-zone *P. microstauron* abundance rises from 0 to 4.5 % of total diatoms counted. Therefore, while zone 3 is representative of a saline lake state, there is evidence of widely fluctuating salinity. Relative abundances of *Navicula directa* in comparison with *Fragilariopsis cylindrus* and *Pinnularia microstauron* account for the observed fluctuations in salinity within this zone and represent phases of wetter versus drier periods in the recent history of the lake core.

Three radiocarbon dates supplied for this core at 111 cm (8450±210), 104 cm

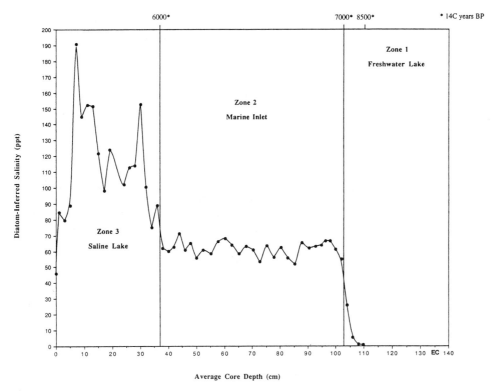

Fig. 3. Reconstructed palaeosalinity for Anderson lake showing radiocarbon dated diatom-salinity inferred zones. EC - end force.

(7110±270) and 40 cm (6730±200) respectively (uncorrected for reservoir effects), reveal the time frame of the inferred salinity changes. Consequently, the Holocene environmental history of Anderson Lake can be reconstructed as follows:

-       8500 - 7000 ^{14}C yr BP fresh water sediment deposited
-       7000 - 6000 ^{14}C yr BP marine sediment deposited
-       6000 ^{14}C yr BP - present hypersaline lake sediment deposited

5   DISCUSSION

The Holocene epoch covers the present interglacial period which started about 10 000 yr BP. The environmental history of Anderson Lake has been recorded in the lake sediments within this time frame. At the beginning of the Holocene, the ice sheet retreated causing isostatic uplift of the bedrock (Lewis 1994) and exposing Anderson basin. Following this period, from approximately 8500 to 7000 ^{14}C yr BP, a freshwater lake existed dominated by benthic diatoms with low salinity tolerance. Relative sea-level rise after this period led to marine inundation of the lake and marine sediments were deposited for approximately 1000 years. The abundance of the planktonic taxa in this zone indicates flushing with seawater. Continued isostatic rebound reisolated the basin from the sea at around 6000 ^{14}C yr BP. The isolated basin developed a negative water balance (evaporation greater than precipitation), which led to a concentration of dissolved salts and the resulting hypersalinity of the present day lake basin dominated by benthic diatoms with high salinity tolerances.

During dry intervals, evaporation exceeds precipitation, lakewater levels in closed basins drop and the concentration of dissolved salts generally increases while, during wetter intervals, precipitation exceeds evaporation, lake levels rise and the lakewater salinity levels often decrease. In Antarctic ice free landscapes increasing evaporation and increasing salinity are related to cold periods, while warm periods tend to be related to rising lake levels and decreasing salinity (Wharton et al. 1992; Doran et al. 1994). These changes are recorded by biological indicators such as diatoms in the lakes sediments. Palaeolimnological analysis of these sediments allows these indicators to be used to reconstruct past variations in lake levels and salinity histories, from which long-term records of past climatic change can be inferred (Zeeb & Smol 1995). Diatoms enumerated from surface sediment assemblages in the lakes of the Vestfold Hills have revealed such indicator species for the hindcasting of salinity history in this region. The application of these indicator taxa via calibration models to lake sediments reveals past environmental conditions of individual lake basins in this lake district.

Interpretation of the diatom-based zones within the Anderson Lake core samples indicates that over the past 8500 ^{14}C years Anderson Lake has passed through distinct stages from a 7000 to 8500 yr old fresh water lake, through a marine inlet phase lasting approximately 1000 years, to finally 6000 years ago where it became the saline lake it is at present. Initially changes in the lakes basin are attributed to glacial retreat and isostatic uplift with relative sea-level changes. These processes have masked climatic cycles. However, within the last 6000 ^{14}C years (Zone 3; Fig. 3), while remaining saline, Anderson Lake has fluctuated between 190 ppt and 50 ppt. Cycles of increasing and decreasing salinity within this zone act as proxies for inferring differing precipitation/evaporation regimes acting on the lake basin.

Lake level research in the McMurdo Dry Valleys (160°-164°E, 76°30'-78°20'S), the largest ice free landscape in Antarctica, reveals similar lake histories. For example, a lake level history of Lake Vanda reveals a warmer period in the Dry Valleys ~2000-3000 years ago followed by a cold period at 1200 years BP (Smith & Friedman 1993). The marked lowering in the water level of Lake Vanda at 1200 years BP corresponds with the highest palaeosalinity estimate and therefore the coldest/driest period in Anderson Lakes basin. Lake levels in the Dry Valleys have been rising consistently in the past two decades, a results of increasing summer time surface air temperatures (Wharton et al. 1992; Doran et al. 1994). Recent salinity decreases in Anderson Lake (Zone 3; Fig. 3) are suggestive of increasing precipitation in this lake's basin, perhaps the result of a similar increase in summertime surface temperatures in the Vestfold Hills.

Consequently, based solely on diatom-inferred salinity changes, it can be suggested that Anderson Lake has gone from a polar desert freshwater lake, through a transitional marine phase, to its present hypersaline state. Recent salinity inference from this lake suggests that, while still saline, it is currently in a wet cycle. This is suggestive of an increase in precipitation in the Vestfold Hills region.

The diatom-salinity model developed here will be applied to several more cores from this region to allow a more complete environmental reconstruction of this area.

6   CONCLUSIONS

Palaeolimnological sediments provide an excellent record of past diatom assemblages. The Vestfold Hills diatom-salinity transfer function developed herein has shown that it is possible to use these diatom assemblages to reconstruct water column (epilimnetic) salinity reliably.

7   ACKNOWLEDGEMENTS

We would like to thank Steve Juggins and Cajo Ter Braak for a beta test version of

CALIBRATE, John Gibson and Mark Wapstra for field assistance, and Dom Hodgson, Steve Juggins and Wim Vyverman for valuable statistical advice.

REFERENCES

Adamson, D.A. & J. Pickard, 1986. Cainozoic history of the Vestfold Hills. In: Pickard, J. (Ed.), *Antarctic Oasis: terrestrial environments and history of the Vestfold Hills*. Academic Press, Sydney.

Battarbee, R.W., 1986. Diatom Analysis. In: Berglund, B.E. (Ed.), *Handbook of Holocene Palaeoecology and Palaeohydrology*. John Wiley & Sons Ltd., Chichester. pp. 527-570.

Bennion, H., 1994. A diatom-phosphorus transfer function for shallow, eutrophic ponds in southeast England. *Hydrobiologia* 275/276:391-410.

Bird, M.I., A.R. Chivas, C.J. Radnell & H.R. Burton, 1991. Sedimentological and stable-isotope evolution of lakes in the Vestfold Hills, Antarctica. *Palaeogeography, Palaeoclimatology, Palaeoecology* 84:109-130.

Birks, H.J.B., J.M. Line, S. Juggins, A.C. Stevenson & C.J.F. Ter Braak, 1990. Diatoms and pH reconstruction. *Philosophical Transactions of the Royal Society of London, B* 327:263-278.

Björck, S., H. Håkansson, R. Zale, W. Karlén & B.L. Jönsson, 1991. A late Holocene lake sediment sequence from Livingston Island, South Shetland Islands, with palaeoclimatic implications. *Antarctic Science* 3(1):61-72.

Collerson, K.D. & J.W. Sheraton, 1986. Bedrock geology and crustal evolution of the Vestfold Hills. In: Pickard, J. (Ed.), *Antarctic Oasis: terrestrial environments and history of the Vestfold Hills*. Academic Press, Sydney. pp. 21-62.

Dixit, S.S., A.S. Dixit & J.P. Smol, 1991. Multivariable environmental inferences based on diatom assemblages from Sudbury (Canada) lakes. *Freshwater Biology* 26:251-266.

Dixon, P.M., 1993. The Bootstrap and the Jackknife: Describing the Precision of Ecological Indicies. In: Scheiner, S.M. & Gurevitch, J. (Eds.), *Design and Analysis of Ecological Experiments*. Chapman & Hall Inc., New York. pp. 290-310.

Doran, P.T., R.A. Wharton Jr., & W.B. Lyons, 1994. Paleolimnology of the McMurdo Dry Valleys, Antarctica. *Journal of Paleolimnology* 10:85-114.

Evans, M.S., 1993. Paleolimnological studies of saline lakes. *Journal of Paleolimnology* 8:97-101.

Fofonoff, N.P. & R.C. Millard, 1983. Algorithms for computation of fundamental properties of seawater. *Technical Papers in Marine Science* 44:1-53.

Fritz, S.C., D.R. Engstrom & B.J. Haskell, 1994. "Little Ice Age" aridity in the North American Great Plains: a high resolution reconstruction of salinity fluctuations from Devils Lake, North Dakota, USA. *The Holocene* 4(1):69-73.

Fritz, S.C., S. Juggins, R.W. Battarbee & D.R. Engstrom, 1991. Reconstruction of past changes in salinity and climate using a diatom-based transfer function. *Nature* 352:706-708.

Fritz, S.C., S. Juggins & R.W. Battarbee, 1993. Diatom assemblages and ionic characterization of lakes of the Northern Great Plains, North America: a tool for reconstructing past salinity and climate fluctuations. *Canadian Journal of Fishing and Aquacultural Science* 50(9):1844-1855.

Gasse, F. & F. Tekaia, 1983. Transfer functions for estimating paleoecological conditions (pH) from East Africa. *Hydrobiologia* 103: 85-90.

Glew, J.R., 1989. A new type of mechanism for sediment samples. *Journal of. Paleolimnology* 2:241-243.

Gore, D.B., 1992. Ice-damming and fluvial erosion in the Vestfold Hills, East Antarctica. *Antarctic Science* 4(2):227-234.

Hall, R.I. & J.P. Smol, 1992. A weighted-averaging regression and calibration model for inferring total phosphorous concentration from diatoms in British Columbia (Canada) lakes. *Freshwater Biology* 27:417-434.

Hammer, U.T., 1986. *Saline Lake Ecosystems of the World*. Dr. Junk Publishers, Dordrecht. 616 p.

Horowitz, N.H., R.E. Cameron & J.S. Hubbard, 1971. Microbiology of the Dry Valleys of Antarctica. *Science* 176:242-245.

Jones, V.J., S. Juggins & J.C. Ellis-Evans, 1993. The relationship between chemistry and surface sediment diatom assemblages in maritime Antarctic lakes. *Antarctic Science* 5(4):339-348.

Jones, V.J. & S. Juggins, 1995. The construction of a diatom-based chlorophyll *a* transfer function and its application at three lakes on Signy Island (maritime Antarctic) subject to differing degrees of nutrient enrichment. *Freshwater Biology* 34:433-445.

Juggins, S. & C. Ter Braak, 1994. *CALIBRATE - a program for species-environment calibration by [weighted averaging] partial least squares regression*. Beta test version 0.54, supplied by S. Juggins, University of Newcastle, United Kingdom.

Last, W.M. & T.H. Schweyen, 1983. Sedimentology and geochemistry of saline lakes of the Great Plains. *Hydrobiologia* 105:245-263.

Lewis, A.H., 1994. *An environmental history of Anderson Lake (Vestfold Hills, Antarctica) derived from lake sediments*. MSc (Polar and Oceanic Science) Thesis, University of Tasmania, Australia.

Mackereth, F.J.H., J. Heron & J.F. Talling, 1978. Water Analysis: Some Revised Methods for Limnologists. *Freshwater Biological Association, Scientific Publication No. 36*:50-68.

Medlin, L.K. & J. Priddle, 1990. *Polar Marine Diatoms*. British Antarctic Survey, Natural Environment Research Council, Cambridge. Cambridge University Press. 214 p.

Peterson, J.A., B.L. Finlayson & Z. Qingsong, 1988. Changing distribution of late Quaternary terrestrial lacustrine and littoral environments in the Vestfold Hills, Antarctica. *Hydrobiologia* 165:221-226.

Pickard, J., 1986. *Antarctic Oasis: terrestrial environments and history of the Vestfold Hills*. Academic Press, Sydney. 351 p.

Pickard, J., D.A. Adamson, & C.W. Heath, 1986. The evolution of Watts Lake, Vestfold Hills, East Antarctica, from marine inlet to freshwater lake. *Palaeogeography Palaeoclimatology, Palaeoecology* 53:271-288.

Priddle, J. & G. Fryxell, 1985. *Handbook of the Common Plankton Diatoms of the Southern Ocean: Centrales except the Genus Thalassiosira*. British Antarctic Survey, Natural Environment Research Council. Cambridge University Press, Cambridge. 159 p.

Roberts, D. & A. McMinn, 1996. Relationships between surface sediment diatom assemblages and water chemistry gradients in saline lakes of the Vestfold Hills, Antarctica. *Antarctic Science* 8(4):331-341.

Round, F.E., R.M. Crawford & D.G. Mann, 1990. *The Diatoms: Biology & Morphology of the Genera*. Cambridge University Press, Cambridge. 747 p.

Seppelt, R.D., P.A. Broady, J. Pickard & D.A. Adamson, 1988. Plants and landscape in the Vestfold Hills, Antarctica. *Hydrobiologia* 165:185-196.

Smith, G.I. & I. Friedman, 1993. Lithology and paleoclimatic implications of lacustrine deposits around Lake Vanda and Don Juan Pond, Antarctica. In: Green, W.J. & E.I. Friedmann (Eds.), *Physical and Biogeochemical Processes in Antarctic Lakes. Antarctic Research Series* 59. American Geophysical Union, Washinigton D.C. pp. 83-94.

Streten, N.A., 1986. Climate of the Vestfold Hills. In: Pickard, J. (Ed.), *Antarctic Oasis: terrestrial environments and history of the Vestfold Hills*. Academic Press, Sydney. pp. 141-164.

Ter Braak, C.J.F., 1986. Canonical correspondence analysis: a new eigenvector technique for multivariate direct gradient analysis. *Ecology* 67(5):1167-1179.

Ter Braak, C.J.F., 1988. *CANOCO - a Fortran program for Canonical Community Ordination*. Microcomputer Power, Ithaca, New York.

Ter Braak, C.J.F., 1990. *Update Notes: CANOCO Version 3.10*. Microcomputer Power, Ithaara, New York. 35 p.

Ter Braak, C.J.F. & L.G. Barendregt, 1986. Weighted averaging of species indicator values: Its efficiency in environmental calibration. *Mathematical Biosciences* 78:57-72.

Ter Braak, C.J.F. & I.C. Prentice, 1988. A theory of gradient analysis. *Advances in Ecological Research* 18:271-313.

Ter Braak, C.J.F. & H. Van Dam, 1989. Inferring pH from diatoms: a comparison of old and new calibration methods. *Hydrobiologia* 178:209-223.

Wasell, A., 1993. Diatom Stratigraphy of Sediments from Nicholson Lake, Vestfold Hills, East Antarctica. In: Wasell, A. (Ed.), *Diatom stratigraphy and evidence of Holocene environmental changes in selected lake basins in the Antarctic and South Georgia*. Stockholm University, Department of Quaternary Research, Report 23, Paper II. 16 p.

Wasell, A. & H. Håkansson, 1992. Diatom stratigraphy in a lake on Horseshoe Island, Antarctica: a marine-brackish-freshwater transition with comments on the systematics and ecology of the most common diatoms. *Diatom Research* 7(1):157-194.

Wharton, R.A. Jr., C.P. McKay, G.D. Clow, D.T. Andersen, G.M. Simmons Jr. & F.G. Love, 1992. Changes in ice cover thickness and lake level of Lake Hoare, Antaractica: implications for local climate change. *Journal of Geophysical Research* 97(C3):3503-3513.

Williams, W.D., 1981. Inland Salt Lakes: An Introduction. *Hydrobiologia* 81:1-14.

Zeeb, B.A. & J.P. Smol, 1995. A weighted-averaging regression and calibration model for inferring lakewater salinity using chrysophycean stomatocysts from lakes in western Canada. *International Journal of Salt Lake Research* 4:1-23

Ecosystem Processes in Antarctic Ice-free Landscapes, Lyons, Howard-Williams & Hawes (eds)
© *1997 Balkema, Rotterdam, ISBN 90 5410 925 4*

Trace metal transport and speciation in Lake Wilson: A comparison with Lake Vanda

J.G.Webster & K.S.Webster
Institute of Environmental Science and Research, Auckland, New Zealand

I.Hawes
National Institute of Water and Atmospheric Research Ltd, Christchurch, New Zealand

ABSTRACT: Located in the Darwin Valley (80°S), Lake Wilson is one of the southern-most of the stratified, ice-covered inland lakes of the arid McMurdo Dry Valley region. In January 1993 samples were collected for the determination of acid-soluble and dissolved trace metal (Cu, Pb, Zn, Co, Cd, and Ni) concentrations in the lake and its inflow. Parameters likely to influence trace metal speciation such as DO, pH, major ion concentrations, suspended sediment load, Fe-, Al- and Mn-oxide concentrations and phytoplankton activity, were also quantified.

The inflow to Lake Wilson transfers a significant load of suspended sediment into the upper/oxic layers of the lake, and this was reflected in the high concentrations of suspended sediment, Fe-oxide and Al-oxide in the upper water column. In the inflow, trace metals were predominantly bound to suspended particulates. In the lake, metals could be broadly classified into two groups on the basis of observed behaviour: those predominantly bound to particulates in the oxic lake water, showing high concentrations in the upper/oxic layer where suspended sediment levels were high (Pb and Co), and those predominantly or partially present in dissolved form in the oxic lake water (Cu, Zn, Ni and Cd). The concentrations of the latter increased with increasing salinity in the lower lake waters. Trace metal concentrations increased or remained relatively high in the anoxic zone below 95 m depth.

The principal process controlling trace metal speciation in the inflow and upper layers of Lake Wilson is considered to be metal adsorption onto, and desorption from, oxide surfaces in the suspended sediment, as has previously been proposed for Lake Vanda. However, the vertical distribution of trace metals in Lake Wilson differed from that of metals in Lake Vanda, due to differences in the density gradients (affecting suspended sediment distribution), pH gradients (affecting the degree of metal adsorption), biological activity and H_2S concentrations (affecting metal solubility in and just above the anoxic zone). Upward diffusion of trace metals from the lower lake appeared to be an important factor influencing metal distribution in the water column of Lake Wilson.

1 INTRODUCTION

Lake Wilson is one of the southern-most of the ice-covered inland lakes of the arid McMurdo Dry Valley region, and is located in the valley of the Darwin Glacier, at 80°S (see Fig. 1). The lake lies at the base of a secondary valley; a pro-glacial lake formed where the Darwin Glacier merges with the McMurdo ice shelf and blocks the valley exit. The main tributary draining into Lake Wilson is a very turbid stream (referred to as site W200) flowing northwards along the west margin of the Darwin Glacier, entering the lake at the south east corner. There is no surface outlet from the lake and, even in mid summer, the ice cover (4.1 m thick) remains practically complete with only minor moat development at a few sites at the margin. In 1993, the maximum depth of the lake was

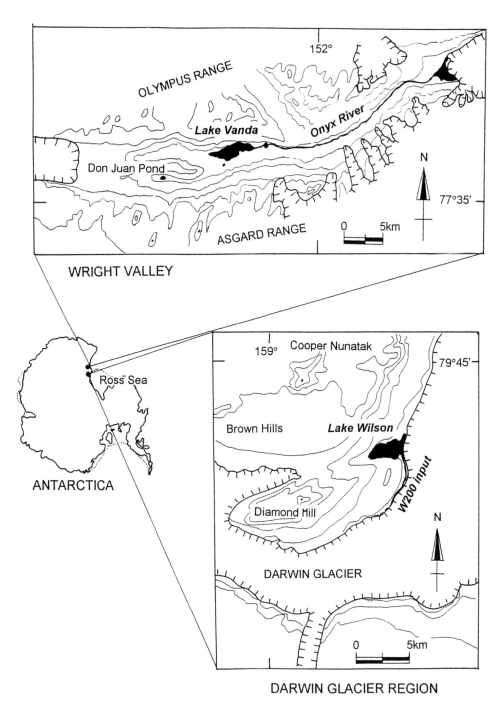

Fig. 1. Site locality for Lakes Wilson and Vanda.

estimated at 120 m, having increased 25 m since it was measured in 1975 (Hendy 1975). The sampling profile used in this study was only 105 m deep, and was clearly not at the deepest part of the lake.

The bathymetry, chemistry and physical structure of Lake Wilson were reported for the first time in Hendy (1975). However, only shallow chemical profiles were determined, due to the unexpected depth of the lake. Full profiles for major ions, pH, temperature, nutrients, redox conditions and biomass were measured on the samples collected in January 1993, together with trace metal concentrations for the lake and its inflow. This paper describes trace metal concentrations, and the factors likely to be controlling metal solubility and speciation, in the inflow and water column of Lake Wilson. A comparison is made with Lake Vanda in the Wright Valley (77°S; Fig. 1), where similar trace metal speciation studies have previously been undertaken (Green *et al.* 1989, 1993; Webster 1994).

2   SAMPLING TECHNIQUES AND ANALYTICAL METHODS

Lake waters were sampled using a 1.5-l capacity "trace metal" sampler, and conductivity, temperature, pH and DO, and H_2S were measured on site using standard portable meters. Major ion concentrations were determined by high performance ion chromatography (HPIC) or by titration against $AgNO_3$ (for Cl) and HCl (for HCO_3). Chlorophyll *a* and suspended sediments were collected onto GF/F filters. Chlorophyll *a* was extracted into 90% acetone, and analysed using a spectrofluorometer callibrated against pure chl *a* standards. These analytical methods are described in more detail in our previous publication on Lake Wilson (Webster *et al* 1996).

For trace metal samples, filtration equipment and bottles were soaked prior to use in redistilled HNO_3, HCl and H_2O_2, and precautions were taken during sampling and sample preservation to prevent any contact with potential sources of trace metal contamination. Up to three 250-ml samples were collected from each depth and acidified to pH ≤ 2 using redistilled HNO_3. The first of these samples remained unfiltered for the determination of "acid-soluble" trace metal concentrations. The second sample was filtered through 0.45-μm Millipore filter membrane prior to acidification, for the determination of "dissolved" trace metal concentrations. For selected sites, a third sample was filtered through a 0.22-μm Millipore filter membrane to determine whether all particulate material was being removed in the 0.45 μm filtration step. Only the upper/oxic waters of Lake Wilson contained acid-soluble suspended particulates of <0.45 μm size. At greater depth in the lake, the waters are more saline and suspended particulates have flocculated under conditions of higher ionic strength.

In the following discussion, "particulate-bound" trace metal concentrations have been calculated as the difference between "acid-soluble" and "dissolved" trace metal concentrations. The "particulate-bound" fraction includes trace metals bound to acid-soluble phases such as amorphous oxides and carbonates, or weakly bound to insoluble silicate surfaces, but does not include trace metal incorporated into silicate and other insoluble phases.

Samples were analysed for the trace metals Cu, Pb, Zn, Cd, Ni and Co by ICP-MS. Ni and Co concentrations have been corrected for Ca interference in the ICP-MS analysis. Field blanks of deionised water were used to provide an indication of likely background contamination levels. Trace metal concentrations in the field blanks were at or near the limits of detection for the analytical methods used (i.e. ≤ 0.05 mg m^{-3} for Co, ≤ 0.1 mg m^{-3} for Cu, Pb, Cd and Ni, ≤ 0.5 mg m^{-3} for Zn). Samples were analysed for oxide components Fe, Mn and Al by ICP-AES. Field blanks again contained suffciently low concentrations of these metals (3 mg m^{-3} for Fe, 0.2 mg m^{-3} for Mn and 5 mg m^{-3} for Al) and no contamination correction was required.

3   RESULTS

Lake Wilson was stratified with respect to salinity, with an upper/oxic layer of low salinity extending to 45 m depth, a moderately saline middle/oxic layer from 50 to 70 m depth, and a more saline lower/oxic layer which extends from 85 m to the redox boundary. All major ion profiles reflected the salinity stratification to some degree (Fig. 2a). The basal waters of the lake were of Na-Cl brine composition, but also contained moderately high concentrations of SO_4. Between the oxic layers there were two diffusion cells: an upper cell from 45 to 50 m and a lower cell from 70 to 85m. The homogeneous temperature and chemistry of the upper/oxic layer suggests that mixing has occurred in these waters, possibly due to thermal convection and/or the turbulence of meltwater inflow.

Temperature and pH profiles in the water column were relatively subdued, and bore little relation to salinity stratification (Fig. 2b). Temperature ranged from 0 to 3.6°C, and pH from 6.62 to 8.02. From the surface to 60 m depth, the water column was supersaturated with respect to DO, with extreme concentrations (>20 g m^{-3}) at 45 to 60 m depth. This feature can not be explained by current photosynthesis and may be a relict layer from a previous lower lake level (Webster et al 1996). Below 60 m, DO concentrations decreased to <2 g m^{-3} at 80 m, thereafter remaining constant until the redox boundary at 95 m. The lake waters were anoxic below 95 m depth and contained measurable but low concentrations of H_2S (maximum: 0.3 g m^{-3}).

The following results are presented in three sections: the first pertaining to trace metal concentrations, and the second and third to components which may influence trace metal solubility and speciation.

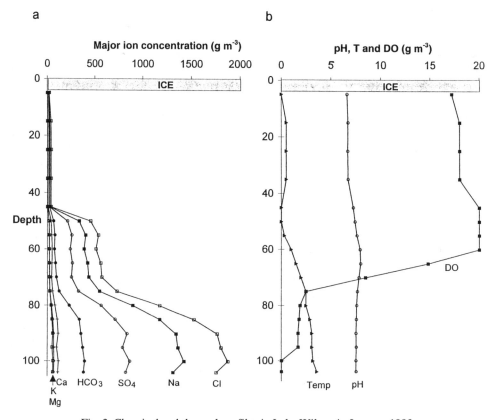

Fig. 2. Chemical and thermal profiles in Lake Wilson in January 1993

3.1   Trace metal concentrations and speciation

(i)   In the inflow (W200)

Acid-soluble metal concentrations in the inflow to Lake Wilson (W200) are shown in Table 1. All trace metals appeared to be predominantly bound to suspended particulate material. Cu, however, may also have been present in either dissolved form, or bound to the fine (<0.45 μm) particulate fraction. This may also have been the case for Cd, however Cd concentrations were barely above the limit of detection, and the error in calculated particulate-bound concentrations is likely to be large.

(ii)   In the water column

Trace metal concentration profiles in the water column of Lake Wilson are shown in Fig. 3. In each case acid-soluble metal concentrations have been plotted as a function of depth, with calculated particulate-bound metal concentrations shown as a secondary, dotted line. On the basis of their distribution profiles, the trace metals were either: (a) predominantly bound to particulates in the oxic lake water (Pb and Co), showing maximum concentrations in the upper/oxic layer and in the anoxic zone, or (b) at least partially present in dissolved form, and showing a direct relationship between metal concentration and salinity (Cu, Zn, Cd and Ni). Consistent values for metal concentrations in non-filtered, 0.45 μm filtered, and 0.22 μm filtered samples confirmed that these metals were indeed dissolved where indicated, and not merely bound to the finer (<0.45 μm) fraction of suspended particulate material which is present in the low salinity, upper lake waters.

Anomalously high concentrations of Zn, Ni, Pb and Cd occured at 5 m depth, just below the ice cover. The possibility of contamination at this depth from the auger, used to drill a hole through the ice cover of the lake, can not be ignored. An analysis of melted ice grinds from the hole drilled through the ice cover also showed high concentrations of these trace metals, particularly Zn, relative to the inflow water (which might otherwise be expected to exert the main influence on lake water chemistry at this level). There was no evidence to suggest, nor any obvious mechanism by which, this possible contamination could affect samples collected at >5 m depth.

Relatively low Cu, Zn and Ni concentrations at 90 m depth also appeared to be inconsistent with general trends in the profiles for these metals, and those of other trace metals and major ions. As there were no significant changes in any of the parameters likely to influence metal solubility at this depth (Figs. 2 & 4), there was no obvious explanation for these anomalously low metal concentrations.

3.2   Suspended sediment and oxide components (Fe, Al & Mn)

The inflow of Lake Wilson was very turbid, with a suspended sediment (TSS) load of 40 g m^{-3} and a relatively small flow, estimated at *ca.* 0.2 to 0.3 m^3 m^{-1} in mid-summer. The inflow to Lake Wilson will clearly have transferred a significant load of suspended sediment into the upper oxic layers of the lake. This was reflected in the concentrations of suspended sediment, and oxide components Fe and Al in the upper water column (Fig. 4). The upper/oxic layer contained between 1.2 and 5.6 g m^{-3} TSS as well as high particulate-bound Fe and Al concentrations (up to 750 and 650 mg m^{-3} respectively).

Particulate-bound Fe and Al concentration maxima at 25 to 45m depth suggest that incoming suspended sediment did not immediately mix with the more saline lake waters in the middle/oxic layer. However, the ratio of particulate-bound Al:Fe in the upper/oxic and middle/oxic layers was unchanged from that of the inflow W200; evidence that the sediment has settled slowly into the middle/oxic water column. The Al:Fe ratio changed only at the redox boundary at 95 m depth. Whereas Al-oxide behaved as a conservative tracer of particulate

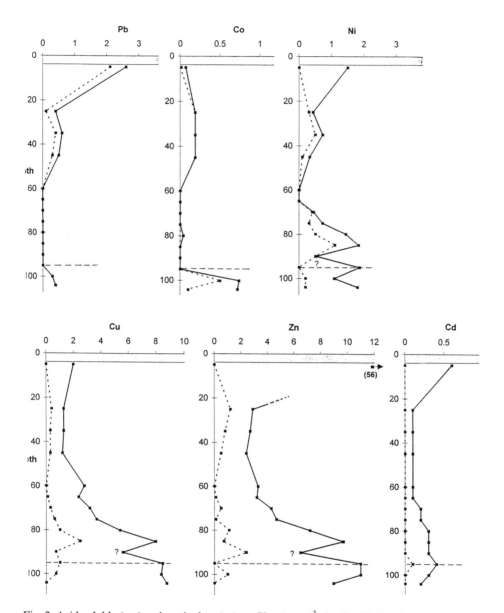

Fig. 3. Acid-soluble (——) and particulate (---) profiles (mg m⁻³) for Cu, Pb, Zn, Cd, Ni, and Co as a function of depth (m) in Lake Wilson. The ice cover is shaded at the top of the profile, and the redox boundary is indicated with a dashed line at 95 m.

movement, Fe-oxide dissolved at the redox boundary, generating higher concentrations of dissolved Fe in the anoxic zone.

In the inflow (see Table 1) and the upper oxic layer of Lake Wilson, Mn-oxides were substantially less abundant than the oxides of Fe and Al. Dissolved Mn concentrations did, however, increase markedly below 75 m depth in the lower/oxic layer of the water column (Fig. 4). This increase corresponded to a decrease in DO concentrations, suggesting that Mn-oxide was reduced at this depth to form the more soluble, divalent manganous ion. Dissolved Mn concentrations remain high in the anoxic zone.

226

Table 1. Acid-soluble metal concentrations (mg m^{-3}) in unfiltered water samples from the inflow into Lake Wilson (W200). The percentage evidently bound to suspended particulates is shown in italics.

	Cu	Pb	Zn	Cd	Co	Ni	Fe	Al	Mn
Lake Wilson inflow	6.1	4.3	24	0.2	3.5	5.7	9600	7970	135
% particulate-bound	*74*	*>98*	*91*	*>50*	*97*	*88*	*99*	*99*	*98*

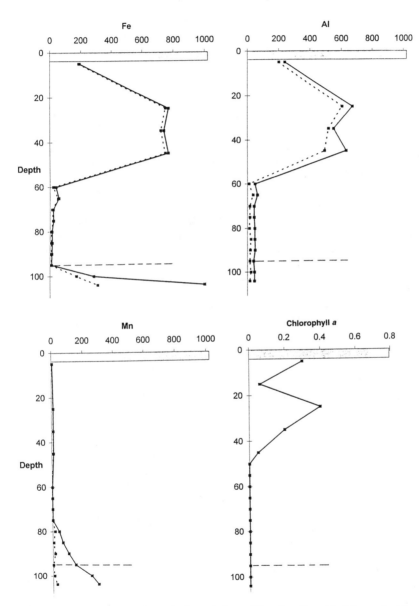

Fig. 4. Acid-soluble (——) and particulate (---) profiles for Fe, Al, Mn and chlorophyll *a* concentrations (mg m^{-3}) as a function of depth (m) in Lake Wilson. The ice cover is shaded at the top of the profile, and the redox boundary is indicated with a dashed line at 95 m.

3.3   Phytoplankton distribution.

High suspended sediment concentrations in the upper/oxic layer of Lake Wilson have significant implications for biological life in the water column. Photoautotrophic production in the Dry Valley lakes depends on PAR through the ice cover and water, and on the ability of phytoplankton to utilise this light. Light penetration through the ice cover of Lake Wilson was low (2.5%), and this was further decreased in the water column by the high turbidity. An estimation of the biovolume of autotrophic planktonic algae and cyanobacteria in the lake showed a maximum in the upper oxic layer, with little evidence of activity below this layer as shown by the distribution of chlorophyll *a* (Fig. 4). The phytoplankton were predominantly cyanobacteria (*Phormidium spp.*) and Chlorophyta (*Chlorella sp.*).

4   DISCUSSION

Oxides of Fe and Al are a significant component of the finely ground rock, or "rock flour", carried in glacially-fed inflow streams, often occurring as oxide-coatings on silicate particles. These oxide surfaces, particularly those of Fe-oxide, have the ability to adsorb selected trace metals under favourable pH conditions. Trace metals in the inflow to Lake Wilson were predominantly bound to the suspended sediment. Preliminary calculations using the geochemical speciation model MINTEQA2 (Allison *et al.* 1991), which has the ability to model adsorption and desorption from hydrous Fe-oxide surfaces, indicated that the pH conditions do favour adsorption of Cu, Pb and Zn, but only partial adsorption of Cd (50%) and Ni (30%). Fe-oxide does not have a high affinity for Ni, and the fact that Ni was apparently predominantly adsorbed on suspended particulate, both in the inflow and upper/oxic lake water, may suggest that the Ni was bound to another mineral phase (e.g. Mn-oxide). MINTEQA2 does not, unfortunately, include a database for Co adsorption.

In the upper/oxic layer of Lake Wilson, conditions were slightly more acidic than in the inflow (pH = 6.6 to 6.7, compared with pH = 7.1). Although small, MINTEQA2 modelling indicated that this pH change would be sufficient to cause desorption of Zn and Cd from an Fe-oxide surface, and at least partial desorption of Cu. The pH change would not be sufficient to cause desorption of Pb, which is relatively strongly adsorbed on Fe-oxide even under low pH conditions. Pb and Co would remain bound to suspended sediment in the turbid upper/oxic layer of Lake Wilson. At greater depth, decreasing Fe-oxide concentrations would discourage readsorption of Zn, Cd and Cu, despite the more favourable pH conditions. Pb and Co were evidently released from Fe-oxide only when the oxide was reduced at the redox boundary.

The distribution profiles of dissolved Cu, Zn, Ni and Cd in the lower lake were not unlike those of the major ions (in Fig. 2a), which were formed by upward diffusion of ions from the deeper, more saline lake water (Webster *et al.* 1996). These trace metals could also be diffusing up through the profile from trace metal-enriched lake water at depth. However, any such diffusion of dissolved Pb and Co from the anoxic zone would be limited by adsorption onto reformed Fe-oxide at the redox boundary. For redox metals such as Mn and Fe, upward diffusion is curtailed by the precipitation of insoluble oxide phases as soon as DO concentrations are sufficiently high. This leads to recycling of Mn and Fe across the redox boundary as the newly-formed oxides once again settle through the water column and redissolve (De Vitre *et al.* 1988). Ferrous ion is oxidised more readily and consequently often closer to the redox boundary (e.g. Fig. 4). In Lake Wilson, for example, the concentration of dissolved Mn diffused up to a depth of 75 m (20 m above the redox boundary), where there was a dramatic increase in DO (Figs. 2 & 4) and Mn-oxide was precipitated.

Finally, the distribution of Al in the water column primarily reflected that of Al-oxide, which was stable under the pH and DO conditions prevailing in Lake Wilson. Unlike the more reactive oxide phases discussed above, the distribution of Al-oxide was affected only by changes in water density, which limit particle settling rates in the middle and lower oxic layers of the lake.

Obvious parallels can be drawn between Lake Wilson and Lake Vanda, the two largest, deepest inland lakes in Victoria Land. Initially these two lakes appear to be very similar, perhaps differing only in the degree to which they have attracted scientific attention. Both lakes have a single inflow over the summer period, have no outlet, and have shown a generally rising trend in lake levels over the past 10 to 20 years. They appear to have been formed from evaporative brine ponds over a similar period (Wilson 1964; Webster *et al.* 1996), and both display gradients of salinity, density, temperature and redox conditions in the water column.

Trace metal distribution profiles for Lake Vanda have previously been determined by Green *et al.* (1989) and Webster (1994), and have been redetermined as part of a current (1995 to 1998) investigation by the authors. Trace metal adsorption and desorption from oxide phases is also considered to be the main process controlling trace metal transport and speciation in the Onyx River and in Lake Vanda. However, there are significant differences between the lakes in terms of density gradients, pH gradients, phytoplankton activity and H_2S concentrations. Individually and collectively, these factors can influence suspended sediment distribution (and therefore adsorbed trace metal distribution) and trace metal adsorption and solubility.

As with the inflow to Lake Wilson, trace metals in the Onyx River flowing into Lake Vanda are predominantly bound to particulate phases. However, trace metal distribution profiles in Lake Vanda are distinctly different, in that:

(a)   There is no peak in particulate Pb, and Co (or Fe, Al and Mn) concentrations in the oxic layer of Lake Vanda. The Onyx River carries a lower suspended sediment load (typically *ca*. 1 to 4 g m^{-3}) into Lake Vanda than the inflow to Lake Wilson, and the Onyx River water does not mix immediately with the upper/oxic layer of the lake to form a suspended sediment-rich layer. Instead the river water remains as a thin layer just under the ice cover.

(b)   Dissolved Cu, Ni and Zn concentrations remain low in the middle layers of Lake Vanda, but peak briefly between 55 m and 60 m depth. The upper/oxic waters of Lake Vanda are relatively alkaline (pH *ca*. 8), favouring trace metal adsorption onto Fe-oxide surfaces, whereas Cu, Ni, Zn and Cd are predominantly desorbed (i.e. dissolved) in the upper waters of Lake Wilson (pH *ca*. 6.5). Below 50 m depth in Lake Vanda, however, pH and salinity changes facilitate metal desorption, corresponding to higher concentrations of dissolved metals at this level (Webster 1994). As there is an algal biomass maximum just above the redox boundary (Vincent & Vincent 1982), the solubility of selected metals may also be influenced by phytoplankton, which are capable of adsorbing metals onto cell walls (Ferris *et al.* 1987). There is no analogous phytoplankton activity in Lake Wilson.

(c)   Sulphide precipitation appears to exert a greater control on Fe and trace metal solubility in the anoxic zone of Lake Vanda, than in that of Lake Wilson. In Lake Wilson, the concentrations of Fe, Pb, Cu, and Ni do not decrease with increasing H_2S concentration and depth in the anoxic zone. Although the H_2S concentrations are lower in Lake Wilson, with a maxiumum of 0.32 g m^{-3} compared to 34 g m^{-3} in Lake Vanda, MINTEQA2 indicated that the anoxic waters of Lake Wilson were indeed supersaturated with respect to the relevant trace metal sulphides, at the time of sampling. Failure to actively precipitate these sulphides may be, at least partly, due to the lack of significant bacterial activity at depths of >50 m (Webster *et al.* 1996). Thiobacteria not only to catalyse the reduction of SO_4 to H_2S, but also the formation of mineral sulphide precipitates.

5   CONCLUSIONS

As in the oxic waters of Lake Vanda, trace metal speciation in Lake Wilson appears to be mainly controlled by adsorption/desorption at oxide surfaces in suspended sediment. Trace metal profiles are, however, quite different from those of Lake Vanda, influenced by the distribution and/or magnitude of suspended sediment load, pH, biological activity and H_2S concentrations. Because trace metal levels are relatively high in the lower lake waters, upward diffusion of metals into the overlying oxic water may be occurring. This is particularly evident for Cu, Ni, Zn and Cd. Pb and Co are more readily adsorbed onto Fe-oxide and do not diffuse above the redox boundary.

6   ACKNOWLEDGMENTS

This study was funded by the Foundation for Research Science and Technology in New Zealand. We acknowledge the assistance of CSIRO Centre for Advanced Chemistry (trace metal analyses) and help of Drs. Mike Timperley, Clive Howard-Williams, Malcolm Downes and Peter Nelson, and other scientists participating in the 1992/93 and 1995/96 Antarctic field seasons. As always, we are indebted to staff of Antarctica New Zealand, and the flight crews of the NZ Air Force and the US Navy VXE-6 squadron for logistic support.

REFERENCES

Allison J.D., D.S. Brown & K.J. Novo-Gradac, 1991. *MINTEQA2/PRODEFA2, a geochemical model for environmental systems.* EPA600/3-91/021.USEPA, Athens, Georgia.

De Vitre R.R., J. Buffle, D. Perret & R. Baudat, 1988. A study of iron and manganese transformations at the O_2/S^{-II} transition layer in a eutrophic lake (Lake Bret, Switzerland): A multimethod approach. *Geochimico et Cosmchimico Acta* 52:1601-1613.

Ferris F.G., W.S. Fyfe & T.J. Beveridge, 1987. Bacteria as nucleation sites for authigenic minerals in a metal-contaminated lake sediment. *Chemical Geology* 63:225-232.

Green W.J., T.G. Ferdleman & D.E. Canfield, 1989. Metal dynamics in Lake Vanda (Wright Valley, Antarctica). *Chemical Geology* 76:85-94.

Green W.J., D.E. Canfield, Y. Shengsong, K.E. Chave, T.G. Ferdelman & G. Delanois, 1993. Metal transport and release processes in Lake Vanda: the role of oxide phases. In: Green, W.J. & E.I. Friedmann (Eds.), *Physical and Biogeochemical Processes in Antarctic Lakes. Antarctic Research Series* 59. American Geophysical Union, Washington D.C. pp. 145-163.

Hendy C.H, 1975. *Report to the University of Waikato Antarctic Research Unit for the 1974/75 field season.* University of Waikato. Unpublished report.

Vincent W.F. & C.L. Vincent, 1982. Factors controlling phytoplankton production in Lake Vanda (77°S). *Canadian Journal of Fisheries & Aquatic Science* 39:1602-1609.

Webster J.G., 1994. Trace metal behaviour in oxic and anoxic Ca-Cl brines of the Wright Valley Drainage, Antarctica. *Chemical Geology* 112:255-274.

Webster J., I. Hawes, M. Downes, M Timperley & C. Howard-Williams, 1996. The recent evolution of Lake Wilson: a pro-glacial lake at 80°S. *Antarctic Science* 8:49-59.

Wilson A.T., 1964. Evidence from chemical diffusion of a climatic change in the McMurdo Dry Valleys 1200 years ago. *Nature* 201: 176-177.

The microbial loop in Antarctic lakes

Johanna Laybourn-Parry
*Department of Physiology and Environmental Science, University of Nottingham, Loughborough,
UK*

ABSTRACT: Antarctic lakes are characterised by a dominance of microbial plankton (bacteria, uni-cellular algae and protozoa). Metazoan plankton are either absent or sparse. Consequently Antarctic lake plankton is made up almost entirely of the so-called microbial loop. This article reviews the present limited data on Antarctic freshwater and saline lakes. Abundances of bacterioplankton and heterotrophic nanoflagellates reported from Antarctic lakes lie at the lower end of the ranges reported for lakes worldwide. Ciliate abundances show much greater variability. They are low in the lakes of the Vestfold and Larsemann Hills (a few hundred l^{-1}) while in the Dry Valley lakes abundances are consistently higher and are comparable to those reported from oligotrophic to mesotrophic lower latitude lakes (1780-7200 l^{-1}). Ciliate communities also show much greater species diversity in the Dry Valley lakes. Overall, however, the species diversity of the protozooplankton is significantly lower in Antarctic lakes. Mixotrophy appears a common nutritional strategy among Antarctic lake protists. It is seen among some phytoflagellate species and in some of the more common ciliates. In the latter it involves both plastid sequestration (in the oligotrichs) and endosymbiotic algae. Limited data on feed rates and specific growth rates indicate temperature and energy availability constraints on physiological functioning. A tentative model of carbon cycling in the plankton of one large ultra-oligotrophic freshwater lake is presented.

1 INTRODUCTION

Despite the small area of the ice-free portion of the Antarctic continent, there are a surprisingly abundant and diverse array of freshwater and saline lakes. Some of these lakes, like those of the Dry Valleys of southern Queen Victoria Land, are ancient systems thought to be hundreds of thousands of years old, while lakes elsewhere are much younger, for example the lakes of the Vestfold Hills and Larsemann Hills, Princess Elizabeth Land (Adamson & Pickard 1986; Burgess *et al.* 1994). Given the diversity of lake types and their huge disparity in age one might expect distinct differences in the microbial plankton inhabiting these water bodies.

Antarctic lakes are usually dominated by microbial plankton. The structure of planktonic food webs as we currently understand them, incorporate a series of microbial trophic levels (Fig. 1). The phytoplankton are known to exude a portion of their photosynthate as dissolved organic carbon (DOC) (Fogg 1958; Chróst & Faust 1983; Sell & Overbeck 1992). Other sources of DOC derived from macrophytes and algal mats, inefficient feeding by zooplankton, and allochthonous inputs may also contribute in varying proportions to the pool. The pool of DOC is exploited by the heterotrophic bacterioplankton as an energy source. There are some heterotrophic nanoflagellates (HNAN) which also appear capable of exploiting the pool of DOC (Marchant & Scott 1993), but the majority are grazers of the bacterioplankton.

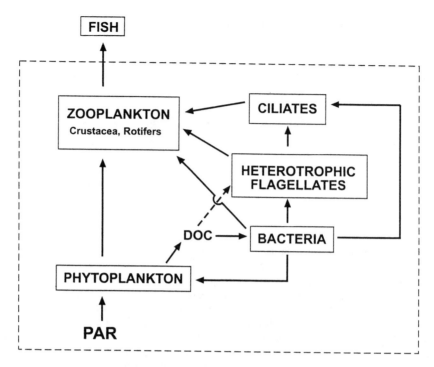

Fig. 1. The trophic structure of typical lake planktonic community. The area in enclosed dashed box is that part of the community found in Antarctic lakes. The Metazoan component is minimal.

Others grazers of the bacterioplankton include some ciliates, especially the scuticociliates and peritrichs, some genera of phototrophic nanoflagellates, some rotifers and members of the Cladocera. In most waters it is the heterotrophic nanoflagellates which are the major grazers of bacterioplankton, but exceptions have been noted (e.g. Sherr & Sherr 1987; Sanders & Porter 1988). In Antarctic lakes Metazoa are either absent or sparse, thus top-down grazing control is lacking and these systems are driven by bottom-up forces.

The microbial plankton (bacteria, protozoa) have only recently become a focus of attention in Antarctic lakes. At present we only have data from lakes in the Vestfold and Larsemann Hills oases (Laybourn-Parry et al. 1991, 1992, 1995; Laybourn-Parry & Marchant 1992a, b; Laybourn-Parry & Bayliss 1996; Laybourn-Parry & Perriss 1995; Perriss et al. 1995; Ellis-Evans et al. 1997, in press) in eastern Antarctica and some tantalising preliminary data from some of the lakes in the Dry Valleys and McMurdo Ice Shelf (Parker et al. 1982; Cathey et al. 1981; James et al. 1995; Laybourn-Parry et al. 1997), together with some comparative information from maritime Antarctic lakes on Signy Island (South Orkneys) (Laybourn-Parry et al. 1996) and from the Antarctic Peninsula (Izaguirre et al. 1993). These studies, some of which encompass the austral winter, have revealed systems with a relatively low species diversity of protists and low biomasses. The most notable feature, however, is the dominance of the microbial plankton.

The species diversity, biomass and community structure of the microbial loop organisms in any lake depends upon a number of factors among which nutrient status, productivity of phytoplankton, sources and levels of DOC and isolation are important.

Heterotrophic bacterioplankton abundances in Antarctic lakes are at the lower end of the spectrum reported for lakes world-wide (Table 1). Across the trophic continuum from oligotrophic to eutrophic, bacterioplankton abundances in cold and warm temperate lakes range between $1.0-118.0 \times 10^8$ l^{-1} (Stockner & Porter 1988; Bloem & Bar-Gilissen 1989; Weisse *et al.* 1990; Laybourn-Parry *et al. 1*994; Vaqué & Pace 1992). In a global sense all Antarctic lakes, except those which have undergone man-induced or animal-induced eutrophication, can be classed as oligotrophic on the basis of their productivity and the abundances and biomasses of microbial plankton supported. In general the higher values seen among the freshwater lakes shown in Table 1 are associated with small shallow water bodies, and the lower values with large deep lakes.

Heterotrophic nanoflagellates, which are very largely dependent on bacteria as an energy resource, have correspondingly low concentrations (Table 2). *Paraphysomonas* has been recorded in virtually all of the freshwater lakes of the Vestfold and Larsemann Hills which have been investigated (Laybourn-Parry *et al.* 1996; Laybourn-Parry & Bayliss 1996; Ellis-Evans *et al.* 1997, in press). Bodonids have also been noted. The abundances seen in Antarctic lakes free from eutrophication overlap with those reported from oligotrophic lakes from lower latitudes. Across the trophic continuum HNAN achieve abundances between $0.12-900 \times 10^5$ l^{-1} (Pick & Caron 1987; Nagata 1988; Bloem & Bar-Gilissen 1989; Bennett *et al.* 1990; Weisse *et al.* 1990; Laybourn-Parry *et al.* 1994). Phototrophic nanoflagellates (PNAN) are relatively more abundant in Lake Fryxell compared with other non-eutrophified Antarctic lakes (25×10^5 l^{-1}). In contrast in Crooked Lake and Lake Druzhby PNAN ranged between 7×10^3 l^{-1} and 1×10^5 l^{-1} (Laybourn-Parry *et al.* 1992; Laybourn-Parry & Bayliss 1996). Among the diverse PNAN

Table 1. Bacterioplankton abundances in Antarctic lakes. Those marked with an asterisk denote annual ranges. Others relate to summer studies only. Data from Laybourn-Parry *et al.* 1995, 1996a, 1996b; Laybourn-Parry & Bayliss 1996; Laybourn-Parry & Marchant 1992; Laybourn-Parry & Perriss 1996; Ellis-Evans *et al.* 1997, in press.

Lake	Bacterial abundance $\times 10^8$ l^{-1}
Frxyell, Southern Queen Victoria Land	1.0-3.8
Crooked Lake, Vestfold Hills, Princess Elizabeth Land	1.19-4.46 *
Lake Druzhby, Vestfold Hills	0.75-2.50 *
Various freshwater lakes, Vestfold Hills	1.3-3.6
Lakes Abraxas, Ace & Highway, Vestfold Hills (saline)	0.54-3.8
Various freshwater lakes in the Larsemann Hills	1.2-17.9
No Worries Lake, Larsemann Hills (eutrophified)	23.0
Heywood and Sombre Lakes (Signy Island)	31.8-80.0 max.

Table 2. Heterotrophic (HNAN) and autotrophic (PNAN) nanoflagellate abundances in Antarctic lakes. Those marked with an askerisk denote annual studies, others relateto the summer only.

Lake	HNAN ($\times 10^5$ l^{-1})	PNAN ($\times 10^6$ l^{-1})
Frxyell, Southern Victoria Land	0.28-7.39	189-2530
Crooked, Vestfold Hills	0-5.09	0.004-0.686
Druzhby, Vestfold Hills	2.0-14.0	0.001-0.10
Abraxas, Ace and Highway, Vestfold Hills	1.0-31.0	0.02-2.8
Various freshwater, Larsemann Hills	2.3-84.0	0.01-0.8
No Worries, Larsemann Hills	15.0	2.16
Heywood and Sombre, Signy Island	2.0-175.0	0.5-105.0

found in Lake Fryxell are several *Chlamydomonas, Cryptomonas, Ochromonas* species as well as *Dinobryon* and *Pyraminomas* (Spaulding *et al.* 1994; Laybourn-Parry *et al.*, in press). Species diversity is much poorer in the Vestfold and Larsemann Hills. *Ochromonas* is found there, and this genus together with *Dinobryon* are known to be mixotrophic (Sanders & Porter 1988).

Compared with lower latitude lakes the species diversity among the ciliated protozoa is low. Table 3 shows species found in lakes of Eastern Antarctic and the Dry Valleys. One feature which is immediately apparent is the much greater diversity found in the more ancient Dry Valley lakes. This is probably related to two factors, firstly the higher nutrient status of some of these lakes, and secondly their greater age, which has allowed a longer period for colonisation. For example, levels of dissolved reactive phosphorus (DRP) and nitrate in Lake Fryxell during summer are typically <3 mg m^{-3} in the oxygenated upper water and ammonia between 10 to 20 mg m^{-3} (Priscu *et al.* 1989). Crooked Lake (Vestfold Hills) typically has DRP levels of <5 mg m^{-3} and nitrogen levels below the level of detection (Laybourn-Parry *et al.* 1992). Shallower neighbouring Lake Druzhby had DRP levels below 5 to 8 mg m^{-3}, nitrate levels around 1 to 3 mg m^{-3} and ammonia concentrations below 5 mg m^{-3} (Laybourn-Parry & Bayliss 1996). The evidence suggests that all of these lakes are nitrogen limited (Vincent 1988). Typical chlorophyll *a* levels in the top oxygenated waters of Lake Fryxell range between <1 to 5 mg m^{-3} (Spaulding *et al.* 1994) whereas those in both Crooked Lake and Lake Druzhby were below 1.1mg m^{-3} during an entire year (Laybourn-Parry *et al.* 1992; Laybourn-Parry & Bayliss 1996).

Not only is there greater species diversity in the Dry Valley lakes, they also support greater numbers of ciliated protozoa. Concentrations of ciliated protozoans reached up to 7200 l^{-1} in Lake Fryxell, 6540 l^{-1} in Lake Vanda, 1780 l^{-1} in Lake Bonney and 1225 l^{-1} in Lake Hoare during the summer (Laybourn-Parry *et al.* 1997; James *et al.*, in press; Roberts & Laybourn-

Table 3. Ciliated protozoa from Continental Antarctic lakes. Lakes Bonney, Fryxell, Hoare and Vanda - Southern Victora Land, Crooked lake and Lake Druzhby - Vestfold Hills, Princess Elizabeth Land, Lake No Worries - Larsemann Hills - Princess Elizabeth Land. Data from Cathey *et al.* 1981; Laybourn-Parry *et al.* 1992, 1995, 1996; Laybourn-Parry & Marchant 1992; Laybourn-Parry & Bayliss 1996; Laybourn-Parry & Perriss 1995; Parker *et al.* 1982; Roberets & Laybourn-Parry in prep.; Ellis-Evans *et al.* 1997, in press; Perriss *et al.* 1995; James *et al.*, in press.

Species	Lakes
Strombidium viride	Fryxell, Crooked, other Vestfold hills lakes, Larsemann Hills lakes
Strombidium sp. (aplastidic)	Some saline Vestfold Hills, Larsemann Hills lakes.
small unidentified oligotrich	Crooked, Druzhby, No Worries
Askenasia sp.	Bonney, Fryxell, Hoare, Vanda, Crooked, Druzhby, some Larsemann Hills lakes
Didinium sp.	Fryxell, Hoare
Euplotes sp.	Bonney, Fryxell, Hoare, Vanda
Halteria grandinella	Bonney, Fryxell, Hoare
Monodinium 1 large sp., 1 small sp.	Bonney, Fryxell, Hoare, Vanda
Nassula sp.	Fryxell
Bursaria sp.	Fryxell
Bursellopsis sp.	Bonney, Vanda
Chilodonella sp.	Fryxell
Small unidentified scuticociliates (possibly 1-2 species)	Crooked, Druzhby, some Larsemann Hills lakes.
Unidentified prostomatid	Fryxell
Vorticella sp.	Hoare
Vorticella mayeri	Fryxell
Sphaerophyra sp.	Bonney, Fryxell, Hoare
Mesodinium rubrum	Vestfold Hills saline lakes

Parry, unpubl.), while in contrast the Vestfold Hills and Larsemann Hills lakes typically have concentrations of a few hundred or less per litre (Laybourn-Parry & Marchant 1992a; Laybourn-Parry *et al. 1992*; Ellis-Evans *et al.*, (in press). However, the inappropriately named No Worries Lake adjacent to the Chinese Station (Zhong Shan) in the Larsemann Hills, has ciliate abundances in excess of 20 000 l^{-1} (Ellis-Evans *et al.* 1997). This system has suffered the effects of human impact and eutrophication. While numbers have increased, the species diversity has not. There are only 3 species of ciliate. Thus the evidence suggests that it is not depaurate waters which determines the species diversity of Antarctic lakes, but the period of time during which the lakes have existed. Length of ice cover and ice thickness do not, on the evidence, appear to play a significant role. The Dry Valley lakes, which have perennial ice up to 5 m thick have a much higher species diversity of ciliates than the younger lakes of the Vestfold Hills (Table 3).

Other evidence comes from the Maritime Antarctic lakes on Signy Island, which are similar in age to those of the oases in Eastern Antarctica. While their species diversity is higher than the lakes of the Vestfold and Larsemann Hills, they contained only 4 to 5 species during a summer study, even in lakes undergoing enrichment from seal faecal inputs. A number of these species or types were common with Lake Fryxell, for example *Halteria grandinella* and two species of *Monodinium*. A species of the peritrich *Vorticella* also occurred, but not the same one described from Lake Fryxell (Laybourn-Parry *et al.* 1996a). In the relatively rich waters of Heywood Lakes ciliates achieved abundances in excess of 6000 l^{-1}. Thus time for colonisation appears to be the crucial factor in determining the species diversity of the ciliate, and other protistan communities, in Antarctic lakes.

The complexity of the community of ciliates is greater in some of the Dry Valley lakes. Among its ciliates is *Sphaerophrya*, a suctorian ciliate which is a predator of other ciliate species (Table 3). It is an ambush predator dependent on chance encounters with its prey. It can only survive where the concentration of other ciliates permits sufficient capture rates to sustain the population. The concentration of potential prey in most of the Vestfold Hills and Larsemann Hills lakes is probably too low to sustain a suctorian ciliate should one ever be introduced.

The saline lakes of the Vestfold Hills differ from the meromictic lakes in the Dry Valleys in that their microbial plankton reflects their marine origin (Adamson & Pickard 1986). Since their formation, the species diversity has decreased leaving a relatively small number of species able to cope with the conditions offered within meromictic lakes. Nutrient levels are comparable to those in Lake Fryxell which contrasts with the nutrient rich marine waters from which they were derived. Species diversity of ciliates is low (Table 3), and those lakes with upper water salinities between 4 to ~35‰ are dominated by the unique, marine autotrophic ciliate *Mesodinium rubrum* (*Myrionecta rubra*), which contains a cryptophycean endosymbiont. On occasions it can reach mean concentrations in excess of 48 000 l^{-1} in the oxygenated portion of the water column (Perriss *et al.* 1995; Laybourn-Parry & Perriss 1995).

Heliozoans are a common feature of Antarctic freshwater lakes. They occurred through out the year in Crooked Lake and periodically in Lake Druzhby, and in the summer communities of other Vestfold Hill lakes and in Lake Fryxell (Laybourn-Parry *et al.* 1992, 1997; Laybourn-Parry & Bayliss 1996: Laybourn-Parry & Marchant 1992a). They tend to be very sporadic in occurrence in most lower latitude lakes, but appear a typical component of the plankton in oligotrophic temperate lakes (Laybourn-Parry *et al.* 1995). Thus they may be a group which thrive best in lakes with limited competition from other species which feed on flagellates.

Dinoflagellates, both heterotrophic and autotrophic are common in the saline lakes of the Vestfold Hills and some of the lakes in the Larsemann Hills. Colourless heterotrophic forms have also been noted in some of the Dry Valley Lakes (Parker *et al.* 1982). Heterotrophic dinoflagellates may be important consumers in the plankton since they are capable of grazing at rates comparable with ciliates (Lessard & Swift 1985). In the Vestfold Hills saline lakes which have been investigated, heterotrophic species include *Gyrodinium lachryma* and several species

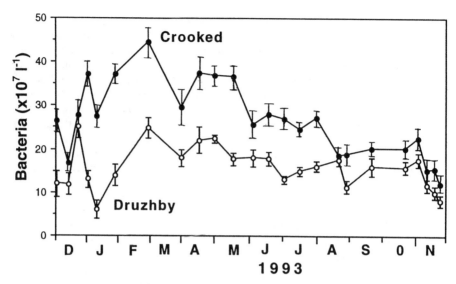

Fig. 2. Bacterioplankton patterns of abundance in Crooked Lake and Lake Druzhby, Vestfold Hills. The values shown are mean values (with SEM) for the water column (n = 7 in each case). Redrawn from Laybourn-Parry *et al.* 1995, Laybourn-Parry & Bayliss 1996).

of *Gymnodinium*. They can reach concentrations in excess of 10 000 cells l^{-1}. From the limited data it appears that heterotrophic forms are more common in the meromictic lakes compared with the brackish lakes (Perriss *et al.*, unpubl.).

3   SEASONAL PATTERNS OF ABUNDANCE

In most cases our data base on the species successions and abundances of bacterio- and protozooplankton are restricted to the summer months. Both Lakes Druzhby and Crooked Lake have been studied for an entire year. These studies have revealed some interesting aspects of the dynamics in these lakes. The bacterioplankton did not show a radical decrease in abundance during the winter (Laybourn-Parry *et al.* 1995; Laybourn-Parry & Bayliss 1996) (Fig. 2). However, a consideration of the mean cell volumes (MCV) of the heterotrophic bacteria revealed that many of them must have been in a starvation phase during winter, because there was a marked decrease in MCV (Fig. 3). With the onset of spring MCV again increased, presumably in relation to increase community metabolism. At this time numbers were still low, but the increase in MCV resulted in a marked increase in biomass. Studies on bacterial production in Crooked Lake during the year showed that community growth was reduced in winter (Laybourn-Parry *et al.* 1995).

HNAN dropped to low levels during the winter in Crooked Lake, presumably reflecting the poor quality of the bacterial community as a food resource (Fig. 4). The HNAN increased with the increase in bacterial biomass in the spring. In contrast the HNAN community in Lake Druzhby remained higher during the winter, which may be related to the larger MCV of the bacterioplankton in this lake (Fig. 3).

4   MIXOTROPHY

The combination of autotrophic and heterotrophic nutrition is a common phenomenon among

236

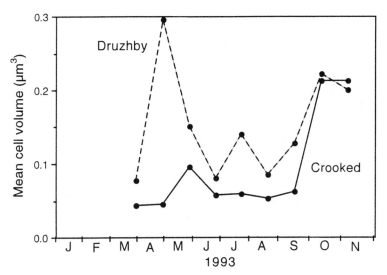

Fig. 3. Mean cell volume (MCV) of bacterioplankton in Crooked Lake and Lake Druzhby. Data from Laybourn-Parry *et al.* 1995 and Laybourn-Parry & Bayliss 1996.

planktonic protozoa. It represents a valuable strategy in the feast and famine environment, particularly in oligotrophic lakes. Some of the autotrophic flagellates, notably the chrysophytes, are capable of feeding on bacteria. *Dinobryon*, for example, can ingest 7 to 24 bacteria cell^{-1} h^{-1} in warm temperate lakes (Sanders & Porter 1988) and is capable of supplementing its carbon budget by as much as 50% by phagotrophy (Bird & Kalff 1986). These organisms are primarily autotrophs but use heterotrophy to obtain nutrients for photosynthesis and to supplement their carbon budget when light levels are low. In Crooked Lake 30% of the PNAN were found to take up bacterial sized fluorescently stained microspheres during the summer, but the uptake rates measured successfully over a 10 h incubation period were very low, only 1 to 2 microspheres cell^{-1} day^{-1} (Laybourn-Parry, unpubl.). Ciliates are essentially heterotrophic and here mixotrophy may be cellular, involving the harbouring of symbiotic zoochlorellae, or organellar where the ciliates sequester the plastids of their phytoflagellate or algal food. The latter form is common among the oligotrich ciliates.

Among the phytoflagellates which occur in Antarctic lakes *Ochromonas* is found in lakes from both the Vestfold Hills and Dry Valleys (Laybourn-Parry *et al.* 1992a 1997; Spaulding *et al.* 1994). In Lake Fryxell there are a wider range of mixotrophic phytoflagellates. In addition to *Ochromonas*, *Dinobryon* and *Chromulina* have been recorded during the summer (Spaulding *et al.* 1994). While no feeding studies have been conducted on these flagellates, it is highly likely that they will be exploiting mixotrophy as a feeding strategy.

The plastidic, mixotrophic ciliate *Strombidium viride* occurs in the Vestfold Hills, Larsemann Hills and Lake Fryxell (Laybourn-Parry *et al.* 1992, 1997; Ellis-Evans *et al.*, in press). *Strombidium* has also been reported from a range of other Dry valley lakes including Vanda, Joyce, Bonney-East and Hoare (Parker *et al.* 1982). Lake Fryxell has at least one other mixotrophic ciliate, possibly a species of *Bursaria*, which was harbouring zoochlorellae (Laybourn-Parry *et al.* 1997).

5   TROPHIC INTERACTIONS

The few data we have reveal little adaptation to cold environments. Antarctic lake protozoa appear to grow very slowly, particularly in ultra-oligotrophic lakes. HNAN specific growth

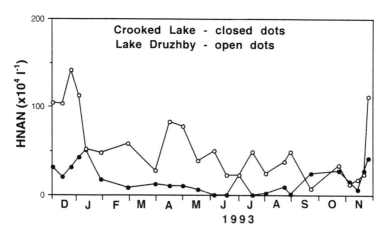

Fig. 4. Mean heterotrophic nanoflagellate (HNAN) seasonal abundances in Crooked Lake and Lake Druzhby. Redrawn from Laybourn-Parry *et al.* 1995 and Laybourn-Parry & Bayliss 1996.

rates in the depaurate waters of Crooked Lake are extremely low (0.00077 h⁻¹) particularly when one compares them with the rates seen in the richer waters of the Signy Islands lakes (0.012 to 0.013 h⁻¹) (Laybourn-Parry *et al.* 1995, 1996). These very low rates are undoubtedly related to energy limitation as well as low temperatures. The grazing rates of HNAN derived from studies using fluorescently labelled bacteria uptake, are correspondingly low, ranging from 0.20 to 0.83 h⁻¹ across the Continental to Maritime Antarctic.

By combining data on bacterial production, derived from the incorporation of tritiated thymidine into DNA and ¹⁴C labelled leucine into protein (Laybourn-Parry *et al.* 1995) and

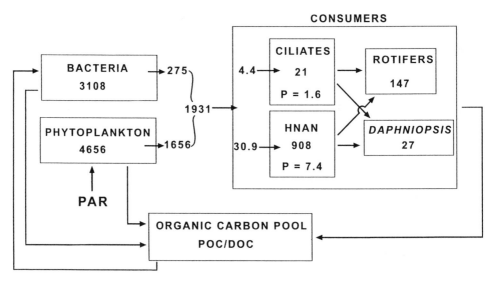

Fig. 5. A simple model of carbon flow in Crooked Lake for the summer (November-January). Biomass (ng carbon l⁻¹) is shown in the compartments. Arrows into Ciliate and HNAN compartments indicate consumed carbon. Figures out of bacteria and phytoplankton indicate production and potential carbon for consumption by the consumer organisms. An unknown portion of primary production enters the DOC pool as exudate. P - production (ng carbon l⁻¹ day⁻¹)

primary production (Bayliss *et al.*, unpubl.) with ecophysiological data on protozoa it is possible to construct a tentative models of carbon flow within the planktonic community of Crooked Lake (Fig. 5). The diagram shown is for the summer months. The biomass of the heterotrophic microbial plankton (bacteria and protozoa) exceeds that of the metazoan plankton, by an order of magnitude.

The productivity of the system is low, with primary production averaging only 1.6 μg l^{-1} day^{-1} in summer. There are no sources of allochthonous carbon and the development of algal mats in the littoral regions of this lake is poor. Thus virtually all the carbon in the pool of organic carbon is derived from photosynthesis undertaken in the plankton. The biomass of protozoan carbon supported by this production via bacteria production is extremely small, but is of the same order of magnitude as that seen in a lower latitude large oligotrophic lake (Laybourn-Parry *et al.* 1994). However, the turnover rate of that carbon is much slower in the Antarctic system, and it is subject to little predation pressure.

What is urgently needed are more long term studies which include an ecophysiological component, on the diverse range of lakes on the continent. At the moment most of the information is descriptive. We need to know much more about the functional dynamics of these extreme, pristine ecosystems, in relation to the size and sources of carbon pools and nutrient dynamics.

REFERENCES

Adamson, D.A. & J. Pickard, 1986. Caenozoic history of the Vestfold Hill. In: Pickard J. (Ed.), *Antarctic Oasis*. Academic Press, Sydney. pp. 63-97.

Bennett, S.J., R.W. Sanders & K.G. Porter, 1990. Heterotrophic, autotrophic and mixotrophic nanoflagellates: seasonal abundances and bacterivory in a eutrophic lake. *Limnology and Oceanography* 35:1821-1832.

Bird, D.F. & J. Kalff, 1986. Bacterial grazing by phytoplanktonic alga. *Science* 231:493-495.

Bloem, J. & M-J.B. Bar-Gilissen, 1989. Bacterial activity and protozoan grazing potential in a stratified lake. *Limnology and Oceanography* 34:279-309.

Burgess, J.S., A.P. Spate & J. Shevin, 1994. The onset of glaciation in the Larsemann Hills, Eastern Antarctica. *Antarctic Science* 6:491-495.

Cathey, D.D., B.C. Parker, G.M. Simmons, W.H. Younge & M.R. Van Brunt, 1981. The microfauna of algal mats and artificial substrates in Southern Victoria Land lakes of Antarctica. *Hydrobiologia* 85:3-15.

Chróst, R.J. & M.A. Faust, 1983. Organic carbon release by phytoplankton: its composition and utilization by aquatic bacterioplankton. *Journal of Plankton Research* 5:477-493.

Ellis-Evans, J.C., J. Laybourn-Parry, P. Bayliss & S.J. Perriss, 1997. Human impact on an oligotrophic lake in the Larsemann Hills. In: *Proceedings of Sixth SCAR Biology Symposium, Venice 1994*.

Ellis-Evans, J.C., J. Laybourn-Parry, P.R. Bayliss & S.J. Perriss, in press. Physical, chemical and microbial community characteristics of lakes in the Larsemann Hills, eastern Antarctica. *Archive fuer Hydrobiology*.

Fogg, G.E., 1958. Extracellular products of phytoplankton and the estimation of primary production. *Rapports et Proces-Verebaux des Reunions Conseil Permanent International pour l'Exploration de la Mer* 144:56-60.

Izaguirre, I., G. Mataloni, A. Vincour & G. Tell, 1993. Temporal and spatial variations of phytoplankton from Boekella Lake (Hope Bay, Antarctic Peninsula). *Antarctic Science* 5:137-141.

James, M.R., R.D. Pridmore & V.J. Cummings, 1995. Planktonic communities of melt ponds on the McMurdo Ice Shelf. *Journal of Plankton Research* 15:555-567.

James, M.R., J.A. Hall, & J. Laybourn-Parry, in press. Protozooplankton and microzooplankton ecology in lakes of the Dry Valleys, Southern Victoria Land. In: Priscu, J. (Ed.), *The McMurdo Dry Valleys, Antarctica: a cold desert ecosystem*. American Geophysical Union, Washington D.C.

Laybourn-Parry, J. & P. Bayliss, 1996. Seasonal dynamics of the planktonic community in Lake Druzhby, Princess Elizabeth Land, Eastern Antarctica. *Freshwater Biology* 35:57-67.

Laybourn-Parry, J. & H.J. Marchant, 1992a. The microbial plankton of freshwater lakes in the Vestfold Hills, Antarctica. *Polar Biology* 12:405-410.

Laybourn-Parry, J. & H.J. Marchant, 1992b. *Daphniopsis studeri* (Crustacea: Cladocera) in lakes of the Vestfold Hills, Antarctica. *Polar Biology* 11:631-635.

Laybourn-Parry, J. & S.J. Perriss, 1995. The role and distribution of the autotrophic ciliate *Mesodinium rubrum* (*Myrionecta rubra*) in three Antarctic saline lakes. *Archiv fuer Hydrobiology* 135:179-194.

Laybourn-Parry, J. M. James, D.M. McKnight, J.C. Priscu, S.A. Spaulding, S. A. & R. Shiel, 1997. The microbial plankton of Lake Fryxell, Southern Victoria Land, Antarctica during the summer of 1992 and 1994. *Polar Biology* 17:62-68.

Ecosystem Processes in Antarctic Ice-free Landscapes, Lyons, Howard-Williams & Hawes (eds)
© *1997 Balkema, Rotterdam, ISBN 90 5410 925 4*

The abundance of planktonic virus-like particles in Antarctic lakes

R.L.Kepner Jr & R.A.Wharton Jr
Desert Research Institute, Reno, Nev., USA

V.Galchenko
Institute of Microbiology, Russian Academy of Sciences, Moscow, Russia

ABSTRACT: We report the first known observations of virus-like particles (VLP) in Antarctic lakes. Mean extracellular VLP densities from Lakes Fryxell and Hoare (McMurdo Dry Valleys, 77°S; 162-163°E), sampled in January of 1996, ranged from 2 to 76×10^6 ml^{-1}. Ratios of VLP to heterotrophic bacteria are higher than those reported from many other aquatic systems. Lake Fryxell had a significantly higher mean density of VLP (3.3×10^7 ml^{-1}) than Lake Hoare (1.8×10^7 ml^{-1}, t-test, p = 0.017, n = 20). Consistent with prior observations, total bacterial densities were significantly higher in the anaerobic zone compared with the aerobic zone of Lake Fryxell. No significant difference in VLP densities was observed between these zones. VLP abundance was significantly correlated with chlorophyll *a* concentration (p <0.05, r = 0.492) in ultra-oligotrophic Lake Hoare, but not in oligotrophic Lake Fryxell. In Lake Hoare, VLP densities increased from the ice-water interface to a point near the chlorophyll *a* maximum (13 m), then decreased consistently with depth. In Fryxell, VLP densities varied greatly with depth. VLP abundance in the Lake Fryxell water column is not significantly correlated with any of the physico-chemical nor microbiological parameters thus far considered. However, in Lake Hoare, VLP abundance is strongly correlated with a variety of parameters including, temperature, pH, dissolved oxygen, specific conductance, salinity, total bacterial density and chlorophyll *a* concentration. Based upon our initial data, we speculate that viruses may constitute an important food web component in these lakes.

1 INTRODUCTION

The traditional view of the "microbial loop" (Azam *et al.* 1983; Pomeroy & Wiebe 1988) suggests a linear system in which protozoa are the major grazers of bacteria which have exploited nutrients and dissolved organic matter (DOM). Previous investigations into the fate of bacteria in aquatic systems have focused on several direct regulators of bacterial numbers and productivity including: available inorganic nutrients (primarily N and P), DOM availability, microprotozoan grazers, and larger, metazoan bactivores. Recent studies have indicated that viruses may contribute significantly to bacterial mortality in planktonic systems, and that mortality attributable to viruses may equal that due to protozoan grazing (Fuhrman & Noble 1995). Roughly 20% of marine heterotrophic bacteria are infected by viruses and 10 to 25% of the bacterial community is lysed daily (Suttle 1994; Steward *et al.* 1996). Viral lysis of phytoplankton may also be significant, with up to 10% of certain populations being destroyed each day (Cottrell & Suttle 1995). With improved methods for enumerating and tracking viruses and virus-like particles (VLP) now developed (Steward *et al.* 1992a; Hennes & Suttle 1995; Fuhrman & Noble 1995), and increasing attention to viral abundance and viral-mediated

Table 1. Correlation matrix for water column parameters (font size indicative of level of significance for biological parameters [ATP, bacteria, chl *a*, and VLP]).

	Depth (m)	T (°C)	pH	DO (mg l⁻¹)	Conductivity	ORP (mV)	Salinity	Light (%)	ATP (ng l⁻¹)	Bacteria (cells ml⁻¹)	chl *a* (mg m⁻³)
Lake Fryxell (n = 12)											
T (°C)	0.745										
pH (std. units)	-0.928	-0.805									
DO (mg l⁻¹)	-0.826	-0.689	0.947								
Cond. (mmhos cm⁻¹)	0.984	0.747	-0.958	-0.892							
ORP (mV)	-0.992	-0.731	0.972	0.975	-0.966						
Salinity	0.985	0.740	-0.956	-0.892	1.000	-0.966					
Light (% incident)	-0.811	-0.974	0.825	0.709	09.799	0.766	-0.794				
ATP (ng l⁻¹)	ns	ns	ns	-0.540	ns	ns	ns	ns			
Bacteria (cells ml⁻¹)	ns	ns	ns	ns	ns	ns	ns	ns	0.744		
chl *a* (mg m⁻³)	ns	ns	ns	ns	ns	ns	ns	ns	ns	ns	
VLP (number ml⁻¹)	ns	ns	ns	ns	ns	ns	ns	ns	ns	ns	ns
Lake Hoare (n = 15)											
T (°C)	-0.694										
pH (std. units)	-0.919	0.836									
DO (mg l⁻¹)	-0.950	0.817	0.892								
Cond. (mmhos cm⁻¹)	0.926	-0.807	-0.970	-0.901							
ORP (mV)	ns	ns	ns	0.447							
Salinity	0.906	-0.823	-0.899	-0.969	0.920	-0.444					
Light (% incident)	-0.728	ns	0.699	0.546	-0.753	ns	-0.591				
ATP (ng l⁻¹)	ns	ns	ns	ns	ns	ns	ns	ns			
Bacteria (cells ml⁻¹)	ns	0.512	ns	0.456	ns	ns	ns	ns	ns		
chl *a* (mg m⁻³)	ns	ns	ns	ns	ns	ns	ns	ns	ns	0.837	
VLP (number ml⁻¹)	-0.502	0.698	0.467	0.648	-0.484	ns	-0.638	ns	ns	0.687	0.492

Significance legend:
$0.05 < p < 0.1$
$0.01 < p < 0.05$
$0.005 < p < 0.01$
$p < 0.005$

242

processes in natural environments (e.g. Suttle 1994; Maranger & Bird 1995), it is crucial to explore the potential role of viruses as regulators of microbial populations in both planktonic and benthic habitats.

The relatively ice-free McMurdo Dry Valleys of Antarctica are characterized by microbially-dominated communities (e.g. Vincent 1988). Aquatic habitats of these cold Antarctic deserts are low in taxonomic richness and total biomass, and thus, ideal for studies of food web structure, trophic transfer and community stability.

Despite attention newly directed at the abundance and impact of aquatic viruses (e.g. Bergh *et al.* 1989), no information on the role of viruses (particularly bacteriophage, cyanophage and other phycoviruses) in high-latitude freshwater systems currently exists. In this paper, we report the first observations and quantification of planktonic VLP in perennially ice-covered Antarctic lakes. We also compare the abundance of VLP with bacerial densities and other physico-chemical parameters.

2   STUDY SITE AND METHODS

Data reported here were collected at Lakes Hoare and Fryxell during the 1995-96 field season. These lakes are separated by the Canada Glacier and located at the eastern end of Taylor Valley in southern Victoria Land. Lake Hoare (77°38'S, 162°53'E) is 58 m above sea-level, has a surface area of 1.9 km^2, a mean depth of 14.2 m, and maximum depth of 34 m. Lake Fryxell (77°37'S, 163°07'E) is approximately 20 m above sea-level, has a surface area of 7.0 km^2, a mean depth of 7.6 m, and maximum depth of 20.5 m. Detailed descriptions of these lakes can be found in the literature (e.g. Lawrence & Hendy 1985, Wharton *et al.* 1986, Vincent 1988, Green *et al.* 1988, Green *et al.* 1989, Wharton *et al.* 1989; Chinn 1993, Wharton *et al.* 1993).

Water samples were collected in early January of 1996 from two sites each in both lakes Hoare and Fryxell using a 1.0-l glass, Kemmerer-type sampler. One shallow and one deep site were sampled at each lake, with the distance between sites being approximately 100 and 500 m for Hoare and Fryxell, respectively. Virus-like particles were enumerated using the method of Hennes and Suttle (1995). Freshly collected samples were filtered onto 0.02-μm pore-size Anodisc 25 membrane filters (Whatman) and stained with the cyanine-based dye, Yo-Pro-1 (4-[3-methyl-2,3-dihydro-(benzo-1,3-oxazole)-2-methylmethyledene]-1-(3'-trimethyl ammoniumpropyl)-quinolinium diiodide; Molecular Probes, Eugene, OR). Virus densities were then estimated using epifluorescent microscopic direct counts (EMC) at a magnification of ×1250 using a Zeiss Standard 16 microscope with 100W Hg lamp and Zeiss P/N 09 combination filter set. Total bacteria were also enumerated by EMC following DAPI (4',6-diamidino-2-phenylindole) staining (Porter & Feig 1980, as modified by Kepner & Pratt 1994) using an Olympus BX60 microscope with U-MWU filter block. At least 400 VLP or bacterial cells were counted on each membrane filter.

Water samples were collected in the same fashion for estimation of chlorophyll *a*. These were filtered (0.45-μm pore-size, 47-mm diam. Gelman cellulose-acetate filters) at <380 mm Hg vacuum and extracted for 12 to 16 h in 90% MgCO$_3$-buffered acetone at 4°C. Measurements were made using a Turner Designs AU10 fluorometer and corrected for phaeophytin by addition of HCl (Holm-Hansen *et al.* 1965). Adenosine 5'-triphosphate (ATP) was measured as an indicator of microbial biomass using an ATP bioluminescent assay kit (Sigma Chemical Co., St. Louis, MO) and ATP photometer. Profiles of physico-chemical parameters (temperature, pH, dissolved oxygen [DO], specific conductance, oxygen reduction potential [ORP], and salinity) were obtained using a HydroLab Surveyor 2 sensor array. HydroLab profile data were collected roughly three weeks prior to sampling for VLPs, bacteria, and chlorophyll *a* concentrations. Light penetration was measured using a LiCor LI-189 meter and LI-193SA underwater spherical irradiance sensor.

3   RESULTS

3.1   Physico-chemical profiles

Observations of temperature, pH, specific conductance, DO and light penetration profiles were consistent with those previously obtained for these lakes (e.g. see Figs. 1-4 in Wharton *et al.* 1983). No major differences were observed between physico-chemical profiles obtained at the shallow and deep sampling holes. Major patterns of interest in Lake Hoare include: (1) a double oxycline with a sharp decreases in DO at 11 to 14 m and then again below 25 m, (2) a minor increase in specific conductance (9 to 14 m) which then continues to increase more gradually with depth, (3) decrease from pH 9 to 7 between 9 and 14 m, and (4) a narrow range of temperatures throughout the water column (0 to 0.5°C), with a thermocline at 11 to 14 m. Characteristics of the Lake Fryxell profile include: (1) a strong oxycline (8.5 to 10 m), with anaerobic (DO <1 mg l^{-1}, negative ORP, high H_2S) waters below 10 m, (2) a constant but gradual increase in conductivity and salinity below 7 m, (3) a slight decrease in pH (7.7 to 6.9) at 7.5 to 9.5 m, and (4) a gradual increase in temperature between 5 and 10 m (from 0 to 2.5°C). Despite similar readings directly below the ice, light attenuated much more rapidly with increasing depth in Lake Fryxell than in Hoare (e.g. % incident solar radiation at 11 m was 0.4 and 2.1% in Fryxell and Hoare, respectively).

3.2   Virus-like particles

Mean extracellular virus-like particle (VLP) densities were higher than expected, ranging from 2 to 76×10^6 ml^{-1}. In Lake Hoare, VLP densities increased from the ice-water interface to a point

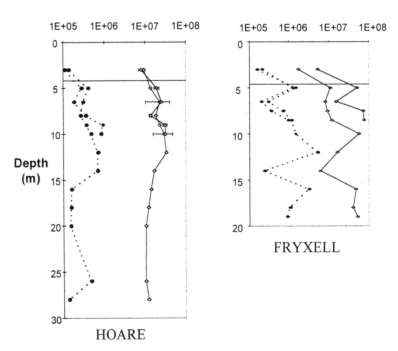

Fig. 1. Water column virus-like particle (solid lines) and total bacterial (dotted lines) densities (numbers ml^{-1}) at paired sites in Lakes Hoare and Fryxell. (Error bars on Lake Hoare VLP represent standard deviations of duplicate sample preparations).

near the chlorophyll *a* maximum (13 m), then decreased consistently with depth (Fig. 1). Little variability was observed between duplicate preparations of samples for VLP enumeration, and changes in VLP abundance mirrored changes in total bacterial abundances throughout the water column.

In Fryxell, densities varied greatly with depth, but generally increased below the oxycline in concert with bacterial densities. Both VLP and bacterial densities decreased markedly at 14 m, but returned to densities more typical of the hypolimnetic waters at a depth of 16 m (Fig. 1). Variability in VLP density with depth at a particular hole, and variabilty between the shallow and deeper sampling holes was greater in Fryxell than in Hoare. Lake Fryxell had a significantly higher mean density of VLP (3.3×10^7 ml^{-1}) than Lake Hoare (1.8×10^7 ml^{-1}, t-test, p = 0.017, n = 20).

Ratios of VLP to heterotrophic bacteria (so-called virus:bacteria ratios; VBR) ranged from 3.2 to 196.7. VBR were higher in Lake Hoare than Fryxell (mean = 62.5 and 42.9, respectively), though this difference was not highly significant (t-test, p = 0.084). However, near the chlorophyll *a* maximum in Fryxell, mean VBR exceeded those found at a similar depth in Hoare (Fig. 2). There was no significant difference in VBR between aerobic and anaerobic zones in Fryxell. In general, VBR appeared to be quite variable both within and between lakes.

3.3   Pattern and correlations with VLP abundance

Lakes Fryxell and Hoare were strikingly different in terms of correlations between extracellular VLP densities and other variables of interest. In Lake Hoare, 8 of 11 measured variables were significantly correlated (at the p <0.10 level, n = 15) with VLP abundance (Table 1). Strongest among these were positive correlations with temperature, DO, and total bacterial density (all p <0.05). Additionally, VLP abundance was significantly correlated with chlorophyll *a*

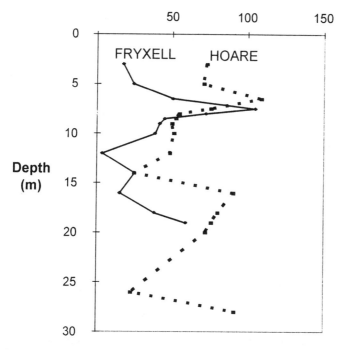

Fig. 2. Mean VLP:bacteria ratios (VBR) in the water columns of lakes Hoare and Fryxell.

concentration (p <0.01) in ultra-oligotrophic Lake Hoare, where distributions of VLP appeared to be closely related to autotrophic biomass (Fig. 3).

Bacterial densities in Lake Hoare were positively correlated with both temperature and DO, but, unexpectedly, not with ATP concentration. In addition, we observed extremely high positive correlations between bacterial counts and both chlorophyll a and VLP concentrations (Table 1). Slightly less than two orders of magnitude more abundant than bacteria, VLP distributions closely mirrored those of bacteria in Hoare. The overall mean density of total bacteria for Hoare and Fryxell were 3.7×10^5 and 1.2×10^6 ml^{-1}, respectively.

Despite an apparent linkage between chlorophyll a concentrations and both VLP and bacterial abundance in Lake Hoare, it should be noted that chlorophyll a levels are seasonally variable (Fig. 4). Whether relationships between phytoplankton biomass, production and VLP densities exist at other times of the year remains unknown. Highest chlorophyll a levels were found later in the season in the more productive Lake Fryxell (Vincent 1981), and chlorophyll a maxima were deeper in Lake Hoare, perhaps as a result of greater light penetration. Overall patterns in pigment distribution were consistent with those previously reported for these lakes (e.g. Lizotte & Priscu 1992; Priscu 1995).

In contrast to findings at Lake Hoare, Fryxell VLP abundance estimates were not correlated with a single water quality parameter considered in this study. Although strong correlations between the various depth-related physico-chemical features (e.g. DO, specific conductance, light penetration) were not surprising to find in either Hoare or Fryxell, the lack of consistent relationships between biological characteristics (e.g. ATP, bacteria, VLP and chlorophyll a concentrations) in Fryxell was somewhat unexpected. Certainly there are many significant hydrologic, chemical and biological differences between these two lakes.

In Lake Fryxell, total bacterial densities were significantly higher in the anaerobic zone (combined data from two sampling sites, t-test, p = 0.040, n = 16). Bacterial densities were

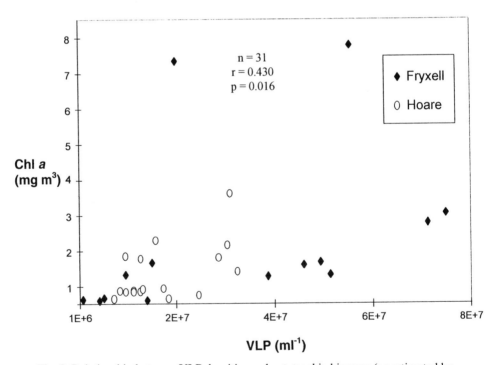

Fig. 3. Relationship between VLP densities and autotrophic biomass (as estimated by chlorophyll a concentration) in lakes Fryxell and Hoare.

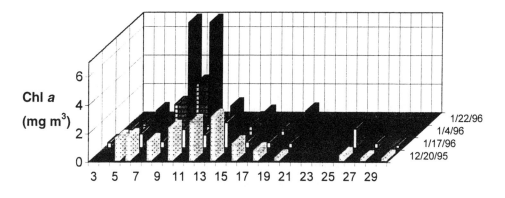

Fig. 4. Chlorophyll *a* concentrations with depth in Lake Hoare (foreground) and Lake Fryxell (background) on two dates each during the 1995/96 field season.

significantly correlated with ATP concentrations, but were not strongly related to any of the other measured variables. Neither bacteria nor VLP densities were significantly correlated with phytoplankton biomass as indicated by chlorophyll *a* concentrations in Fryxell.

4   DISCUSSION AND FUTURE RESEARCH

Viruses are increasingly recognized as important components of aquatic microbial communities (e.g. Borsheim 1993; Fuhrman & Suttle 1993; Suttle 1994). However, little information exists regarding the role of viruses in structuring microbial assemblages in natural environments. This is particularly true in the case of freshwater lakes. Data on VLP abundances from the McMurdo Dry Valleys lakes represent a first effort at examining the potential significance of viruses in these high-latitude, microbially-dominated systems.

Of major interest is the relatively high density of VLPs encountered in these lakes compared to other aquatic systems (Table 2). In all cases but one (Hennes & Suttle 1995) these densities were estimated using transmission electron microscopy (TEM). Total bacterial densities, however, are no higher than those typically found in pelagic marine systems and oligotrophic temperate lakes. The resulting Antarctic lake VBR values are at the high end of the spectrum of values thus far reported in the literature (Table 2). Whether this is due to high rates of viral production, low rates of viral decay, or a combination of the two, remains unknown. Based upon literature surveys as well as their own data from 22 lakes in Quebec, Maranger & Bird (1995) report mean VBR ranging from 20 to 25 and 1 to 5 in fresh water and marine systems respectively.

Only a few prior studies have documented significant relationships between viral abundance and planktonic bacterial densities in either freshwater (Maranger & Bird 1996) or marine (Cochlan *et al.* 1993; Steward *et al.* 1996) systems. As in Antarctic lakes, viral abundance typically parallels bacterial abundance, with numbers of viruses often exceeding bacteria by roughly 5 to 10-fold (Fuhrman & Suttle 1993, and citations therein). Maranger & Bird (1996) report a strong positive correlation between virus abundance and chlorophyll *a* concentration, and, similar to our findings in Fryxell, an increase in viral and bacterial abundance in the anoxic hypolimnion of the small, dimictic, Lac Gilbert in Quebec. Relatively high densities of bacteria and VLP in anoxic waters may reflect a correspondingly low biomass of bactivores. Interestingly, the high variability in VBR with depth observed in the Antarctic

247

Table 2. Extracellular virus abundance and virus:bacteria ratio (VBR) values obtained from previous studies of virus densities in a variety of aquatic habitats.

Mean Virus or VLP Density ($\times 10^6$ ml^{-1})	VBR	System	Reference
10.1	3.2	Chesapeake Bay	Bergh et al. 1989
6.1	5.5	Korsfjorden	Bergh et al. 1989
4.9	12.2	Raunefjorden	Bergh et al. 1989
14.9	49.7	North Atlantic	Bergh et al. 1989
0.1	3.0	Barents Sea	Bergh et al. 1989
158.0	37.1	Southern California Bight (nearshore stations)	Steward et al. 1992a
9.8	20.7	Southern California Bight (offshore stations)	Steward et al. 1992a
25.0	11.9	Chesapeake Bay	Wommack et al. 1992
2.5-36.0	10.0	Bering and Chukchi Seas	Steward et al. 1996
5.5*	12*	Mission Bay, Southern California	Steward et al. 1992b
7.9	10.8	Japanese coastal waters (Osaka & Otsuchi Bays)	Hara et al. 1991
5.3	3.4	Japanese offshore waters	Hara et al. 1991
251.5	171.2	Gulf of Mexico, nearshore at Port Aransas, Texas	Hennes & Suttle 1995
254.0	39.0	Plussee, Germany	Bergh et al. 1989
3.4-8.4	12.0-61.0	Danube River backwater	Mathias et al. 1995
10-40	10*	Lake Constance, Germany	Hennes & Simon 1995
36*	13*	Lac Gilbert, Quebec	Maranger & Bird 1996
18.3	62.5	Lake Hoare, McMurdo Dry Valleys, Antarctica	This study
33.4	42.9	Lake Fryxell, McMurdo Dry Valleys, Antarctica	This study

*value estimated from figures

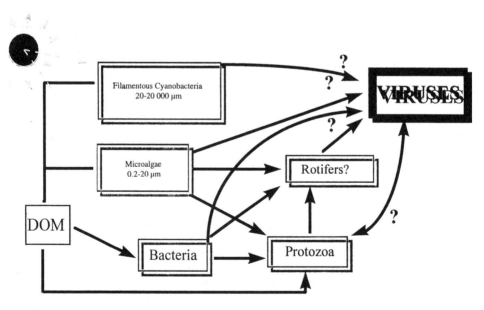

Fig. 5. Schematic linking of conceptual viral and microbial loop models. Do viruses substitute for bactivores and algivores in antarctic ecosystems where multicellular organisms are rare or absent?

lakes was not observed in Lac Gilbert, despite a similar range of VLP and bacterial densities encountered.

Our research is the first step in a program designed to address the question, "to what extent do viruses account for microbial mortality in Antarctic lakes?" In order to address this question we must determine naturally occurring densities of both intra- and extracellular viruses, types and diversity of viruses, rates of viral production, frequency of infected cells, and ultimately, the mortality rates within the various microbial populations attributable to viruses. As suggested by Pesan *et al.* (1994), viruses should be considered as potential regulators of both heterotrophic and autotrophic protozoan, algal, and bacterial community successions (Fig. 5).

In conclusion, our VLP data suggest that viruses may constitute a potentially important component of microbially-dominated food webs in these Antarctic lakes. Future work should be directed at examining viral-mediated mechanisms potentially responsible for the regulation of aquatic microbial community composition.

5   ACKNOWLEDGEMENTS

This research was supported by National Science Foundation grant OPP 92-11773. Much thanks to Jeanette Ward and Chris Moore for laboratory assistance with bacterial counts and to the helpful comments of two anonymous reviewers.

REFERENCES

Azam, F., T. Fenchel, J. Gray, L. Meyer-Reil & F. Thingstad, 1983. The ecological role of water-column microbes in the sea. *Marine Ecology Progress Series* 10:257-263.

Bergh, O., K.Y. Borsheim, G. Bratbak & M. Heldal, 1989. High abundance of viruses found in aquatic environments. *Nature* 340:467-468.

Borsheim, K.Y., 1993. Native marine bacteriophages. *FEMS Microbioial Ecology* 102:141-159.

Chinn,T.J.H., 1993. Physical hydrology of dry valley lakes. In: W. Green & E.I. Friedman (Eds.), *Physical and Biogeochemical Processes in Antarctic Lakes. Antarctic Research Series* 59. American Geophysical Union, Washington, D.C. pp. 1-52.

Cochlan, W.P., J. Wikner, G.F. Stewart, D.C. Smith & F. Azam, 1993. Spatial distribution of viruses, bacteria and chlorophyll *a* in neritic, oceanic and estuarine environments. *Marine Ecology Progress Series* 92:77-87.

Cottrell, M.T. & C.A. Suttle, 1995. Dynamics of a lytic virus infecting the photosynthetic marine picoflagellate *Micromonas pusilla. Limnolology Oceanography* 40(5):730-739.

Fuhrman, J.A. & R.T. Noble, 1995. Viruses and protists cause similar bacterial mortality in coastal seawater. *Limnology and Oceanography* 40(7):1236-1242.

Fuhrman, J.A. & C.A. Suttle, 1993. Viruses in marine planktonic systems. *Oceanography* 6:50-62.

Green, W.J., M.P. Angle & K.E. Chave, 1988. The geochemistry of antarctic streams and their role in the evolution of our lakes in the McMurdo Dry Valleys. *Geochimica et Cosmochimica Acta* 52:1265-1274.

Green, W.J., T.J. Gardner, T.G. Ferdelman, M.P. Angle, L.C. Varner & P. Nixon, 1989. Geochemical processes in the Lake Fryxell Basin (Victoria Land, Antarctica). *Hydrobiology* 172:129-148.

Hennes, K.P. & M. Simon, 1995. Significance of bacteriophages for controlling bacterioplankton growth in a mesotrophic lake. *Applied Environmental Microbiology* 61(1):333-340.

Hennes, K.P. & C.A. Suttle, 1995. Direct counts of viruses in natural waters and laboratory cultures by epifluorescence microscopy. *Limnology and Oceanography* 40(6):1050-1055.

Holm-Hansen, O., C.J. Lorenzen, R.W. Holms & J. Strickland, 1965. Fluorometric determination of chlorophyll. *Journal du Conseil. Conseil Permanent International pour l'Exploration de la Mer* 30:169-204.

Kepner, R.L., Jr. & J.R. Pratt, 1994. Use of fluorochromes for direct enumeration of total bacteria in environmental samples: past and present. *Microbial Reviews* 58:603-615.

Lawrence, M.J.F. & C.H. Hendy, 1985. Water column and sediment characteristics of Lake Fryxell, Taylor Valley, Antarctica. *New Zealand Journal of Geology and Geophysics* 28:543-552.

Lizotte, M.P. & J.C. Priscu, 1992. Spectral irradiance and bio-optical properties in perenially ice-covered lakes of the dry valleys (McMurdo Sound, Antarctica). In: Elliot, D.H. (Ed*.), Contributions to Antarctic Research. Antarctica Research Series* 57. American Geophysical Union, Washington D.C. pp. 1-14.

Maranger, R. & D.F. Bird, 1995. Viral abundance in aquatic systems: a comparison between marine and fresh waters. *Marine Ecology Progress Series* 121:217-226.

Maranger, R. & D.F. Bird, 1996. High concentrations of viruses in the sediments of Lac Gilbert, Quebec. *Microbial Ecology* 31:141-151.

Mathias, C.B., A.K.T. Kirschner & B. Velmirov, 1995. Seasonal variations of virus abundance and viral control of the bacterial production in a backwater system of the Danube River. *Applied and Environmental Microbiology* 61(10):3734-3740.

Pesan, B.F., M.G. Weinbauer & P. Peduzzi, 1994. Significance of the virus-rich 2-200-nm size fraction of seawater for heterotrophic flagellates: I. Impact on growth. *Marine Ecology 15:*281-290.

Pomeroy, L.R. & W.J. Wiebe, 1988. Energetics of microbial food webs. *Hydrobiology 159:*18-37.

Porter, K.G., & Y.S. Feig, 1980. The use of DAPI for identifying and counting aquatic microflora. *Limnology and Oceanography* 25:943-948.

Priscu, J.C., 1995. Phytoplankton nutrient deficiency in lakes of the McMurdo dry valleys, Antarctica. *Freshwater Biology 34:*215-227.

Steward, G.F., D.C. Smith & F. Azam, 1996. Abundance and production of bacteria and viruses in the Bering and Chukchi Seas. *Marine Ecology Progress Series* 131:287-300.

Steward, G.F., J. Wikner, D.C. Smith, W.P. Cochlan & F. Azam, 1992a. Estimation of virus production in the sea: I. Method development. *Marine Microbial Food Webs* 6:57-78.

Steward, G.F., J. Wikner, W.P. Cochlan, D.C. Smith & F. Azam, 1992b. Estimation of virus production in the sea: II. Field results. *Marine Microbial Food Webs* 6(2):79-90.

Suttle, C.A., 1994. The significance of viruses to mortality in aquatic microbial communities. *Microbial Ecology* 28:237-244.

Vincent, W.F., 1981. Production strategies in antarctic inland waters: Phytoplankton eco-physiology in a permanently ice-covered lake. *Ecology* 62:1215-1224.

Vincent, W.F., 1988. *Microbial Ecosystems of Antarctica*. Cambridge University Press, Cambridge. 304 p.

Wharton, R.A., Jr., C.P. McKay, G.M. Simmons, Jr. & B.C. Parker, 1986. Oxygen budget of a perenially ice-covered Antarctic lake. *Limnology Oceanography* 31:437-443.

Wharton, R.A., Jr., C.P. McKay, G.D. Clow & D.T. Andersen, 1993. Perennial ice covers and their influence on Antarctic lake ecosystems. In: W. Green & E.I. Friedman (Eds.), *Physical and Biogeochemical Processes in Antarctic Lakes. Antarctic Research Series* 59. American Geophysical Union, Washington, D.C. pp. 53-72.

Wharton, R.A., Jr., B.C. Parker & G.M. Simmons Jr., 1983. Distribution, species composition and morphology of algal mats in Antarctic dry valley lakes. *Phycologia* 22(4):355-365.

Wharton, R.A., Jr., G.M. Simmons, Jr., & C.P. McKay, 1989. Perennially ice-covered Lake Hoare, Antarctica: Physical environment, biology and sedimentation. *Hydrobiologia* 172:306-320.

Wommack, K.E., R.T. Hill, M. Kessell, E. Russek-Cohen & R.R. Colwell, 1992. Distribution of viruses in the Chesapeake Bay. *Applied and Environmental Microbiology* 58:2965-2970.

4 Human impacts

Hydrocarbon-degrading bacteria in oil-contaminated soils near Scott Base, Antarctica

Jackie Aislabie
Manaaki Whenua, Landcare Research, Hamilton, New Zealand

ABSTRACT: Oil is spilt in the Antarctic when fuel oils such as JP8 jet fuel and MoGas are moved or stored. Hydrocarbons, both *n*-alkanes and aromatic compounds, have been detected in soils around Scott Base. In such areas hydrocarbon-degrading microbes could be used to clean up the oil spills. Soil samples from oil-impacted and control sites were analysed for hydrocarbon-degrading microbes and a range of parameters known to limit biodegradative activity. Soils were analysed for water content, pH, electrical conductivity, and concentrations of nutrients (N and P). Some of the oil-contaminated samples were enriched with culturable heterotrophic bacteria and hydrocarbon degraders. Bacteria able to utilise JP8 jet fuel were isolated, and most used *n*-alkanes as their sole source of carbon. Preliminary results indicate that low levels of nitrogen may limit biodegradation of oil at some of the contaminated sites.

1 INTRODUCTION

Oil pollution in Antarctica poses a threat to marine life and the limited terrestrial biota concentrated in ice-free regions along the coast line. These regions are located mostly in the Ross Dependency and on the Antarctic Peninsula (Gregory *et al.* 1984), where many scientific stations are based and field operations, such as storage and refuelling of aircraft and vehicles, can result in oil spills (Cripps & Priddle 1991). Applying hydrocarbon-degrading microbes for *in situ* bioremediation of oil spills in the Antarctic has been proposed by Kerry (1993) and Wardell (1995). The success of this technology is dependent not only on the presence of appropriate biodegradative microbes but also on environmental parameters.

There have been a few studies of microbial oil degradation in Antarctic soils, hydrocarbon-degrading microbes have been detected (Konlechner 1985; Kerry 1990; Tumeo & Wolk 1994) and it has been established that bacteria, not fungi, are the major colonisers of oil-contaminated soils (Kerry 1990). Environmental parameters that affect the biodegradation of oil in cold climates have been studied extensively in the Arctic (Atlas 1981). Although oil degradation rates are sensitive to temperature extremes (Atlas 1981), at Scott Base, soil temperatures from 0°C to 20°C have been measured during summer (Balk *et al.* 1995). As hydrocarbon degradation activity can occur within this range, oil degradation should not be limited by temperature at least in summer. Atlas (1981) also observed that it was available water not temperature that was the major limiting factor. To enhance oil biodegradation rates *in situ* common strategies employed include the addition of nutrients, lime to modify pH, and moisture (Wardell 1995).

If hydrocarbon-degrading microbes are present in Antarctic soil, and environmental parameters do not inhibit their action, then bioremediation has potential as a cleanup technology for oil-contaminated soils in this environment. We therefore set out to determine the numbers and types of hydrocarbon-degrading microbes in oil-contaminated and non-contaminated soils from around Scott Base. These soils were also analysed for water content, pH, electrical conductivity, and levels of nutrients, parameters known to affect the rate at which hydrocarbons are biodegraded. This paper describes preliminary results from this study.

2 MATERIALS AND METHODS

2.1 Soil collection and characterisation

Soils, both oil-contaminated and non-contaminated, were collected from sites around Scott Base (Table 1). Shallow pits were dug down to the ice-cemented surface and, unless stated otherwise, samples were collected from the 0 to 5 cm depth. Aseptic soil samples for microbial analysis were collected into sterile plastic bags using ethanol sterilised tools. Sufficient material was obtained for both chemical and physical analyses of soil. Frozen soil samples were transported to New Zealand for analysis.

2.2 Microbial analyses

Numbers of culturable heterotrophs were determined by plating soil dilutions onto R2A (Difco)

Table 1. Levels of total petroleum hydrocarbons (TPH) and numbers of culturable heterotrophic bacteria and the most probable number (MPN) of hydrocarbon degraders in oil-contaminated and control sites from the vicinity of Scott Base, Antarctica, collected January 1996.

Location	Sample No.	TPH $\mu g\ g^{-1}$ dry weight	Number of culturable heterotrophs g^{-1} dry weight	MPN of hydrocarbon degraders g^{-1} dry weight
Former site for storage of Mogas. Sample 1 was taken from within the ice-cemented layer and Sample 2 from the upper layer at 3-5 cm.	1	5928	1.11×10^3	<10
	2	12271	3.7×10^2	<10
Below the generator plant.	4	1955	1.08×10^5	<10
Site of fuel leak near the Scott Base kitchen.	5	<1	2.52×10^7	2.3×10^2
All samples were collected from beneath the kitchen along a 2 m transect. Sample 5 was furtherest from the contamination. 0-2 cm depth.	6	10228	5.78×10^7	2.5×10^4
	7	17488	2.12×10^7	2.6×10^4
Site of a leaking pipeline carrying JP8 jet fuel from McMurdo Base to Williams Field. Sample 8 was taken from the contaminated site. Sample 9 is from a site 30 m east of Sample 8.	8	1048	6.62×10^6	8.6×10^1
	9	<1	5.79×10^5	3.5×10^1
Control site on Observation Hill.	3	<1	9.69×10^4	<10
Control site above Scott Base away from vehicle tracks.	10	<1	1.96×10^5	<10

agar plates. Soil samples (10 g wet weight) were shaken for 1 hour in 90 ml of 0.1% sterile sodium pyrophosphate solution containing 30 g of glass beads. The resulting suspension was diluted in 0.1% sodium pyrophosphate and each dilution vortexed for one minute to separate bacteria from soil particles. Portions (0.1 ml) of each dilution were spread onto plates and incubated at 16°C for four weeks.

Relative differences in numbers of hydrocarbon-degrading microbes in the samples were determined by a most probable number technique, using the five-tube method. Soil samples were diluted to extinction in sterile Bushnell Haas Medium (BH), then 10 ml placed in sterile tubes and amended with 50 µl of JP8 jet fuel as sole carbon and energy source. All tubes were sealed and incubated at 8°C for at least two months. Poisoned controls were prepared by adding 0.2 ml of cHCl to each tube. To determine if growth had occurred in the tubes they were compared with the controls and those that were both turbid and showed disruption to the film of oil on the surface of the medium were scored as positive.

2.3   Isolation of jet fuel-degrading bacteria

Bacteria capable of degrading JP8 jet fuel (commonly used in the Antarctic) were isolated directly from the soil samples. Soil dilutions were prepared in 0.1% sterile sodium pyrophosphate solution as described above. Portions (0.1 ml) of each dilution were spread onto Bushnell Haas medium (BH) (Difco) solidified with 1.5% purified agar (Oxoid) and supplied with JP8 jet fuel as sole carbon and energy source by placing it in a vapour tube in the lid of the plate. Control plates without substrate were also inoculated. All plates were prepared in triplicate for each dilution and incubated at 16°C for at least two months.

Jet fuel-degrading bacteria were removed from the isolation plates and purified on BH plates supplied with jet fuel as sole source of carbon. Colonies larger and different from those on the substrate-free control plates were selected and 58 colonies were purified for further investigation. Representative strains were characterised using standard microbiological tests (Gerhardt 1994) and some were sent to Microbial Id. Inc. (Newark, Del.) for presumptive identification by whole cell fatty acid analysis.

2.4   Screening of bacterial isolates for ability to degrade hydrocarbons

The jet-fuel degrading bacteria were screened on minimal agar plates for their ability to degrade both aromatic and aliphatic hydrocarbons. Toluene was supplied as a vapour, by placing 2 ml of the substrate in an open beaker in an air-tight container along with inoculated plates. Naphthalene was added as crystals to the lid of the petri dish. Hexadecane and dodecane (100 µl) were added in an overlayer of BH medium solidified with 1% purified agar. All plates were incubated at 16°C for at least one month. To determine whether growth had occurred, substrate-free control plates were used for comparison.

2.5   Chemical and physical analyses of soil

Soils were analysed for parameters known to influence the rate at which oil is biodegraded, including water content, pH, electrical conductivity, total carbon, and concentrations of the nutrients nitrogen and phosphorus (Blakemore et al. 1987). Total carbons were determined using wet samples to reduce loss of hydrocarbons on drying. Hydrocarbon levels were determined by extracting the soil samples in methylene chloride then analysing the extracts by capillary gas chromatography with a flame ionisation detector as outlined in the USEPA method 418.1. Those samples contaminated with hydrocarbons were subsequently analysed using capillary gas chromatography with mass spectrometry (GC-MS) to speciate the contaminants.

3   RESULTS

Ten soil samples were collected from around Scott Base (Table 1). Hydrocarbons were detected in samples from the former site for storage of MoGas (1 & 2), beneath the generator (4), the fuel leak near the kitchen (6 & 7) and the leaking pipeline (8). Samples from beneath the generator (4), the fuel leak near the kitchen (6 & 7), and the leaking pipeline (8) were contaminated mainly with *n*-alkanes with chain length from C9 to C14, whereas aromatic compounds, both methyl benzenes and naphthalene derivatives, dominated samples from the former MoGas storage site (1 & 2). No hydrocarbons were detected in the control samples (3 & 10) nor the two samples collected furthest from the fuel spills near the kitchen (5) and the pipeline (9).

Numbers of culturable heterotrophic bacteria and hydrocarbon degraders were enriched in samples from the fuel leak near the kitchen (5, 6 & 7) and the leaking pipeline (8) when compared to those of the control samples (3 & 10) (Table 1). Hydrocarbon-degrading microbes were detected in all samples except those from the site of the former MoGas storage area (1 & 2) and one of the control sites (10). Numbers of heterotrophs and hydrocarbon degraders in the soil sample from beneath the generator (4) were comparable with those in the control site (3). Lowest numbers of heterotrophic bacteria and hydrocarbon degraders were detected in samples from the former MoGas storage site (1 & 2).

Bacteria able to utilise JP8 jet fuel for growth were isolated directly from the oil-contaminated samples from near the kitchen (5, 6 & 7) and the leaking pipeline (8). Bacterial colonies on plates inoculated with soil from samples 6 and 8 were typically large, white and mucoid. In contrast, those originating from samples 5 and 7 were more diverse. They contained a number of small yellow pigmented or non-pigmented colonies. Of the representative colony types presumptively identified, all but one were Gram-positive rod shaped bacteria. The white, mucoid isolate was identified as *Rhodococcus erythropolis* by fatty acid analysis, and the small yellow pigmented or non-pigmented isolates as *Nocardia asteroides* and *Rhodococcus luteus* respectively. All the Gram-positive bacteria utilised the *n*-alkanes, hexadecane and dodecane, for growth, but only one degraded the aromatic compound toluene. The single Gram-negative isolate grew on toluene but not on the *n*-alkanes and was presumptively identified as a *Pseudomonas* sp.

Chemical and physical characteristics of the soil samples are shown in Table 2. The water content of the samples was low. Although the pH values of the soils were moderately too highly

Table 2. Soil parameters of Antarctic samples that could limit microbial activity *in situ*.

Sample No.	% Soil water	pH	Total C (%)	Total N (%)	Total P (%)	EC (mS/cm)
1	12.8	9.2	0.57	0.01	0.22	0.27
2	11.1	9.1	0.97	0.01	0.20	0.31
3	7.2	9.3	0.07	0.02	0.21	0.46
4	2.2	8.5	2.99	0.01	0.18	0.80
5	6.2	9.3	0.17	0.01	0.17	0.13
6	9.4	9.5	0.79	0.01	0.21	0.23
7	13.5	9.3	1.63	0.01	0.23	0.17
8	8.9	9.4	0.26	0.01	0.20	0.22
9	5.7	8.3	0.09	0.01	0.20	2.31
10	11.7	8.9	0.18	0.01	0.19	0.65

alkaline and ranged from 8.3 to 9.5, these values are typical for this area. Total % carbon of the soils ranged from 0.07% for a pristine soil to 2.99% for sample 4. Enhanced levels of carbon at sites 1, 2, 6, 7 and 8 were attributed to hydrocarbon contamination. Sample 4 contained wood chips, presumably from building activity. Nitrogen content of the soil samples was very low, whereas the phosphorus contents were very high. The EC values for the soil were variable ranging from a low of 0.17 mS cm^{-1} in sample 7 from the spill site near the kitchen, to a very high value of 2.31 mScm^{-1} for sample 9 from near the pipeline spill.

4   DISCUSSION

For bioremediation of the Antarctic soils investigated, microbes with the ability to degrade n-alkanes from C9 to C14, methyl benzenes and methyl naphthalene are required. Fortunately these hydrocarbons are readily biodegradable and support the growth of many different microbes (Atlas 1981). Alkane-degrading bacteria were readily isolated, and although only one aromatic degrader was obtained, we have isolated them previously when naphthalene or 1-methyl naphthalene was supplied as carbon source rather than JP8.

Increased numbers of hydrocarbon degrading microbes in soil from the oil spill sites near the kitchen and the pipeline, provides indirect evidence for their activity and growth under *in situ* conditions (Madsen 1991), the numbers detected were comparable to those in samples from Tank Farm 2 at the nearby McMurdo Station (Tumeo & Wolk 1994). The low numbers of culturable heterotrophic bacteria and the absence of hydrocarbon degraders in samples from the former MoGas storage site indicate that this site may be contaminated with some compound(s) that limits microbial growth and activity, for example toxic levels of organic lead, and is not therefore amenable to bioremediation.

The hydrocarbon-degrading bacteria isolated are probably indigenous to the Antarctic rather than introduced with the oil, as they are predominantly Gram-positive rod shaped bacteria and belong to the Nocardiaceae family. Members of this family are widely distributed in Antarctic soils, and have been reported previously to degrade alkanes (Kerry 1990).

Soil analyses indicate that a number of parameters are likely to limit oil biodegradation rates *in* situ. They are the availability of nitrogen, and possibly the alkaline pH and low water content of the soil. For optimum oil biodegradation rates C:N:P ratios of 120:10:1 are recommended (Wilson & Jones 1993). Kerry (1993) was able to enhance biodegradation of n-alkanes at an experimental spill site near Davis Station by the addition of nitrogen, phosphorus and potassium. Experiments are ongoing to investigate the influence of pH and nitrogen on alkane mineralisation rates in soil from the site of the leaking pipeline. Most hydrocarbon degraders grow optimally in soils at neutral pH with a water holding capacity of 50 to 80% (Wilson & Jones 1993). To predict if the soil water contents do limit oil biodegradation rates we plan to generate a water release curve using Antarctic soil from around Scott Base (Parr *et al.* 1981).

Preliminary results provide evidence for the presence of indigenous hydrocarbon-degrading bacteria in surface soil samples from the Antarctic. However, oil deposited on soil, particularly light fuel oils, migrates down and along the ice-cemented surface of the permafrost (Balk & Campbell 1995). Hence, we plan to collect samples down a soil profile to the permafrost to determine both the numbers of hydrocarbon-degrading microbes present in the subsurface layers, and the potential rates of oil degradation.

In this study conventional plating methods have been used for the isolation and enumeration of bacteria. These methods provide indirect evidence for the effect of oil contamination on indigenous microbes. In the future we plan to use the bacteria isolated to develop molecular techniques which will allow us to study hydrocarbon-degrading bacteria under "*in situ*" conditions in Antarctic soil.

5   ACKNOWLEDGEMENTS

This work was supported by funding from the Foundation for Research, Science and Technology, New Zealand. I thank Malcolm McLeod and Gareth Lloyd-Jones, who collected the soil samples, Brian Daly who did the chemical and physical analyses of the soil, and Jason Simpson who did the total petroleum hydrocarbon analyses.

REFERENCES

Atlas, R.M., 1981. Microbial degradation of petroleum hydrocarbons: an environmental perspective. *Microbiological Review* 45:180-209.

Balk, M.R. & I.B. Campbell, 1995. Fuel contamination in antarctic soils. *New Zealand Soil News* 43. 235 p.

Balk, M.R.; D.I. Campbell, I.B. Campbell & G.G.C. Claridge, 1995. *Interim results of 1993/94 soil climate, active layer and permafrost investigations at Scott Base, Vanda and Beacon Heights, Antarctica.* University of Waikato, Antarctic Research Unit. Special Report No. 1. 64 p.

Blakemore, L.C., P.L. Searle & B.K. Daly, 1987. Methods for chemical analysis of soils. *New Zealand Soil Bureau Scientific Report* 80.

Cripps, G.C. & J. Priddle, 1991. Hydrocarbons in the Antarctic environment. *Antarctic Science* 3:233-250.

Gerhardt, P., 1994. *Methods for general and molecular bacteriology.* American Society for Microbiology Publications, Washington D.C. 791 p.

Gregory, M.R., R.M. Kirk & M.C.G. Mabin, 1984. Shore types of Victoria Land, Ross Dependency, Antarctica. *New Zealand Antarctic Record* 5:23-40.

Kerry, E., 1990. Microorganisms colonizing plants and soil subjected to different degrees of human activity, including petroleum contamination in the Vestfold Hills and Mac Robertson Land Antarctica. *Polar Biology* 10:423-430.

Kerry, E., 1993. Bioremediation of experimental petroleum spills on mineral soils in the Vestfold Hills, Antarctica. *Polar Biology* 13:163-170.

Konlechner, J.C., 1985. An investigation of the fate and effects of a paraffin based crude oil in an Antarctic terrestrial ecosystem. *New Zealand Antarctic Record* 6:40-46.

Madsen, E.L., 1991. Determining *in situ* biodegradation: facts and challenges. *Environmental Science and Technology* 25:1663-73.

Parr, J.F. W.R. Gardner, & L.F. Elliot, 1981. *Water potential relations in soil* microbiology. Soil Science Society of America, Special Publication No. 9. Soil Science Society of America, Madison. 151 p.

Tumeo, M.A. & A.E. Wolk, 1994. Assessment of the presence of oil-degrading microbes at McMurdo Station. *Antarctic Journal of the United States* 29:375:377

Wilson, S.C. & K.C. Wilson, 1993. Bioremediation of soil contaminated with polynuclear aromatic hydrocarbons (PAHs): a review. *Enviromental Pollution* 81:229-249.

Wardell, L.J., 1995. Potential for bioremediation of fuel-contaminated soil in Antarctica. *Journal of Soil Contamination* 4:111-121.

Ecosystem Processes in Antarctic Ice-free Landscapes, Lyons, Howard-Williams & Hawes (eds)
© *1997 Balkema, Rotterdam, ISBN 90 5410 925 4*

Some aspects of human impact on lakes in the Larsemann Hills, Princess Elizabeth Land, Eastern Antarctica

J.S. Burgess
Department of Geography and Oceanography, ADFA, University of New South Wales, A.C.T., Australia

E. Kaup
Institute of Geology, Estonian Academy of Sciences, Tallin, Estonia

ABSTRACT: Human impacts on lake systems in the Larsemann Hills, Eastern Antarctica since 1987, when the first significant human occupation occurred are outlined. It is suggested that most impact has been confined to the areas in the immediate vicinity of the bases and the interlinking road networks. Storness and most of Broknes remain in the pre-1987 state. Previously reported elevated phosphorus levels in some lakes that were attributed to human activity is questioned.

1 INTRODUCTION

The Larsemann Hills is an ice-free area of approximately 50 km² on the Ingrid Christensen Coast of Princess Elizabeth Land, Eastern Antarctica (69°30'S, 76°20'E) (Fig. 1). The area was not charted until 1935 when Captain Klarius Mikkelsen in the *Thorshaun* led an expedition for the Norwegian whaling magnate Lars Christensen. The expedition flew over the area and took air photographs but it is not certain whether they actually landed. Prior to that both Sir Douglas Mawson and Lars Christensen had sailed along the coast in 1930/31.

Despite that early interest in the area few visits were made until 1986, and until that date the area could be considered pristine. In February 1986 ANARE began to establish Law Base, and later that year USSR began to establish Progress I Station. That station was subsequently abandoned and Progress II was built. In 1987/88 the Peoples Republic of China established Zhong Shan Station. As a result of station development the number of visitors to the Larsemann Hills has increased very substantially (Table 1). The increased activity in such a small area has raised concerns that irreversible adverse environmental impacts may be occurring. Given that the history and extent of occupation is well documented the area provides a good example of the possible human impacts that can occur in an ice-free area of Antarctica. For example simple maps of the distribution of rubbish/litter provide obvious evidence of the probable extent of impact (Fig. 2) (also see Burgess *et al.* 1992 for an earlier map). Further concerns have been voiced by Burgess *et al.* 1993; Ellis-Evans *et al.* 1994 and Wang *et*

Table 1. Larsemann Hills vistiation 1986-1994.

Year	Summer population	Winter population	Totals visitors
1987	40	12	71
1988	43	16	?
1989	103	38	245
1990	103	18	375
1991	122	40	445
1992	?	?	?
1993	55	18	238
1994	17	12	114
1995	22	20	90?

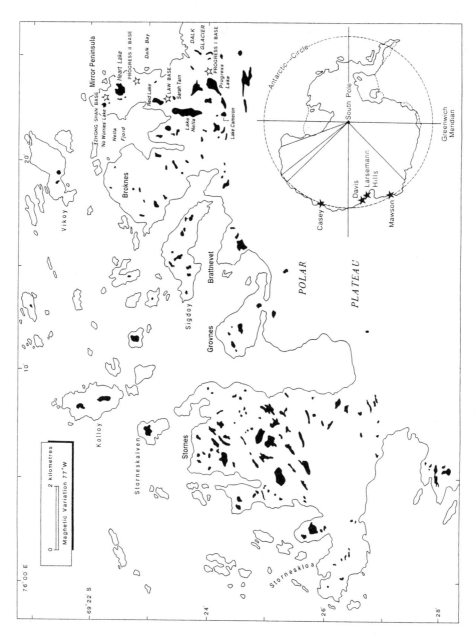

Fig. 1. Location of the Larsemann Hills

al. 1996. Of considerable interest is possible adverse impact on the lake systems. The purpose of this paper is to outline the general status of the lakes prior to major human occupation, indicate how the lakes may be affected, and comment on the significance of possible eutrophication of the systems.

2   LAKE SYSTEMS

In early 1987 a baseline limnological survey of 74 of the 150+ lakes was made (Gillieson *et al.* 1990). That survey indicated that the lakes are generally shallow ponds or ice-deepened basins that thaw, or at least partially thaw, between December and February. Some of the shallower lakes reached temperatures of up to +8°C in early January (Burgess *et al.* 1988). Conductivity, pH, turbidity and ionic composition of the waters confirm their freshwater status (Table 2). Conductivity of lake water ranged from 14 mS cm^{-1} in a small lake on Stornes to 3340 mS in Sarah Tarn close to Law Base. Turbidity was low (usually less than 20 NTU) and the waters contained negligible amounts of silt and clay. Most lakes were characterised by cyanobacterial mats, some up to a metre thick. Chemical analyses indicated that the ionic order was:

$$Na^+ > Mg^{2+} > Ca^{2+} > K^+.$$

The lakes were considered oligotrophic with very low concentrations of P and N. Considerable evidence exists to suggest that lake water levels have varied considerably; higher shorelines of close to five metres were not uncommon.

3   ENVIRONMENTAL IMPACTS

Most impacts in the Larsemann Hills have been related to the building of the Stations. Land transport has been by walking, wheeled vehicles, skidoos (during winter) or by tracked vehicles. With the need to resupply bases and transport people and equipment, a network of roads has been established (Burgess *et al.* 1992) (Fig. 2). Construction of the roads has involved either the informal creation of a track, or compaction by use or bulldozing cuttings. Little attention has been paid to terrain/relief, and no consideration given to drainage or side cuttings. Roads therefore usually take the most direct course, run up and down steep slopes and pass directly through water courses. The roads run through predominantly gneiss that breaks down under loading to fine sand and silt (with little or no clay) that is very easily mobilised by water. Snow and exposed permafrost melt results in the road becoming a 'quasi' watercourse that channelises flow, and in a number of cases, alters the nature of lakes and natural watercourses. For example, Heart Lake on Mirror Peninsula only 20 m from the ocean has in the last few years become progressively less saline, and in January 1996 was 30cm higher than in 1994. This is attributed to the road system in the vicinity of the lake channelling snow melt from an enlarged catchment into the lake. Measurements of height at Lake Nella, Reid Lake and Sarah Tarn showed different trends in lake levels. Lake Nella had a similar mean height to 1994, Reid Lake and Sarah Tarn dropped by 15 cm and 25 cm respectively. The four lakes are, however, very different in their

Table 2. Water Chemistry of the Larsemann Hills lakes

Area	n	Conductivity		pH	Na$^+$	Mg^{2+}	Ca^{2+}	K$^+$
Larsemann Hills total	71	mean	556	6.86	86.7	10.7	10.8	3.2
		sd	705	0.56	111.8	12.4	16.2	4.2
Broknes	20	mean	585	6.83	100.4	10.0	10.3	3.4
		sd	815	0.47	150.1	12.7	14.4	4.7
Stornes	29	mean	288	6.59	45.6	6.7	4.9	2.0
		sd	412	0.52	62.6	8.2	10.3	3.2

Source: Adapted from Gillieson *et al.* 1990.

hydrologic characteristics. Sarah Tarn and Lake Reid are both closed systems and depend totally on snow melt in summer and the extent of evaporation and ablation. Heart Lake was a closed system in 1987 that depended on snow melt in the immediate catchment area. As a result of the changes caused by the road network it is now fed by a small stream during December and January, as well as the original snow melt source. No outlet exists and water loss again depends on evaporation and ablation. Lake Nella has two small input streams, as well as a substantial snow melt source to the east and south. Flows out of the lake are in the vicinity of 20 l s^{-1} for parts of December, January and February and at times the flows have been as high as 1 metre^3 s^{-1}. Lake Nella is therefore subjected to considerably more flushing than the other lake systems. Ellis-Evans et al. (1994) have also reported changes in the hydrologic status of a lake ('No Worries') near Zhong Shan Station which are attributed to human impacts and that are at least partially the result of changes in the flow regime.

Other possible impacts include those from sewage and rubbish disposal, building construction (concrete), helicopter operations and walking. In the case of sewage disposal the three nations have different methods. At Law Base grey water is collected in 200-l drums and returned to Australia while other material is burnt in semi-closed 200-l drums along with other burnable rubbish and returned to Australia or Davis Station. At Zhong Shan sewage waste is disposed after treatment to the ocean and other waste burnt or stockpiled awaiting removal. At Progress 2 waste is burnt or removed to the ocean. The impacts of these activities are unknown although research has begun on Mirror Peninsula and Broknes (Riddle 1996). Similarly the impact of vehicle emissions is also unknown, although preliminary investigations have begun. These activities, however, have been almost solely confined to Mirror Peninsula and Broknes and apart from occasional short term field operations, Storness has been unaffected.

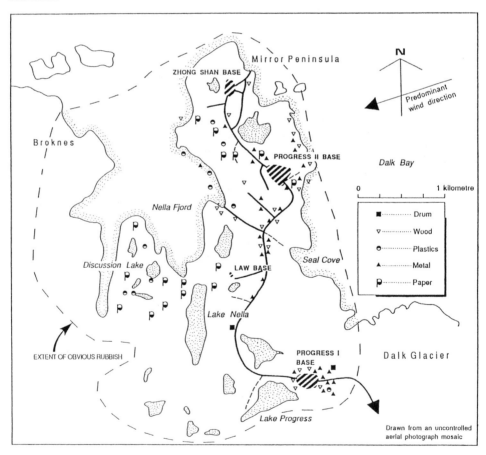

Fig. 2. Distribution of Rubbish and Road Network 1993/94

Of major interest is the possibility that increased nutrient inputs to some lakes have occurred as a result of human impact. Kaup & Burgess (1994) reported that during 1993/94 the water columns of some freshwater lakes revealed P-PO$_4$ concentrations consistently between 30 to 60 mg P m^{-3}. Ellis-Evans et al. (1994) also report relatively high phosphorus levels but the levels in a lake in the vicinity of Zhong Shan were significantly less than those detected in Lake Cameron, a lake distant from human activity and less than one kilometre from the ice plateau. Kaup & Burgess (1994) report similar elevated levels in Lake Cameron during the 1993/94 summer. This leads to the question as to whether elevated nutrient levels are related to human activity, or natural variation related to differences in the nature of the weathered rock in the catchments, or to some other reason. Measurement of total phosphorus for 24 samples of weathered rock from Broknes (collected on a equally spaced grid) showed that values varied considerably (mean 140 g m^{-3}, standard deviation 63 mg m^{-3}). The extent of weathering and backgroud levels of P are considered insufficient to explain the variations in lake concentrations.

Perhaps of more significance are the values of P for inflow streams and for ground water. Measurement of P of inflow streams at Lake Cameron, Lake Nella and Sarah Tarn returned values of 113 mg P m^{-3}, 10 to 83 mg P m^{-3} and 58 to 61 mg P m^{-3}. During the 1993/94 summer a pilot survey of ground water was undertaken. Ground water was collected from ceramic tipped piezometers installed at lakes Cameron, Nella and Sarah. Bedrock at these three sites was similar (yellow gneiss) and runoff was evident at all three. The sub surface stratigraphy of the sampled slopes was typically 50 to 60 cm of coarse weathered sand-sized gneiss underlain by permafrost, with most larger boulders having been sorted to the surface. The depth at which the permafrost was encountered varied. Considerable sub-surface flow of water occurs such that soil pits dug quickly collapsed and infilled. Sediment cores collected from the permafrost revealed varying amounts of organic material, which in the case of Lakes Cameron and Nella was considerable (Burgess et al. 1994). Much of this material has been dated as preceding the last glacial maximum. At this stage it is not known what areas are underlain by this material and hence the size of the 'reservoir' of organic material is conjecture. In the case of Lake Cameron, water drains from a nearby large 'dry' lake bed that contains substantial organic material that includes the remains of benthic cynobacterial mats and moss. Water samples collected from the piezometers during January on an irregular basis showed higher P values than those from the lakes or the input streams (Sarah > 65 mg P m^{-3}, Nella > 130 mg P m^{-3}, Cameron > 180 mg P m^{-3}).

4   DISCUSSION

Environmental impacts in the Larsemann Hills, despite only recent occupation, are not well known. The history of occupation has been relatively well documented and base line information regarding the status of the lake systems collected. Clearly the use of the road network has had an impact on some lake systems and that impact will continue with increasing quantities of fine sediment being transported into some lakes. The problem of vehicle use is not, however, easily solved. There are numerous occasions when tracked vehicles have to be driven during summer. Irresponsible vehicle use can, and should, be avoided (a tracked vehicle was driven through Sarah Tarn in 1988? for no valid reason and the track marks were still noticeable in 1996) and summer use restricted. That elevated P-PO$_4$ levels have been measured in some lakes has caused concern as it possibly indicates cultural eutrophication, which would be a major concern given the pristine state of those ecosystems a decade ago. Clearly this concern should in no way be discounted. However, that high levels have been observed in a lake well distant from human activity seems to indicate that the processes whereby nutrient reaches water bodies is more complicated than at first thought. Some evidence would suggest that ground water is a major contributor of nutrients to the lake systems. This leads to speculation about the role of frozen organic material contained in permafrost. It has been noted that the construction of roads and the operation of vehicles along those roads promotes melting and possibly provides additional contribution of nutrients. Additionally the possibility that global warming may result in the permafrost melting to greater depths and for longer periods of time could consequently result in greater exposure of sub-surface organic material to through flow, that may be of greater impact to lake ecosystems than those resulting from more direct human impacts.

5   ACKNOWLEDGMENTS

This research was funded by ARC, an ASAC grant, the Australian Antarctica Foundation and by the Rector, University College, University of NSW. Considerable assistance was also provided by staff of the Australian Antarctic Division (especially Martin Betts and Rob Easther) and members of 1986-96 ANARE.

REFERENCES

Burgess, J.S., C. Carson, J. Head & A. Spate, in press. *Landform evolution in the Larsmann Hills in the last 25,000 years.* Proceedings of VII International Symposium on Antarctic Earth Sciences, Sienna, September 1995. 69 p.

Burgess, J.S., D.S. Gillieson & A.P. Spate, 1988. On the thermal stratification of freshwater lakes in the Snowy Mountains, Australia, and the Larsemann Hills, Antarctica. *Search* 19:147-149.

Burgess, J.S, A.P. Spate & F.I. Norman, 1992. Environmental impacts of station development in the Larsemann Hills, Princess Elizabeth Land, Antarctica. *Journal of Environmental Management* 36:287-299.

Burgess, J.S., A.P. Spate & J. Shevlin, 1994. The onset of deglaciation in the Larsemann Hills, Eastern Antarctica. *Antarctic Science* 6: 491-495.

Ellis-Evans, J.C., J. Laybourn-Parry, P. Bayliss, & S.T. Perriss, in press. Human impact on an oligotrophic lake in the Larsemann Hills. *Proceedings of the Sixth SCAR Biology Symposium, Venice, 1994.*

Gillieson, D., J. Burgess, A. Spate & A.Cochran, 1990. An Atlas of the Lakes of the Larsemann Hills Princess Elizabeth Land, Antarctica. *ANARE Research Notes* 74. 133 p.

Kaup E. & J. Burgess, 1995. Elevated concentrations of phosphorus in the surface waters of Broknes Peninsula, Larsemann Hills, Antarctica-natural or the result of human impact? *Poster Paper Abstracts, International Association of Theoretical and Applied Limnology, 26th Congress, Sao Paulo, Brazil, 23-29 July 1995.* p. 151

Riddle, M., 1996. Human Impacts Research in the Larsemann Hills. *ANARE News* 76:9.

Wang, Z., F.I. Norman, J.S. Burgess, S.J. Ward, A.P. Spate & C. Carson, 1996. Human influence on variations in numbers of territories and success of South Polar Skuas breeding in the eastern Larsemann Hills, Princess Elizabeth Land, Antarctica. *Polar Record* 13:255-262

Ecosystem Processes in Antarctic Ice-free Landscapes, Lyons, Howard-Williams & Hawes (eds)
© *1997 Balkema, Rotterdam, ISBN 90 5410 925 4*

Implementing the protocol on ice free land: The New Zealand experience at Vanda Station

Emma J.Waterhouse
Antarctica New Zealand, Christchurch, New Zealand

ABSTRACT: The Protocol on Environmental Protection to the Antarctic Treaty was signed in 1991 and provides for the comprehensive protection of the Antarctic environment. Antarctica New Zealand's recent activities at Lake Vanda in the McMurdo Dry Valleys have encompassed a full range of environmental management actions required under the protocol in ice-free areas. In 1994/95, the last remains of a field research station were removed from the shores of the lake, and a new smaller facility resited nearby. Environmental impact assessments have been carried out, and special procedures and plans developed for removal of the station and for the construction and operation of the new facility. Sampling for site contamination was carried out prior to station removal and remediation activities planned on the basis of the results. Specific codes of conduct covering scientists and other visitors were developed and a targeted monitoring programme set up. Monitoring is ongoing and should provide an indication of the success of remediation activities and help to identify the impacts of any contamination on Lake Vanda ecosystems. Implementation of the protocol requirements in this ice-free area required a cooperative effort between scientists, environmental managers and operations staff to ensure that the effects of 26 years of direct human impact at Vanda Station were minimised.

1 INTRODUCTION

Seasonal variations in the climate of the Dry Valleys have resulted in raised levels of Lake Vanda, in the Wright Valley, South Victoria Land. Recent increases had the lake water within two metres of the floor level of the lowest building at the station. Given this immediate threat of inundation, the decision was made to decommission and remove Vanda Station.

Vanda Station was operated as a New Zealand research base for 26 years from 1969-1993. Over this period extensive scientific programmes were carried out. The station was staffed during the summer months throughout this period and for three winters. The occupation of the station and associated activities led to considerable human impacts at the site. Impacts were largely the result of disturbance by trampling and vehicle movement, excavations and erection of buildings, storage of consumables, accidental spills and waste disposal.

Should the site be flooded, contaminants in the soils around the station and from the structures themselves have the potential to enter the lake ecosystem with resulting impacts on the pristine nature of Lake Vanda and its high scientific values (Goguel & Webster 1990; Green & Friedman 1993; Yu *et al.* 1992). A carefully planned process for decommissioning and remediation activities was required. An Initial Environmental Evaluation (IEE) was prepared to cover the activities required to decommission and remove the station. The likely impact of the decommissioning on the lake ecosystem was considered and the requirements set out for remediation activities in order to minimise the potential for impacts on the lake when the site is flooded.

The IEE and all subsequent activities carried out during the decommissioning of Vanda Station were planned and conducted on the basis of the requirements of the Protocol on Environmental Protection to the Antarctic Treaty. Carrying out a decommissioning project of this scale, on ice free land and adjacent to Lake Vanda, presented a number of challenges to the New Zealand managers. The approach taken, and subsequent revisions and modification to the planning and conduct of activities, were centred on the need to minimise or avoid significant impacts in this pristine environment.

2 THE PROTOCOL ON ENVIRONMENTAL PROTECTION TO THE ANTARCTIC TREATY

2.1 General provisions - a natural reserve devoted to peace and science

The protocol establishes Antarctica as a "natural reserve devoted to peace and science" and provides for the comprehensive protection of the Antarctic environment and dependent and associated ecosystems. It strengthens and reiterates the Antarctic Treaty System (ATS), enhancing the protection of the Antarctic environment and recognising the unique scientific and research opportunities the continent offers. The protocol applies to all activities within the Antarctic Treaty area (south of 60°), providing general rules for environmental protection under Article 3, "Environmental Principles", and specific rules under a number of annexes.

The environmental principles provide an over-arching framework for minimising the impacts of activities in Antarctica. Section 1 of Article 3 (below) sets the tone for the remaining principles.

"The protection of the Antarctic environment and dependent and associated ecosystems and the aesthetic value of Antarctica, including its wilderness and intrinsic values and its value as an area for the conduct of scientific research, in particular research essential to understanding the global environment, shall be fundamental considerations in the planning and conduct of all activities in the Antarctic Treaty area." (Section 1, Article 3)

Article 3 goes on to state four additional principles which all parties must adhere to. These include requirements for: all activities to be planned and conducted so as to limit adverse impacts on the Antarctic environment and dependent and associated ecosystems including on biological, scientific, aesthetic and wilderness values; all activities to be planned and conducted on the basis of sufficient information to assess impacts; and, for monitoring to take place. Section 3 of Article 3 accords priority to scientific research when planning and conducting activities and to preserving Antarctica as an area to conduct research. Notable is the inclusion in section 4 of tourism and all other governmental and non governmental activities in the treaty area (together with scientific research programmes) as activities to which the principles apply.

The annexes to the protocol build on these principles by providing additional specific rules for implementation and include provisions for environmental impact assessment, conservation of Antarctic flora and fauna, waste disposal and waste management, prevention of marine pollution and area protection and management. An annex on liability under the protocol is currently being negotiated. This annex, once agreed, should provide a mechanism by which parties could be held liable for environmental damage.

The protocol does not come into force until it has been ratified in domestic legislation by all 26 Antarctic Treaty consultative parties, a process that is expected to be completed by 1997/98. However all countries did undertake, at the signing of the protocol in 1991, to act, as far a practicable, as if the protocol was already in force.

Antarctica New Zealand, as New Zealand's government Antarctic agency, has undertaken to operate in accordance with the protocol. Throughout the project, from the planning and environmental assessment stages to the final clean up and monitoring, the requirements of the protocol have been incorporated into all actions.

2.2   The protocol and ice-free land

The protocol refers specifically to ice-free areas under Annex III Waste Disposal and Waste Management. Under Article 4 of that annex, the protocol sets out the requirements for disposal of waste on ice-free land.

"Wastes not removed or disposed of in accordance with Articles 2 and 3 [requirements for specific hazardous wastes] *shall not be disposed of onto ice-free areas or into fresh water systems".*

The major potential contaminating source from activities at Vanda Station were the result of disposal of domestic waste onto the land and spillage of oil in the vicinity of the station. The introduction of the protocol in 1991, resulted in a halt to disposal of waste water onto the ground at Vanda Station in 1992/93. The prevention of any waste disposal (solid or liquid), onto either the land or into the lake, was a high priority throughout the decommissioning phase.

Other fundamental provisions of the protocol that were incorporated into the planning process included the requirement to produce an environmental impact assessment, in this case an IEE, to avoid any significant impacts on terrestrial and aquatic environments, to provide sufficient information to allow adequate assessment of potential impacts and the requirement under Annex III, Article 1(5) to clean up past and present waste disposal sites. The protocol also requires that clean up and remediation be carried out following the decommissioning or abandonment of a facility or work site.

3   VANDA STATION - 26 YEARS OF OPERATION

3.1   History of the station

Vanda Station was established in the summer of 1968/69 to support New Zealand science activities in the region. The station consisted of eight buildings and could accommodate up to 14 people. Since establishment, until the 1991/93 summer, the station was occupied for the period November to late January (Hayward *et al.* 1994). The station was staffed each year with two support personnel and two to three science personnel. In 1969, 1970 and 1974 a team of four wintered at the station. During periods of occupation, extensive scientific programmes were carried out on the physical sciences, including meteorology, seismology and geomagnetics. In the summer months, Vanda Station acted as a support base for field parties working in the immediate area.

Since 1990, there has been a marked decrease in the amount of science requiring support at Vanda. Operations decreased to a point where in 1992/93 the station was not staffed permanently over summer although scientists working in the area utilised the facility.

3.2   Rising lake levels - the need to remove the station

Increases in the level of Lake Vanda led directly to the decision to decommission and remove the station. Because the lake has no outfall its level is determined by the balance between annual inflows, principally the Onyx River, and ablation plus evaporation. Climate and lake level measurements have been taken at the station since establishment. In the last two decades the lake has risen to seven metres above its 1970 level. Predictions were for lake levels to continue to rise, up to two metres a year. If the lake continued to rise at this rate, station buildings would have begun to be inundated by the summer of 1993/94. A rise of one metre would result in the station site becoming an island.

Lake Vanda is one of the world's most pristine ecosystems and is one of the world's most ultra-oligotrophic lakes. The lake has been extensively studied for over 35 years and continues to have high scientific values. The inundation of Vanda Station represented a significant threat to these values and the pristine qualities of the lake. Although no major contamination events were reported or noted around the station, 26 years of occupation and associated activities meant that cumulative effects were potentially significant. With 1992 lake levels just two metres below the floor level of the lowest building, the decision was made to remove the station. In accordance with protocol requirements, an environmental impact assessment was prepared to cover the activities required to decommission the station.

4 THE ENVIRONMENTAL IMPACT ASSESSMENT PROCESS

4.1 The objectives and process

An Initial Environmental Evaluation (IEE) was prepared to cover the decommissioning activities. This level of assessment was chosen to reflect the high science values of Lake Vanda and the scale and nature of activities that would be required to remove the station. There was considerable concern among scientists and managers about the removal process, and in particular, the need to minimise the entry of contaminants from around the station environment into the lake system. The main objectives in carrying out the evaluation on the decommissioning are stated in the IEE (Hayward *et al.* 1994) and are listed below:

1. *To ensure that that in the event of the waters of Lake Vanda inundating the Station, there will be an absolute minimum of human induced environmental change to the lake system.*
2. *To ensure that if change is unavoidable the nature of that change is understood and documented.*
3. *To submit proposals for protecting Lake Vanda to national and international review.*

The draft IEE was completed in October 1993 and circulated for comment to interested parties in New Zealand. The document was also reviewed by New Zealand's independent Antarctic environmental assessment panel. Comments and recommendations were taken into account in preparing the final IEE, which was completed in March 1994. The IEE concluded that the impacts of the decommissioning would have a minor or transitory impact on the Antarctic environment and outlined the process to be followed in the decommissioning of Vanda Station. The IEE was presented as an information paper to the XVIII Antarctic Treaty Consultative Meeting in April 1994.

4.2 Gathering information on impacts and contamination

Two main investigations were carried out as part of the IEE process to determine the extent of possible soil contamination at, and adjacent to, Vanda Station. A survey was carried out of all past leaders of Vanda Station in 1992/93 to acquire information on the location of disposal sites and spills, the nature and amount of any materials disposed of and the year of disposal. A high response rate to the survey was achieved and no major contamination of the station site was reported. Minor fuel spills were identified but the largest volume of contaminant was determined to have derived from domestic waste water. All solid toilet wastes had been removed from Vanda, and after 1970, all liquid toilet wastes were also removed although some ongoing urine contamination probably occurred. Strained grey water continued to be dumped on the ground until the 1992/93 season. Contamination from other sources is likely to have been small (for example chemicals, batteries and explosives). Some dry wastes were also burnt on site in the first years of operation.

A study of soil contamination was carried out in the 1992/93 season (Sheppard *et al.* 1993). The study analysed up to 23 samples collected from nine soil pits. Sampling sites were determined based on either known or predicted sites of contamination events (based on the leaders survey). Control sites were also included. Analysis concentrated on a range of substances and chemicals likely to have been introduced from the various activities during the station's history. The results of the study were subject to an independent review aimed at assessing the impact on Lake Vanda should contaminated soils become flooded by rising lake levels (Howard-Williams 1993). The review concluded that, should the whole area be flooded and all contaminants leach into the lake in a single year, the contribution of all elements (except total phosphorus, copper and lead) to the natural annual load to the lake from the Onyx River or other meltwater streams, would be negligible. Contaminants P, Cu and Pb would provide very small additional input to the lake. The report concluded that these elements would have a negligible effect on the Lake Vanda ecosystem.

While heavy metals and nutrients such as nitrogen and phosphorus are natural compounds in Lake Vanda, organic compounds associated with human contamination are not naturally present. These contaminants include condensed poly-phosphates, oil residues, soot, animal fats and oils. In addition, organic carbon, present at the main disposal site, Greywater Gully, is not in a form natural to the area. Consequently it was advised to remove the most heavily impacted soils, particularly in Greywater Gully where most domestic waste was disposed of.

4.3   Decommissioning approach - removing the threat of contaminating Lake Vanda

The information enabled the development of an effective and targeted decommissioning process. The IEE set out a programme for the decommissioning of all structures on the station site and the removal of soil and grey water from significantly contaminated areas. The following activities were proposed as necessary to decommission the station and prevent contamination of the pristine waters of Lake Vanda.

(i)   Removal of direct hazards for lake water, sediment, biota and ice. These were all facilities and equipment above ground, i.e. buildings, antennae, towers, and associated equipment.

(ii)   Identification of contaminated sites and remedial action required to minimise leaching and/or diffusion of soil contaminants into the lake system.

(iii)   Identification of potential hazards to the lake's scientific values.

(iv)   Removal of sub-surface installations as far as practical, i.e. buried power and communication cables and building anchors, that may pose future hazards to the lake system.

(v)   Restoration of the natural surface as close as possible to the original form.

(vi)   Monitoring of any ongoing impacts and the effectiveness of decommissioning actions.

5 DECOMMISSIONING AND REMOVAL

5.1 Time frame and resources

During the 1993/94 season some preliminary decommissioning work was carried out. However the majority of work involved in the decommissioning was completed in the 1994/95 summer. Facilities decommissioned at Vanda during the 1994/95 season included four buildings, antennas and associated towers, anchors and underground electrical wiring. These facilities were decommissioned over a period from November 1994 to January 1995 by two teams dedicated to this work plus personnel from Scott Base. Approximately 180 person days were required to complete decommissioning and removal of the station. In addition, removal of materials and contaminated soil from the site to Scott Base required over 70 helicopter hours to complete.

5.2 Decommissioning process

During decommissioning activities all reasonable precautions were undertaken to minimise the release of foreign items into the environment. All obvious surface litter was collected, including particles originating from food, building materials, crates, packaging and wrapping material. The use of saws was kept to minimum and where possible were used inside the workshop so saw dust could be easily collected. Asbestos was present in the buildings and required special handling and precautions. All fuels stored at the station were returned to Scott Base. All painted rocks were also collected and removed from the site or relocated to the new Vanda Huts to mark the helicopter pad.

Helicopter visits to the station were minimised to reduce the blowing of sand over any small items on the ground or the blowing away of light objects. Loads were moved 50 to 100 m away for pick up. The explosives cache at Vanda Station was also removed. Several anchors which could not be practically removed (buried in permafrost up to two metres) were cut off half a metre below the ground surface and remain buried.

6 REMEDIATION

6.1 Remediation activities

Contaminated soil and water was removed from the most heavily contaminated sites around the station. Key aspects of the remediation programme were based on an evaluation of the potential contamination load on the lake from station soils.

Soils known to contain compounds not naturally occurring in the lake were given priority for removal especially at sites under immediate threat from inundation of lake waters. Some remediation work was necessary in 1993/94 in low lying areas. Two 209-l drums of contaminated water and approximately 5000 kg of contaminated soil were removed from Greywater Gully in January 1994.

Further extensive remediation was carried out in 1994/95 following site visits by environmental staff. This work included the following:

(i)   Removal of painted rocks from around the station. Most paint used during the operation of the station was lead based.

(ii)   Removal of soil showing obvious signs of hydrocarbon contamination in the immediate station area, helicopter pads and at other sites identified by the 1992/93 survey. Sites were then covered with fines from the track between the station and helicopter pad and surface rocks replaced.

(iii)   Fat and grease coated pebbles and contaminated fines were removed from an old grey water disposal site in grey water gully. Approximately 2000 kg of rocks and soil was removed from an area of 8 m x 4 m x 150 mm. All visibly contaminated material (dark and non friable) was removed. The excavation was covered using fines and gravel from the helicopter pad track and surface rocks were replaced.

(iv)   Soil was removed from an early (1969) incinerator/dump site. Excavated soil contained elevated levels of copper, lead and zinc as well as macro waste items such as packaging, paper and dry cell batteries.

(v)   Disused science equipment was removed and surface areas remediated. Sites included an old dust trap, a seismic vault and a water staff gauge.

(vi)   All tracks and rock walls were remediated. Rocks were replaced right way up where ever possible (wind worn surface uppermost). Granite was replaced where granite occurs naturally, dolerite in dyke formations, i.e. as "naturally" as possible.

(vii)   All cairns were dismantled and rocks replaced "naturally" in the surrounding area taking care of rock type and orientation.

(viii)   Old rock lines used to hold down wires for various science experiments were randomised. Rocks were replaced "naturally" in the surrounding area taking care of rock type and orientation.

(ix)   A daily inspection collected exposed rubbish (nails, wood chips, splinters, etc).

An estimated total of 15 000 kg of contaminated soil and rocks were removed from Vanda Station and vicinity over the two years of decommissioning. Soil returned to Scott Base has been deposited at a designated site. The site will not be reworked by base activity in the future, and records kept of all Vanda soil deposited on this site.

6.2   What level of remediation?

Several factors determined the level of remediation that was carried out at the site. The major concern was to avoid as far as practicable the release of non-naturally occurring contaminants into lake waters. Consequently the bulk of remediation action was concentrated on sites in Greywater Gully where domestic waste water was disposed of. Contaminated soil and water was removed from this site. Fuel contaminated soils were also removed from the immediate station environs and the helicopter pads to avoid contamination of lake waters with oil products. These actions ensured that the major sources of contamination identified in previous studies were removed.

Solid structures such as anchors and buildings presented relatively few problems to remove, with the greatest difficulty being containment of decommissioning debris and minimising the amount of litter produced. Additional remediation activities were determined on the basis of aesthetic values and the commitment in the IEE to return the site to as natural a state as possible. Consequently, considerable effort was put into repositioning rocks across tracks, dismantling rock walls and cairns, and removing rocks with even the slightest paint residue.

The result of remediation activity is a site where an attempt has been made to remove all major contamination sources, and where almost all visual signs of human presence have also been removed.

The reactions of visitors to the site since remediation was completed has reinforced the value of continuing the remediation process to this extent. The perception of visitors is of an effective and successful decommissioning and removal process based largely on what they see, that is, the aesthetics of the site. Their knowledge of activities and actions taken to prevent serious contamination of the lake is often limited. While considerable extra effort was put into restoring aesthetic values at the site, these activities were considered to be an essential element in the remediation process.

7   MONITORING

7.1   The aim of monitoring - IEE requirements

The IEE recognised that all contaminating substances in the soils of the Vanda Station environs would not be successfully removed. One problem for example was the nature of the terrain, which comprises shattered bedrock at shallow depth. Contaminating substances which reached this layer would have been impractical to remove. Therefore the monitoring programme is aimed primarily at assessing the rate of release and quantity of any contaminants entering the lake waters from contaminated areas at the station site.

7.2   Monitoring programme

The potential contamination load on Lake Vanda from flooded sites at the station site was calculated based on the 1992/93 soil contamination study. Calculations showed that if no remediation was carried out, contaminant phosphorus, copper and lead (found in high levels in station soils) would be unlikely to provide significant additional input to the lake. The hypothesis was put forward that the extensive benthic algal mats in Lake Vanda could serve as effective traps for phosphorus and dissolved metals, including copper and lead, being circulated across their surfaces. Therefore sampling of both algal mats and lake water were included as integral components of the monitoring programme.

Monitoring activities commenced in December 1994. The main components of the monitoring programme include the following:

(i)   Sampling benthic algal mats (cyanobacteria) and water at the lake edge near the most heavily impacted area (Greywater Gully) and at a control site. Samples are analysed for nutrients and heavy metals.
(ii)   Sampling water from snow bank run-off ponding over contaminated areas in Greywater Gully and a control site. Samples are analysed for nutrients and heavy metals.

Conductivity is also measured at each water sample collection point.

It is expected that the chemistry of the shallow waters near the station site and the control site will vary through the season depending on the Onyx River flow, whilst the chemistry of the microbial mats will not vary in this way. Thus water samples will provide short time scale indications of water quality, and mats will provide biological indications on a longer time scale of water and sediment quality. The programme will then be reviewed based on these results and in consultation with scientists assisting with the sample collection and analysis.

7.3   Results

Results available to date only cover sampling work carried out in the 1994/95 and 1995/96 summers. Samples were taken from Vanda Bay (on the lake shore at the mouth of Greywater Gully) and at a control site, as well as two meltwater ponds in the gully. The results show that within Greywater Gully there are water bodies with grossly elevated nutrient concentrations. Unusual algal growths were also recorded. However, nutrient contamination of lake waters is barely detectable (Hawes & Howard-Williams 1996). Further conclusions about the impacts on the lake and the effectiveness of remediation activities will need to await the results of heavy metal analysis and sampling over future seasons.

7.4   Future work

It is intended that the monitoring programme continues for at least two more years. Lake levels in the last two seasons have remained at approximately their 1993/94 heights so that only a very limited area of the station site has been flooded, including the lower helicopter pad. The monitoring programme will be reassessed based on lake levels, that is, whether the main station site is flooded and the results of previous seasons sampling.

The Vanda Station site offers a number of opportunities for research into the impacts of human activities on the Antarctic environment, both in the terrestrial and aquatic environments. The history of activities at the site is relatively well documented, as is the remediation activities that were carried out. The likely flooding of the Vanda Station site has the potential to provide valuable scientific, as well as practical, information directly applicable to the development of environmental management practices elsewhere in the Dry Valleys and at other similar sites in the Antarctic.

Research is planned at the site for the 1995/96 summer concentrating on understanding how cyanobacteria obtain, process and accumulate, or restrict the mobility of nutrients, heavy metals and hydrocarbons derived from contaminated soils (Webster 1995). Further research opportunities may also arise as the site is flooded, in particular looking at the rate of release of contaminants into lake waters, and more work on the role of benthic algal mats in the uptake of nutrients and heavy metals. On land, the extensive physical disturbance at and around the site will begin to recover and offers further opportunities for research into recovery rates in the Antarctic terrestrial environment.

8   CONCLUSIONS - INTERPRETING THE PROTOCOL REQUIREMENTS AT VANDA STATION

The threat of flooding of Vanda Station and the potential for contamination of the lake's water required that urgent action be taken. The decommissioning and remediation activities that were carried out by Antarctica New Zealand were to a standard that was determined by the flooding potential (absent at most other decommissioned facilities), the general requirements of the protocol, and more specific, rules for ice-free areas and for the protection of scientific values.

Together these factors led to a decommissioning programme involving dedicated teams of well trained staff working over two summers. The programe was based on adequate information about contamination sites and likely impacts on Lake Vanda. Remediation requirements were reviewed throughout the project and changes made where necessary.

While decommissioning and remediation activities were extensive, Antarctica New Zealand believed they were necessary in order to fulfil the requirements of the protocol in this environment, and to return the site to as natural a state as possible. Of particular concern, especially from scientists, was the level of contamination in some low lying areas and the potential for release of contaminants into one of the world's most pristine lakes. The effectiveness of remediation measures is yet to be determined. The monitoring programme will continue for at least two more seasons and should provide an indication of the success of the decommissioning in minimising the environmental impact on the lake.

9 ACKNOWLEDGMENTS

The author thanks the staff of Antarctica New Zealand for their assistance in the preparation of this paper.

REFERENCES

Goguel, R.L. & J.G. Webster, 1990. Trace element concentrations in Lake Vanda, the Onyx River and Don Juan Pond, Wright Valley, Antarctica. *New Zealand Antarctic Record* 10:2-8.

Green, W.J. & I.E. Friedmann (Eds.), 1993. *Physical and Biogeochemical Processes in Antarctic Lakes. Antarctic Research Series* 59. American Geophysical Union, Washington D.C. 216 p.

Hayward, J., M.J. Macfarlane, J.R. Keys & I.B. Campbell, 1994. Decommissioning Vanda Station, Wright Valley Antarctica. *Initial Environmental Evaluation. March 1994.*

Hawes, I. & C. Howard-Williams, 1996. Lake Vanda monitoring: Report No. 1 (May 1996). *NIWA Christchurch Consultancy Report, No. NZA60501. May 1996.*

Howard-Williams, C., 1993. Review of the Initial Environmental Evaluation on: Decommissioning Vanda Station, Wright Valley, Antarctica. *New Zealand Freshwater Report* 136.

Sheppard, D.S., I.B. Campbell & G.G.C. Claridge, 1993. Contamination in the soils at Vanda Station. Client report to NZAP. August 1993.

Webster, J., 1995. Predicting contaminant impacts in Antarctica. Application for Research Funding 1996/97, Foundation for Research, Science and Technology and the Ross Dependency Research Committee.

Yu, S.S., W.J. Green, & P.A. Nixon, 1992. Trace metals in Vanda Lake in Antarctica. *Science in China, Series B, November 1992,* 35:1397-1408.

Synthesis: Polar deserts as indicators of change

David H. Walton – *British Antarctic Survey, Cambridge, UK*

Warwick F. Vincent – *Département de Biologie et Centre d'Etudes Nordiques, Université Laval, Que., Canada*

Michael H. Timperley – *National Institute of Water and Atmospheric Research Ltd, Auckland, New Zealand*

Ian Hawes & Clive Howard-Williams – *National Institute of Water and Atmospheric Research Ltd, Christchurch, New Zealand*

1 INTRODUCTION

Much has been written in recent years concerning the prediction and identification of effects of climate change on ecosystems. This debate has been fuelled by the potential for anthropogenically induced changes to interact with natural climatic trends. Hence global scale impacts can be accentuated with consequent economic and political implications. Antarctica has become a focus for aspects of this work, particularly in the context of the observed effect of reduced atmospheric ozone on penetration of UV-B radiation to the Earth's surface, and the potential consequences to sea level of climate-induced changes to the continental ice sheets.

It is particularly evident that, although changes in the UV radiation climate have attracted significant interest, it is changes in water balance and water availability which appear to have had, and be having, most impact on Antarctic ice-free landscapes. As pointed out by Bodeker (this volume), while proportional rises in ground level UV-B have been large in Antarctica, absolute levels remain low compared to lower latitudes. Indeed at an ecosystem level, papers in this volume have rather emphasised the sensitivity of inland Antarctic sites to climate changes which affect the ice/water equilibria of these systems.

Temporal changes or spatial differences in water balance have been shown to affect a wide range of processes. These include soil processes (Ellis-Evans; Campbell *et al*; Wynn-Williams, this volume), hydrology and biogeochemistry of waters (Priscu 1997; Lyons *et al.*, Timperley; Howard-Williams *et al.*, this volume) and terrestrial plant productivity (Schroeter *et al.*, this volume). The sensitivity of these ecosystems to *climate change* is reinforced by the palaeoecological record which shows that lakes in particular respond to changing climate with changes to depth, water quality and hence community composition (Lyons *et al.*; Roberts & McMinn, this volume). Indeed, it is important to recognise that ecosystems in Antarctic ice-free landscapes can be highly dynamic, with, for example, Lake Vanda changing from a lake over 125 m deep to a 3.5 m deep hypersaline pond and back to its current 75 m depth within the past 3000 years. While Antarctic landscapes may be very old, systems within them may be relatively ephemeral.

There are a number of aspects of Antarctic ice-free landscapes, and indeed polar desert ecosystems in general, which make them sensitive indicators of regional climate change, and therefore models of change on a global scale. A concomitant of using these ecosystems as locations for measuring change must be a concern about the effects of human impacts on them and the approaches needed to minimise these effects which could otherwise confuse the signals of global change. In this short chapter we summarise these aspects using examples drawn from the workshop discussions and papers published in this volume. Three questions may be posed:

- What are the key issues and questions regarding global change which should be considered priorities for research on polar desert ecosystems and how should these best be approached.

- Why are polar desert ecosystems appropriate locations to study global change, and how can their unique attributes help us better detect and understand global change processes.

- What are the scientific, logistical and environmental management concerns and constraints for such research.

2   KEY ISSUES FOR RESEARCH

Global change can be interpreted in a broad sense to mean any long term shift in the planetary environment. Clearly, in this context, global change is a continuous, ongoing process. Within Antarctica this is reflected on a number of time scales, from the consequences of continental drift, to the waxing and waning of ice sheets in response to ice age cycles, and in the decadal variability in climate. Ongoing change in Antarctic ecosystems can be divided into two components: the regional component of global change (which may include the effects of anthropogenic activities elsewhere) and local anthropogenic effects caused by human presence in these systems. The former includes greenhouse gas accumulation in the atmosphere and its influence on radiation balance, wind, humidity and precipitation. Also within this category are stratospheric ozone depletion with its effects on the ground-level flux and spectral composition of UV-B radiation, and other environmental phenomena such as long range transport of contaminants. The evidence for and potential impact of greenhouse gas accumulation and increasing UV-B radiation have been summarised in the Intergovernmental Panel on Climate Change (Houghton *et al.* 1995).

The second component is related primarily to activities within the last 90 years of human exploration and settlement. It includes redistribution of the world's biota as well as local changes that have resulted from the general expansion of human activities such as, transport, tourism, technology development and scientific research (Vincent 1996; SCAR/ COMNAP 1996).

Certain generic questions are common to all of these climate-related as well as direct anthropogenic effects. For example, we ultimately require knowledge about the direction, amplitude, frequency and duration of change; i.e., the shape of the overall 'global change function'. To develop a predictive understanding of this change function will require detailed information about the mechanisms of change including feedback controls, critical thresholds, buffer capacities, and the resistance to change over certain parameter ranges. Different components of an ecosystem, as well as different ecosystems, are likely to vary in their resistance, sensitivity, and qualitative response to global change, and also in their ability to recover from varying levels of change (resilience).

Three questions are of central importance in current and future research in these areas:

- How did global climate change proceed in the past? New research findings from ice cores, lake cores, soil profiles and a variety of other historical records are providing new insights into this question over a range of time scales (Lyons *et al.*, this volume).

- How can we differentiate any long term global change signal from interannual variability, spatial variations, and local anthropogenic effects? New monitoring initiatives will provide an important step forward in this area (SCAR/COMNAP 1996).

- How do regional climatic changes translate into the microscale features of habitat which directly affect organisms. In polar deserts, this particularly relates to moisture availability (Wynn-Williams *et al.*; Schroeter *et al.* this volume).

3 POLAR DESERT ECOSYSTEMS AS SITES FOR GLOBAL CHANGE RESEARCH

Arctic and Antarctic desert ecosystems offer a number of advantages as sites for detecting global environmental change. These include:

- Global climate models predict that the polar regions are likely to be more sensitive than regions at other latitudes to climate changes related to increasing greenhouse gas concentrations. This effect is likely to be significant in terms of temperature; changes in the amount and timing of precipitation are still inadequately modelled and may also turn out to be very important.
- The relative increase in ground-level UV-B radiation is greatest in the polar regions. The low UV-screening by dissolved organic carbon content of many high latitude lakes (especially in the Antarctic) may make them particularly vulnerable to rising UV-B (see Vincent, this volume), only partially offset by snow and ice-cover. Polar deserts generally lack snow accumulation making terrestrial organisms exposed to high springtime UV radiation.
- The baseline levels of contaminants is low, especially in Antarctica, and small absolute increases of introduced chemicals are therefore more likely to be detectable than at lower latitudes. Atmospheric circulation patterns result in the transfer of airborne contaminants from low to high latitudes, hence Antarctica provides an opportunity to identify truly global pollutants. Organisms growing in this environment under temperature or water stress may be more severely impaired by xenobiotic materials than organisms growing under less stressful conditions at lower latitudes.
- The low species diversity and absence of certain functional groups (e.g. angiosperms and aquatic macroinvertebrates) of polar desert ecosystems including the lakes means that loss or gain of even a single species as a result of environmental change may have major implications for food web integrity and productivity.
- Physical properties of polar desert ecosystems render them highly sensitive to small variations in the physical environment. For example, small changes in temperature could have a major impact on the balance between freezing and melting (e.g. on glacier surfaces and on the depth of the active layer in the soil). Similarly, the highly desiccated and often saline soils of polar deserts may be highly responsive to changes in precipitation.
- Polar desert regions are becoming increasingly accessible to tourism and other human activities (Vincent 1996; SCAR/COMNAP 1996). Global change research in these environments will help identify environmental management options for the long term protection and conservation of these sites.
- The environmental extremes encountered in polar systems provide unique opportunities for ecosystem research; for example, in providing 'end-member' systems for the study of geochemical processes; the development of microbial communities under severe environmental stress; the provision of useful models for research in ecophysiology and exobiology.

4 CONSTRAINTS ON POLAR DESERT RESEARCH

Polar deserts and ice free landscapes may be attractive research sites for addressing global change questions but there are also a number of constraints and management concerns which need to carefully considered during the research planning stage:

- The high logistic costs and potential environmental impact of polar research need to be balanced against the value of the anticipated research findings.

- The paucity of long term environmental data for many locations and subject areas may make it difficult to detect changes in the environment. Long term data sets, however, are gradually becoming available in both polar regions.
- The absence of certain kinds of palaeo-environmental indicators (e.g. absence of a pollen record in Antarctica) is a limit on the availability of some types of long term data.
- The high sensitivity to climatic forcing may make it difficult to discern long term trends in ecosystem processes from interannual variability.
- Conversely, the presence of generalist species in polar desert communities may provide a buffer against change.
- The scarcity of sites may pose an impediment to certain types of research. There are a small number of potential research sites, and most sites are limited in areal extent.
- Polar desert ecosystems are likely to be highly sensitive to the human activities associated with research; for example, desert pavement soils are seriously disturbed by vehicle movements; the low biotic diversity of these systems could be perturbed in major ways by accidental transfers.

5   MANAGEMENT OF POLAR DESERT ECOSYSTEMS

The sensitivity of these ecosystems makes them especially vulnerable to the effects of local human activities. Management needs to be on an ecosystem basis in order to protect the processes controlling activities in the system as well as its particular biotic components. This in itself poses difficulties since there has been little evidence so far of the necessary interdisciplinary integration to ensure that all key site values are identified. The assessment matrices developed for environmental impact assessment (e.g. Vincent 1996; SCAR/COMNAP 1996) offer some hope of identifying the interactions between the temporal and spatial components, an apparently crucial feature of assessing impacts at a site specific level.
    A number of management tools already exist to protect ice-free landscapes and minimise impacts. Parts of these landscapes are already designated as protected areas, with management plans and permitting regulations which limit both access to and activities within them. It is not clear that all the key areas have yet been afforded this level of protection. For the ice free areas of Victoria Land an agreed Code of Conduct is now in use for most researchers working there. This Code attempts to minimise sampling, avoid cross-contamination of sites, ensure all wastes removal, and restrict travel and camping. In addition there is considerable scientific interest in extending and developing the existing informal management rules to provide the framework for an Antarctic Specially Managed Area covering, for example, all of the McMurdo Dry Valleys. This is recognised to be a complex undertaking but almost certainly essential if the damage to these sensitive systems is to be kept to an acceptable level. This will be a complex but essential undertaking to minimise damage to these sensitive environments.

6   THE FUTURE USE OF POLAR DESERTS

The papers at this workshop clearly indicate the value of polar deserts for a wide variety of scientific investigations. In looking forward to their future use it is important to recognise the major gaps in our understanding. These can be broadly classified into three main fields; ecosystem structure, variability and control processes. For ecosystem structure these include the geochemical characteristics, the structure and dynamics of periglacial features, the hydrology of glacier meltwater streams, the bio-optical characteristics of ice-covered lakes and the connectivity between ecosystem sub-components. At a biological level it includes topics such as the biodiversity and genetic characteristics of the desert ecosystem communities, the productivity and balance of phototrophy versus heterotrophy, the biological linkages between terrestrial and aquatic biological processes, and the trophic organisation of these desert systems.

There is a need to improve our understanding of ecosystem variability at all timescales, from less than 24 h to interannual, in order to define the limits of resolution for detecting long term trends. A focus upon historical variations in these systems (e.g., from palaeo-limnological records) will be extremely helpful in this regard. The development of sensitive remote sensing techniques is required to provide data with minimal site impact by investigators and maximum spatial and temporal coverage.

A broad range of controlling factors in the polar desert ecosystem have been identified including temperature, radiation and nutrient supply. As in any desert system, however, the critical limiting factor is the availability of liquid water. Water supply in the Arctic and Antarctic desert environment acts as a master control on a great variety of ecosystem properties and processes, for example geochemical weathering, substrate stability, transport of materials, terrestrial productivity etc. Despite this key role there is only a poor understanding of the temporal and spatial availability of water and of the most appropriate measure of each process involving water (e.g., vapour pressure, osmotic potential etc).

Precipitation is difficult to measure in polar environments, and there are few reliable long term records. Moisture availability itself is a poorly defined concept, and sensor technology severely limits both the type and the domain size of measurements. What is more, there is at present little understanding of the processes which determine water availability at ecologically relevant scales. A better quantification of 'biologically available water' in different habitats is required for most climatic regimes but it becomes especially difficult in the polar regions where repeated freezing and thawing may occur. Little is known about the non-linearities, feedback effects and thresholds associated with changes in water supply. This field offers perhaps the greatest single research challenge to all environmental and biological scientists working in polar deserts.

Global climate change is likely to strongly manifest its effects on polar desert ecosystems through temperature-dependent (freezing-melting) and direct (precipitation) changes in water availability. By using this as a thematic focus for inter-disciplinary research it might prove possible to bridge the gaps between disciplines and encourage the holistic approach which we consider to be essential for the long term management and protection of these sensitive areas.

7   ACKNOWLEDGEMENTS

The authors acknowledge the input of the participants of the International Workshop on Polar Desert Ecosystems. Particular thanks are due to Dr Julie Hall and Anne-Maree Schwarz, rapporteurs for the individual workshops from which this discussion paper is condensed.

REFERENCES

Houghton, J.T., L.G. Meira Filho, B.A. Callander, N. Harris, A. Kattenberg, & K. Maskell, (Eds.) 1996. Climate Change 1995 - The science of climate change. Contribution of Working Group I to the Second Assessment Report of the Intergovernmental Panel on Climate Change. Cambridge University Press. 572 p.

Priscu, J.C., (Ed.) 1997. The McMurdo Dry Valleys: A Cold Desert Ecosystem. American Geophysical Union Antarctic Research Series, Washington D.C. in press.

Vincent, W.F., (Ed.) 1996. Environmental management of a cold desert ecosystem: the McMurdo Dry Valleys. *Desert Research Institute, University of Nevada, USA, Special Publication.* 57 p.

SCAR/COMNAP 1996. Monitoring of environmental impacts from science and operations in Antarctica. Published by the Scientific Committee on Antarctic Research, Cambridge, UK. 40 p.

Author index

281